Galois' Theory
of Algebraic Equations

Galois' Theory
of Algebraic Equations

Jean-Pierre Tignol
Université Catholique de Louvain, Belgium

 World Scientific

NEW JERSEY · LONDON · SINGAPORE · BEIJING · SHANGHAI · HONG KONG · TAIPEI · CHENNAI

Published by

World Scientific Publishing Co. Pte. Ltd.

5 Toh Tuck Link, Singapore 596224

USA office: 27 Warren Street, Suite 401-402, Hackensack, NJ 07601

UK office: 57 Shelton Street, Covent Garden, London WC2H 9HE

Library of Congress Cataloging-in-Publication Data
Tignol, Jean-Pierre
  Galois' theory of algebraic equations / by Jean-Pierre Tignol.
  p. cm.
  Includes bibliographical references and index.
  ISBN-13: 978-981-02-4541-2 (pbk)
  ISBN-10: 981-02-4541-6 (pbk)
  1. Equations, Theory of. 2. Galois theory. I. Title.
  QA211.T5413 2001
  512'.3--dc21

                                        2001276236

British Library Cataloguing-in-Publication Data
A catalogue record for this book is available from the British Library.

First published 2001
Reprinted 2004, 2006, 2011

Printed in Singapore.

à Paul

For inquire, I pray thee, of the former age,
and prepare thyself to the search of their fathers:
For we are but of yesterday, and know nothing,
because our days upon earth are a shadow.

Job 8, 8–9.

Preface

In spite of the title, the main subject of these lectures is not algebra, even less history, as one could conclude from a glance over the table of contents, but methodology. Their aim is to convey to the audience, which originally consisted of undergraduate students in mathematics, an idea of how mathematics is made. For such an ambitious project, the individual experience of any but the greatest mathematicians seems of little value, so I thought it appropriate to rely instead on the collective experience of generations of mathematicians, on the premise that there is a close analogy between collective and individual experience: the problems over which past mathematicians have stumbled are most likely to cause confusion to modern learners, and the methods which have been tried in the past are those which should come to mind naturally to the (gifted) students of today. The way in which mathematics is made is best learned from the way mathematics has been made, and that premise accounts for the historical perspective on which this work is based.

The theme used as an illustration for general methodology is the theory of equations. The main stages of its evolution, from its origins in ancient times to its completion by Galois around 1830 will be reviewed and discussed. For the purpose of these lectures, the theory of equations seemed like an ideal topic in several respects: first, it is completely elementary, requiring virtually no mathematical background for the statement of its problems, and yet it leads to profound ideas and to fundamental concepts of modern algebra. Secondly, it underwent a very long and eventful evolution, and several gems lie along the road, like Lagrange's 1770 paper, which brought order and method to the theory in a masterly way, and Vandermonde's visionary glimpse of the solution of certain equations of high degree, which hardly unveiled the principles of Galois theory sixty years before Galois' memoir. Also instructive from a methodological point of view is the relationship between the general theory, as developed by Cardano, Tschirnhaus, Lagrange and Abel, and the attempts by Viète, de Moivre, Vandermonde and Gauss at significant examples, namely the so-called cyclotomic equations, which arise from the division of the circle into equal parts. Works in these two directions are closely intertwined like themes in a counterpoint, until their resolution in Galois' memoir. Finally, the algebraic theory of equations is now a closed subject, which reached complete maturity a long time ago; it is therefore possible to give a fair assessment of its various aspects. This is of course not true of Galois theory, which still provides inspiration for original research in numerous directions, but these lectures are concerned with the theory of equations and not with Galois theory of fields. The evolution from Galois' theory to modern Galois theory falls beyond the scope of this work; it would certainly fill another book like this one.

As a consequence of emphasis on historical evolution, the exposition of mathematical facts in these lectures is genetic rather than systematic, which means that it aims to retrace the concatenation of ideas by following (roughly) their chronological order of occurrence. Therefore, results which are logically close to each other may be scattered in different chapters, and some topics are discussed several times, by little touches, instead of being given a unique definitive account. The expected reward for these circumlocutions is that the reader could hopefully gain a better insight into the inner workings of the theory, which prompted it to evolve the way it did.

Of course, in order to avoid discussions that are too circuitous, the works of mathematicians of the past—especially the distant past—have been somewhat modernized as regards notation and terminology. Although considering sets of numbers and properties of such sets was clearly alien to the patterns of thinking until the nineteenth century, it would be futile to ignore the fact that (naive) set theory has now pervaded all levels of mathematical education. Therefore, free use will be made of the definitions of some basic algebraic structures such as field and group, at the expense of lessening some of the most original discoveries of Gauss, Abel and Galois. Except for those definitions and some elementary facts of linear algebra which are needed to clarify some proofs, the exposition is completely self-contained, as can be expected from a genetic treatment of an elementary topic.

It is fortunate to those who want to study the theory of equations that its long evolution is well documented: original works by Cardano, Viète, Descartes, Newton, Lagrange, Waring, Gauss, Ruffini, Abel, Galois are readily available through modern publications, some even in English translations. Besides these original works and those of Girard, Cotes, Tschirnhaus and Vandermonde, I relied on several sources, mainly on Bourbaki's Note historique [6] for the general outline, on Van der Waerden's "Science Awakening" [62] for the ancient times and on Edwards' "Galois theory" [20] for the proofs of some propositions in Galois' memoir. For systematic expositions of Galois theory, with applications to the solution of algebraic equations by radicals, the reader can be referred to any of the fine existing accounts, such as Artin's classical booklet [2], Kaplansky's monograph [35], the books by Morandi [44], Rotman [50] or Stewart [56], or the relevant chapters of algebra textbooks by Cohn [14], Jacobson [33], [34] or Van der Waerden [61], and presumably to many others I am not aware of. In the present lectures, however, the reader will find a thorough treatment of cyclotomic equations after Gauss, of Abel's theorem on the impossibility of solving the general equation of degree 5 by radicals, and of the conditions for solvability of algebraic equations after Galois, with complete proofs. The point of view differs from the one in the quoted references in that it is strictly utilitarian, focusing (albeit to a lesser extent than the original papers) on the concrete problem at hand, which is to solve equations. Incidentally, it is striking to observe, in comparison, what kind of acrobatic tricks are needed to apply modern Galois theory to the solution of algebraic equations.

The exercises at the end of some chapters point to some extensions of the theory and occasionally provide the proof of some technical fact which is alluded to in the text. They are never indispensable for a good understanding of the text. Solutions to selected exercises are given at the end of the book.

This monograph is based on a course taught at the Université catholique de Louvain from 1978 to 1989, and was first published by Longman Scientific & Technical in 1988. It is a much expanded and completely revised version of my "Leçons sur la théorie des équations" published in 1980 by the (now vanished) Cabay editions in Louvain-la-Neuve. The wording of the Longman edition has been recast in a few places, but no major alteration has been made to the text.

I am greatly indebted to Francis Borceux, who invited me to give my first lectures in 1978, to the many students who endured them over the years, and to the readers who shared with me their views on the 1988 edition. Their valuable criticism and encouraging comments were all-important in my decision to prepare this new edition for publication. Through the various versions of this text, I was privileged to receive help from quite a few friends, in particular from Pasquale Mammone and Nicole Vast, who read parts of the manuscript, and from Murray Schacher and David Saltman, for advice on (American-) English usage. Hearty thanks to all of them. I owe special thanks also to T.S. Blyth, who edited the manuscript of the Longman edition, to the staffs of the Centre général de Documentation (Université catholique de Louvain) and of the Bibliothèque Royale Albert 1er (Brussels) for their helpfulness and for allowing me to reproduce parts of their books, and to Nicolas Rouche, who gave me access to the riches of his private library.

On the TEXnical side, I am grateful to Suzanne D'Addato (who also typed the 1988 edition) and to Béatrice Van den Haute, and also to Camille Debième for his help in drawing the figures.

Finally, my warmest thanks to Céline, Paul, Ève and Jean for their infectious joy of living and to Astrid for her patience and constant encouragement. The preparation of the 1988 edition for publication spanned the whole life of our little Paul. I wish to dedicate this book to his memory.

Contents

Chapter 1

Quadratic Equations

1.1 Introduction

Since the solution of a linear equation $aX = b$ does not use anything more than a division, it hardly belongs to the algebraic theory of equations; it is therefore appropriate to begin these lectures with quadratic equations

$$aX^2 + bX + c = 0 \qquad (a \neq 0).$$

Dividing both sides by a, we may reduce this to

$$X^2 + pX + q = 0.$$

The solution of this equation is well-known: when $\left(\frac{p}{2}\right)^2$ is added to each side, the square of $X + \frac{p}{2}$ appears and the equation can be written

$$\left(X + \frac{p}{2}\right)^2 + q = \left(\frac{p}{2}\right)^2.$$

(This procedure is called "completion of the square"). The values of X easily follow:

$$X = -\frac{p}{2} \pm \sqrt{\left(\frac{p}{2}\right)^2 - q}.$$

This formula is so well-known that it may be rather surprising to note that the solution of quadratic equations could not have been written in this form before the seventeenth century.* Nevertheless, mathematicians had been solving quadratic

---

*The first uniform solution for quadratic equations (regardless of the signs of coefficients) is due to Simon Stevin in "L'Arithmetique" [55, p. 595], published in 1585. However, Stevin does not use literal coefficients, which were introduced some years later by François Viète: see chapter 4, §4.1.

1

equations for about 40 centuries before. The purpose of this first chapter is to give a brief outline of this "prehistory" of the theory of quadratic equations.

1.2 Babylonian algebra

The first known solution of a quadratic equation dates from about 2000 B.C.; on a Babylonian tablet, one reads (see Van der Waerden [62, p. 69])

> I have subtracted from the area the side of my square: 14.30. Take 1, the coefficient. Divide 1 into two parts: 30. Multiply 30 and 30: 15. You add to 14.30, and 14.30.15 has the root 29.30. You add to 29.30 the 30 which you have multiplied by itself: 30, and this is the side of the square.

This text obviously provides a procedure for finding the side of a square (say x) when the difference between the area and the side (i.e. $x^2 - x$) is given; in other words, it gives the solution of $x^2 - x = b$.

However, one may be puzzled by the strange arithmetic used by Babylonians. It can be explained by the fact that their base for numeration is 60; therefore 14.30 really means $14 \cdot 60 + 30$, i.e. 870. Moreover, they had no symbol to indicate the absence of a number or to indicate that certain numbers are intended as fractions. For instance, when 1 is divided by 2, the result which is indicated as 30 really means $30 \cdot 60^{-1}$, i.e. 0.5. The square of this 30 is then 15 which means 0.25, and this explains why the sum of 14.30 and 15 is written as 14.30.15: in modern notations, the operation is $870 + 0.25 = 870.25$.

After clearing the notational ambiguities, it appears that the author correctly solves the equation $x^2 - x = 870$, and gets $x = 30$. The other solution $x = -29$ is neglected, since the Babylonians had no negative numbers.

This lack of negative numbers prompted Babylonians to consider various types of quadratic equations, depending on the signs of coefficients. There are three types in all:

$$X^2 + aX = b, \quad X^2 - aX = b, \quad \text{and} \quad X^2 + b = aX,$$

where a, b stand for positive numbers. (The fourth type $X^2 + aX + b = 0$ obviously has no (positive) solution.)

Babylonians could not have written these various types in this form, since they did not use letters in place of numbers, but from the example above and from other numerical examples contained on the same tablet, it clearly appears that the

Babylonians knew the solution of

$$X^2 + aX = b \quad \text{as} \quad X = \sqrt{\left(\frac{a}{2}\right)^2 + b} - \frac{a}{2}$$

and of

$$X^2 - aX = b \quad \text{as} \quad X = \sqrt{\left(\frac{a}{2}\right)^2 + b} + \frac{a}{2}.$$

How they argued to get these solutions is not known, since in every extant example, only the procedure to find the solution is described, as in the example above. It is very likely that they had previously found the solution of geometric problems, such as to find the length and the breadth of a rectangle, when the excess of the length on the breadth and the area are given. Letting x and y respectively denote the length and the breadth of the rectangle, this problem amounts to solving the system

$$\begin{cases} x - y = a \\ xy = b. \end{cases} \tag{1.1}$$

By elimination of y, this system yields the following equation for x:

$$x^2 - ax = b. \tag{1.2}$$

If x is eliminated instead of y, we get

$$y^2 + ay = b. \tag{1.3}$$

Conversely, equations (1.2) and (1.3) are equivalent to system (1.1) after setting $y = x - a$ or $x = y + a$.

They probably deduced their solution for quadratic equations (1.2) and (1.3) from their solution of the corresponding system (1.1), which could be obtained as follows: let z be the arithmetic mean of x and y.

In other words, z is the side of the square which has the same perimeter as the given rectangle:

$$z = x - \frac{a}{2} = y + \frac{a}{2}.$$

Compare then the area of the square (i.e. z^2) to the area of the rectangle ($xy = b$). We have

$$xy = \left(z + \frac{a}{2}\right)\left(z - \frac{a}{2}\right)$$

whence $b = z^2 - \left(\frac{a}{2}\right)^2$. Therefore, $z = \sqrt{\left(\frac{a}{2}\right)^2 + b}$ and it follows that

$$x = \sqrt{\left(\frac{a}{2}\right)^2 + b} + \frac{a}{2} \quad \text{and} \quad y = \sqrt{\left(\frac{a}{2}\right)^2 + b} - \frac{a}{2}.$$

This solves at once the quadratic equations $x^2 - ax = b$ and $y^2 + ay = b$.

Looking at the various examples of quadratic equations solved by Babylonians, one notices a curious fact: the third type $x^2 + b = ax$ does not explicitly appear. This is even more puzzling in view of the frequent occurrence in Babylonian tablets of problems such as to find the length and the breadth of a rectangle when the perimeter and the area of the rectangle are given; which amounts to the solution of

$$\begin{cases} x + y = a \\ xy = b. \end{cases} \tag{1.4}$$

By elimination of y, this system leads to $x^2 + b = ax$. So, why did Babylonians solve the system (1.4) and never consider equations like $x^2 + b = ax$?

A clue can be discovered in their solution of system (1.4), which is probably obtained by comparing the rectangle with sides x, y to the square with perimeter $\frac{a}{2}$:

One then sets $x = \frac{a}{2} + z$, whence $y = \frac{a}{2} - z$, and finishes as before.

Whatever their method, the solution they get is

$$x = \frac{a}{2} + \sqrt{\left(\frac{a}{2}\right)^2 - b},$$

$$y = \frac{a}{2} - \sqrt{\left(\frac{a}{2}\right)^2 - b},$$

thus assigning one value for x and one value for y, while it is clear to us that x and y are interchangeable in the system (1.4): we would have given two values for each one of the unknown quantities, and found

$$x = \frac{a}{2} \pm \sqrt{\left(\frac{a}{2}\right)^2 - b}, \qquad y = \frac{a}{2} \mp \sqrt{\left(\frac{a}{2}\right)^2 - b}.$$

In the Babylonian phrasing, however, x and y are not interchangeable: they are the length and the breadth of a rectangle, so there is an implicit condition that $x \geq y$. According to S. Gandz [22, §9], the type $X^2 + b = aX$ was systematically and purposely avoided by Babylonians because, unlike the two other types, it has *two* positive solutions (which are the length x and the breadth y of the rectangle). The idea of two values for one quantity was probably very embarrassing to them, it would have struck Babylonians as an illogical absurdity, as sheer nonsense.

However, this observation that algebraic equations of degree higher than 1 have several interchangeable solutions is of fundamental importance: it is the corner-stone of Galois theory, and we shall have the opportunity to see to what clever use it will be put by Lagrange and later mathematicians.

As André Weil commented in relation to another topic [69, p. 104]

> This is very characteristic in the history of mathematics. When there is something that is really puzzling and cannot be understood, it usually deserves the closest attention because some time or other some big theory will emerge from it.

1.3 Greek algebra

The Greeks deserve a prominent place in the history of mathematics, for being the first to perceive the usefulness of proofs. Before them, mathematics were rather empirical. Using deductive reasoning, they built a huge mathematical monument, which is remarkably illustrated by Euclid's celebrated masterwork "The Elements" (c. 300 B.C.).

The Greeks' major contribution to algebra during this classical period is foundational. They discovered that the naive idea of number (i.e. integer or rational number) is not sufficient to account for geometric magnitudes. For instance, there is no line segment which could be used as a length unit to measure the diagonal and the side of a square by integers: the ratio of the diagonal to the side (i.e. $\sqrt{2}$) is not a rational number, or in other words, the diagonal and the side are incommensurable.

The discovery of irrational numbers was made among followers of Pythagoras, probably between 430 and 410 B.C. (see Knorr [39, p. 49]). It is often credited to Hippasus of Metapontum, who was reportedly drowned at sea for producing a downright counterexample to the Pythagoreans' doctrine that "all things are numbers." However, no direct account is extant, and how the discovery was made is still a matter of conjecture. It is widely believed that the first magnitudes which were shown to be incommensurable are the diagonal and the side of a square, and the following reconstruction of the proof has been proposed by Knorr [39, p. 27]:

Assume the side AB and the diagonal AC of the square $ABCD$ are both measured by a common segment; then AB and AC both represent numbers (= integers) and the squares on them, which are $ABCD$ and $EFGH$, represent square numbers. From the figure, it is clear (by counting triangles) that $EFGH$ is the double of $ABCD$, so $EFGH$ is an even square number and its side EF is therefore even. It follows that EB also represents a number, whence $EBKA$ is a square number.

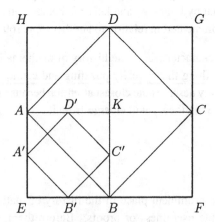

Since the square $ABCD$ clearly is the double of the square $EBKA$, the same arguments show that AB is even, whence $A'B'$ represents a number.

We now see that $A'B'$ and $A'C'$ $(= EB)$, which are the halves of AB and AC, both represent numbers; but $A'B'$ and $A'C'$ are the side and the diagonal of a new (smaller) square, so we may repeat the same arguments as above.

Iterating this process, we see that the numbers represented by AB and AC are indefinitely divisible by 2. This is obviously impossible, and this contradiction proves that AB and AC are incommensurable.

This result obviously shows that integers are not sufficient to measure lengths of segments. The right level of generality is that of ratios of lengths. Prompted by this discovery, the Greeks developed new techniques to operate with ratios of geometric magnitudes in a logically coherent way, avoiding the problem of assigning numerical values to these magnitudes. They thus created a "geometric algebra," which is methodically taught by Euclid in "The Elements."

By contrast, Babylonians seem not to have been aware of the theoretical difficulties arising from irrational numbers, although these numbers were of course unavoidable in the treatment of geometric problems: they simply replaced them by rational approximations. For instance, the following approximation of $\sqrt{2}$ has been found on some Babylonian tablet: $1.24.51.10$, i.e. $1 + 24 \cdot 60^{-1} + 51 \cdot 60^{-2} + 10 \cdot 60^{-3}$ or $1.41421296296296\ldots$, which is accurate up to the fifth place.

Although Euclid does not explicitly deal with quadratic equations, the solution of these equations can be detected under a geometric garb in some propositions of the Elements. For instance, Proposition 5 of Book II states [30, v. I, p. 382]:

> If a straight line be cut into equal and unequal segments, the rectangle contained by the unequal segments of the whole together with the square on the straight line between the points of section is equal to the square on the half.

On the figure above, the straight line AB has been cut into equal segments at C and unequal segments at D, and the proposition asserts that the rectangle AH together with the square LG (which is equal to the square on CD) is equal to the square CF. (This is clear from the figure, since the rectangle AL is equal to the rectangle DF).

If we understand that the unequal segments in which the given straight line $AB = a$ is cut are unknown, it appears that this proposition provides us with the core of the solution of the system

$$\left\{ \begin{array}{l} x + y = a \\ xy = b. \end{array} \right.$$

Indeed, setting $z = x - \frac{a}{2}$ "the straight line between the points of section," it states that $b + z^2 = \left(\frac{a}{2}\right)^2$. It then readily follows that

$$z = \sqrt{\left(\frac{a}{2}\right)^2 - b}$$

whence

$$x = \frac{a}{2} + \sqrt{\left(\frac{a}{2}\right)^2 - b} \quad \text{and} \quad y = \frac{a}{2} - \sqrt{\left(\frac{a}{2}\right)^2 - b},$$

as in Babylonian algebra. In subsequent propositions, Euclid also teaches the solution of

$$\left\{ \begin{array}{l} x - y = a \\ xy = b \end{array} \right.$$

which amounts to $x^2 - ax = b$ or $y^2 + ay = b$. He returns to the same type of problems, but in a more elaborate form, in propositions 28 and 29 of book VI (Compare Kline [38, pp. 76–77] and Van der Waerden [62, p. 121].)

The Greek mathematicians of the classical period thus reached a very high level of generality in the solution of quadratic equations, since they considered equations with (positive) *real* coefficients. However, geometric algebra, which was the only rigorous method of operating with real numbers before the XIXth century, is very difficult. It imposes tight limitations which are not natural from the point of view of algebra; for instance, a great skill in the handling of proportions is required to go beyond degree three.

To progress in the theory of equations, it was necessary to think more about formalism and less about the nature of coefficients. Although later Greek mathematicians such as Hero and Diophantus took some steps in that direction, the really new advances were brought by other civilizations. Hindus, and Arabs later, developed techniques of calculation with irrational numbers, which they treated unconcernedly, without worrying about their irrationality. For instance, they were familiar with formulas like

$$\sqrt{a} + \sqrt{b} = \sqrt{a + b + 2\sqrt{ab}}$$

which they obtained from $(u + v)^2 = u^2 + v^2 + 2uv$ by extracting roots of both sides and replacing u and v by $\sqrt{a}$ and $\sqrt{b}$ respectively. Their notion of mathematical rigor was rather more relaxed than that of Greek mathematicians, but they paved the way to a more formal (or indeed algebraic) approach to quadratic equations (see Kline [38, ch. 9, §2]).

1.4 Arabic algebra

The next landmark in the theory of equations is the book "Al-jabr w' al muqabala" (c. 830 A.D.), due to Mohammed ibn Musa al-Khowarizmi.

The title refers to two basic operations on equations. The first is *al-jabr* (from which the word "algebra" is derived) which means "the restoration" or "making whole." In this context, it stands for the restoration of equality in an equation by adding to one side a negative term which is removed from the other. For instance, the equation

$$x^2 = 40x - 4x^2$$

is converted into

$$5x^2 = 40x$$

by *al-jabr* [36, p. 105]. The second basic operation *al muqabala* means "the opposition" or "balancing"; it is a simplification procedure by which like terms are removed from both sides of an equation. For instance, *al muqabala* changes

$$50 + x^2 = 29 + 10x$$

into

$$21 + x^2 = 10x$$

[36, p. 109].

In this work, al-Khowarizmi initiates what might be called the classical period in the theory of equations, by reducing the old methods for solving equations to a few standardized procedures. For instance, in problems involving several unknowns, the systematically sets up an equation for one of the unknowns, and he solves the three types of quadratic equations

$$X^2 + aX = b, \qquad X^2 + b = aX, \qquad X^2 = aX + b$$

by completion of the square, giving the two (positive) solutions for the type $X^2 + b = aX$.

Al-Khowarizmi first explains the procedure, as a Babylonian would have done:

> The following is an example of squares and roots equal to numbers: a square and 10 roots are equal to 39 units. The question therefore in this type of equation is about as follows: what is the square which combined with ten of its roots will give a sum total of 39? The manner of solving this type of equation is to take one-half of the roots just mentioned. Now the roots in the problem before us are 10. Therefore take 5, which multiplied by itself gives 25, an amount which you add to 39, giving 64.
>
> Having taken then the square root of this which is 8, subtract from it the half of the roots, 5, leaving 3. The number three therefore represents one root of this square, which itself, of course, is 9. Nine therefore gives that square. [36, pp. 71–73]

However, after explaining the procedure for solving each of the six types $mX^2 = aX$, $mX^2 = b$, $aX = b$, $mX^2 + aX = b$, $mX^2 + b = aX$ and $mX^2 = aX + b$, he adds:

> We have said enough, says Al-Khowarizmi, so far as numbers are concerned, about the six types of equations. Now, however, it is necessary that we should demonstrate geometrically the truth of the same problems which we have explained in numbers. [36, p. 77]

He then gives geometric justifications for his rules for the last three types, using completion of the square as in the following example for $x^2 + 10x = 39$:

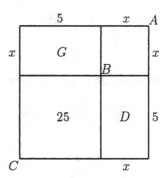

Let x^2 be the square AB. Then $10x$ is divided into two rectangles G and D, each being $5x$ and being applied to the side x of the square AB. By hypothesis, the value of the shape thus produced is $x^2 + 10x = 39$. There remains an empty corner of value $5^2 = 25$ to complete the square AC. Therefore, if 25 is added, the square $(x + 5)^2$ is completed, and its value is $39 + 25 = 64$. It then follows that $(x + 5)^2 = 64$, whence $x + 5 = 8$ and $x = 3$ (see [36, p. 81]).

It should be observed that the geometry behind this construction is much more elementary than in Euclid's Elements, since it is not logically connected by deductive reasoning to a small number of axioms, but relies instead on intuitive geometric evidence. From the point of view of algebra, on the other hand, al-Khowarizmi's work is incommensurately ahead of Euclid's, and it set the stage for the later development of algebra as an independent discipline.

Another remarkable achievement of the Arabs in the theory of equations is a geometric solution of cubic equations due to Omar Khayyam (c. 1079). For instance, the solution of $x^3 + b^2x = b^2c$ is obtained by intersecting the parabola $x^2 = by$ with the circle of diameter c which is tangent to the axis of the parabola at its vertex.

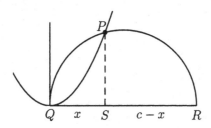

To prove that the segment x as shown on the figure above satisfies $x^3 + b^2x = b^2c$, we start from the relation $x^2 = b \cdot PS$, which yields

$$\frac{b}{x} = \frac{x}{PS}.$$

On the other hand, since the triangles QSP and PSR are similar, we have

$$\frac{x}{PS} = \frac{PS}{c - x},$$

whence

$$\frac{b}{x} = \frac{PS}{c - x}.$$

As $PS = b^{-1}x^2$, this equation yields

$$\frac{b}{x} = \frac{x^2}{b(c - x)},$$

whence $x^3 = b^2c - b^2x$, as required.

Omar Khayyam also gives geometric solutions for the other types of cubic equations by intersection of conics, but these brilliant solutions are of little use for practical purposes, and an algebraic solution was still longed for.

In 1494, Luca Pacioli closes his book "Summa de Arithmetica, Geometria, Proportione e Proportionalita" (one of the first printed books in mathematics) with the remark that the solutions of $x^3 + mx = n$ and $x^3 + n = mx$ (in modern notations) are as impossible as the quadrature of the circle. (See Kline [38, p. 237], Cardano [11, p. 8].) However, unexpected developments were soon to take place.

Chapter 2

Cubic Equations

2.1 Priority disputes on the solution of cubic equations

The algebraic solution of $X^3 + mX = n$ was first obtained around 1515 by Scipione del Ferro, professor of mathematics in Bologna. Not much is known about him nor about his solution as, for some reason, he decided not to publicize his result. After his death in 1526, his method passed to some of his pupils.

The second discovery of the solution is much better known, through the accounts of its author himself, Niccolo Fontana (c. 1500–1557), from Brescia, nicknamed "Tartaglia" ("Stammerer") (see Hankel [28, pp. 360 ff]). In 1535, Tartaglia, who had dealt with some very particular cases of cubic equations, was challenged to a public problem-solving contest by Antonio Maria Fior, a former pupil of Scipione del Ferro. When he heard that Fior had received the solution of cubic equations from his master, Tartaglia threw all his energy and his skill into the struggle. He succeeded in finding the solution just in time to inflict upon Fior a humiliating defeat.

The news that Tartaglia had found the solution of cubic equations reached Girolamo Cardano (1501–1576), a very versatile scientist, who wrote a number of books on a wide variety of subjects, including medicine, astrology, astronomy, philosophy and mathematics. Cardano then asked Tartaglia to give him his solution, so that he could include it in a treatise on arithmetic, but Tartaglia flatly refused, since he was himself planning to write a book on this topic. It turns out that Tartaglia later changed his mind, at least partially, since in 1539 he handed on to Cardano the solution of $X^3 + mX = n$, $X^3 = mX + n$ and a very brief indication on $X^3 + n = mX$ in verses* (see Hankel [28, pp. 364–365]):

*As pointed out by Boorstin (cited by Weeks [66, p. lx]), verses were a useful memorization aid at a

13

> *Quando che'l cubo con le cose appresso*
> *Se agguaglia a qualche numero discreto:*
> *Trovan dui altri, differenti in esso.*

> *Dapoi terrai, questo per consueto,*
> *Che'l lor produtto, sempre sia eguale*
> *Al terzo cubo delle cose neto;*

> *El residuo poi suo generale,*
> *Delli lor lati cubi, bene sottratti*
> *Varra la tua cosa principale.*

> . . .

This excerpt gives the formula for $X^3 + mX = n$. The equation is indicated in the first two verses: the cube and the things equal to a number. *Cosa* (= thing) is the word for the unknown. To express the fact that the unknown is multiplied by a coefficient, Tartaglia simply uses the plural form *le cose*. He then gives the following procedure: find two numbers which differ by the given number and such that their product is equal to the cube of the third of the number of things. Then the difference between the cube roots of these numbers is the unknown.

With modern notations, we would write that, to find the solution of

$$X^3 + mX = n,$$

we only need to find t, u such that

$$t - u = n \qquad \text{and} \qquad tu = \left(\frac{m}{3}\right)^3;$$

then

$$X = \sqrt[3]{t} - \sqrt[3]{u}.$$

The values of t and u are easily found (see the system (1.1), p. 3)

$$t = \sqrt{\left(\frac{n}{2}\right)^2 + \left(\frac{m}{3}\right)^3} + \frac{n}{2}$$

$$u = \sqrt{\left(\frac{n}{2}\right)^2 + \left(\frac{m}{3}\right)^3} - \frac{n}{2}.$$

time when paper was expensive.

Therefore, a solution of $X^3 + mX = n$ is given by the following formula:

$$X = \sqrt[3]{\sqrt{\left(\frac{n}{2}\right)^2 + \left(\frac{m}{3}\right)^3} + \frac{n}{2}} - \sqrt[3]{\sqrt{\left(\frac{n}{2}\right)^2 + \left(\frac{m}{3}\right)^3} - \frac{n}{2}}.$$

However, the poem does not provide any justification for this formula. Of course, it "suffices" to check that the value of X given above satisfies the equation $X^3 + mX = n$, but this was far from obvious to a sixteenth century mathematician. The major difficulty was to figure out that

$$(a - b)^3 = a^3 - 3a^2b + 3ab^2 - b^3,$$

a formula which could be properly proved only by dissection of a cube in three-dimensional space.

Having received Tartaglia's poem, Cardano set to work; he not only found justifications for the formulas but he also solved all the other types of cubics. He then published his results, giving due credit to Tartaglia and to del Ferro, in the epoch-making book "Ars Magna, sive de regulis algebraicis" (The Great Art, or the Rules of Algebra [11]). A bitter quarrel then erupted between Tartaglia and Cardano, the former claiming that Cardano had solemnly sworn never to publish Tartaglia's solution, while the latter countered that there had never been any question of secrecy.

2.2 Cardano's formula

Although Cardano lists 13 types of cubic equations and gives a detailed solution for each of them, we shall use modern notations in this section, and explain Cardano's method for the general cubic equation

$$X^3 + aX^2 + bX + c = 0.$$

First, the change of variable $Y = X + \frac{a}{3}$ converts the equation into one which lacks the second degree term:

$$Y^3 + pY + q = 0 \tag{2.1}$$

where

$$p = b - \frac{a^2}{3} \quad \text{and} \quad q = c - \frac{a}{3}b + 2\left(\frac{a}{3}\right)^3. \tag{2.2}$$

If $Y = \sqrt[3]{t} + \sqrt[3]{u}$, then

$$Y^3 = t + u + 3\sqrt[3]{tu}(\sqrt[3]{t} + \sqrt[3]{u})$$

and equation (2.1) becomes

$$(t + u + q) + (3\sqrt[3]{tu} + p)(\sqrt[3]{t} + \sqrt[3]{u}) = 0.$$

This equation clearly holds if the rational part $t + u + q$ and the irrational part $(\sqrt[3]{t} + \sqrt[3]{u})(3\sqrt[3]{tu} + p)$ both vanish or, in other words, if

$$\begin{cases} t + u = -q \\ tu = -\left(\frac{p}{3}\right)^3. \end{cases}$$

This system has the solution

$$t, u = -\frac{q}{2} \pm \sqrt{\left(\frac{p}{3}\right)^3 + \left(\frac{q}{2}\right)^2}$$

(see (1.4), p. 4); hence a solution for equation (2.1) is

$$Y = \sqrt[3]{-\frac{q}{2} + \sqrt{\left(\frac{p}{3}\right)^3 + \left(\frac{q}{2}\right)^2}} + \sqrt[3]{-\frac{q}{2} - \sqrt{\left(\frac{p}{3}\right)^3 + \left(\frac{q}{2}\right)^2}} \qquad (2.3)$$

and a solution for the initial equation $X^3 + aX^2 + bX + c = 0$ easily follows by substituting for p and q the expressions given by (2.2). Equation (2.3) is known as *Cardano's formula* for the solution of the cubic (2.1).

2.3   Developments arising from Cardano's formula

The solution of cubic equations was a remarkable achievement, but Cardano's formula is far less convenient than the corresponding formula for quadratic equations since it has some drawbacks which undoubtedly baffled XVI-th century mathematicians (to begin with, its discoverers).

(a) First, when some solution is expected, it is not always yielded by Cardano's formula. This could have struck Cardano when he was devising examples for illustrating his rules, such as

$$X^3 + 16 = 12X$$

(see Cardano [11, p. 12]) which is constructed to give 2 as an answer. Cardano's formula yields

$$X = \sqrt[3]{-8} + \sqrt[3]{-8} = -4.$$

Why does it yield -4 and not 2?

It is likely that the above observation had first prompted Cardano to investigate a question much more interesting to him: How many solutions does a cubic equation have? He was thus led to observe that cubic equations may have three solutions (including the negative ones, which Cardano terms "false" or "fictitious," but not the imaginary ones) and to investigate the relations between these solutions (see Cardano [11, Chapter I]).

(b) Next, when there is a rational solution, its expression according to Cardano's formula can be rather awkward. For instance, it is easily seen that 1 is solution of

$$X^3 + X = 2,$$

but Cardano's formula yields

$$X = \sqrt[3]{1 + \frac{2}{3}\sqrt{\frac{7}{3}}} + \sqrt[3]{1 - \frac{2}{3}\sqrt{\frac{7}{3}}}.$$

Now, the equation above has only one real root, since the function $f(X) = X^3 + X$ is monotonically increasing (as it is the sum of two monotonically increasing functions) and, therefore, takes the value 2 only once. We are thus compelled to conclude

$$1 = \sqrt[3]{1 + \frac{2}{3}\sqrt{\frac{7}{3}}} + \sqrt[3]{1 - \frac{2}{3}\sqrt{\frac{7}{3}}}, \tag{2.4}$$

a rather surprising result.

Already in 1540, Tartaglia tried to simplify the irrational expressions arising in his solution of cubic equations (see Hankel [28, p. 373]). More precisely, he tried to determine under which condition an irrational expression like $\sqrt[3]{a + \sqrt{b}}$ could be simplified to $u + \sqrt{v}$. This problem can be solved as follows (in modern notations): starting with

$$\sqrt[3]{a + \sqrt{b}} = u + \sqrt{v} \tag{2.5}$$

and taking the cube of both sides, we obtain

$$a + \sqrt{b} = u^3 + 3uv + (3u^2 + v)\sqrt{v},$$

whence, equating separately the rational and the irrational parts (this is licit if a, b, u and v are rational numbers),

$$\begin{cases} a &= u^3 + 3uv \\ \sqrt{b} &= (3u^2 + v)\sqrt{v}. \end{cases} \tag{2.6}$$

Subtracting the second equation from the first, we then obtain

$$a - \sqrt{b} = (u - \sqrt{v})^3$$

whence

$$\sqrt[3]{a - \sqrt{b}} = u - \sqrt{v}. \tag{2.7}$$

Multiplying (2.5) and (2.7), we obtain

$$\sqrt[3]{a^2 - b} = u^2 - v \tag{2.8}$$

which can be used to eliminate v from the first equation of system (2.6). We thus get

$$a = 4u^3 - 3(\sqrt[3]{a^2 - b})u.$$

Therefore, if a and b are rational numbers such that $\sqrt[3]{a^2 - b}$ is rational and if the equation

$$4u^3 - 3(\sqrt[3]{a^2 - b})u = a \tag{2.9}$$

has a rational solution u, then

$$\sqrt[3]{a + \sqrt{b}} = u + \sqrt{v} \qquad \text{and} \qquad \sqrt[3]{a - \sqrt{b}} = u - \sqrt{v},$$

where v is given by equation (2.8)

$$v = u^2 - \sqrt[3]{a^2 - b}.$$

This effectively provides a simplification in the irrational expressions

$$\sqrt[3]{-\frac{q}{2} \pm \sqrt{\left(\frac{p}{3}\right)^3 + \left(\frac{q}{2}\right)^2}}$$

which appear in Cardano's formula for the solution of $X^3 + pX + q = 0$, but this simplification is useless as far as the solution of cubic equations is concerned.

Indeed, if $a = -\frac{q}{2}$ and $b = \left(\frac{p}{3}\right)^3 + \left(\frac{q}{2}\right)^2$, one has to find a rational solution of equation (2.9):

$$4u^3 + pu = -\frac{q}{2},$$

and this exactly amounts to finding a rational solution of the initial equation $X^3 + pX + q = 0$, since these equations are related by the change of variable $X = 2u$. However, this process can be used to show, for instance, that

$$\sqrt[3]{1 + \frac{2}{3}\sqrt{\frac{7}{3}}} = \frac{1}{2} + \frac{1}{2}\sqrt{\frac{7}{3}} \quad \text{and} \quad \sqrt[3]{1 - \frac{2}{3}\sqrt{\frac{7}{3}}} = \frac{1}{2} - \frac{1}{2}\sqrt{\frac{7}{3}}$$

from which formula (2.4) follows.

(c) The most serious drawback of Cardano's formula appears when one tries to solve an equation like

$$X^3 = 15X + 4.$$

It is easily seen that $X = 4$ is a solution, but Cardano's formula yields a very embarrassing expression:

$$X = \sqrt[3]{2 + \sqrt{-121}} + \sqrt[3]{2 - \sqrt{-121}},$$

in which square roots of negative numbers are extracted. The case where $\left(\frac{p}{3}\right)^3 + \left(\frac{q}{2}\right)^2 < 0$ is known as the "casus irreducibilis" of cubic equations. For a long time, the validity of Cardano's formula in this case had been a matter of debate, but the discussion of this case had a very important by-product: it prompted the use of complex numbers.

Complex numbers had been, up to then, brushed aside as absurd, nonsensical expressions. A remarkably explicit example of this attitude appears in the following excerpt from chapter 37 of the Ars Magna [11, p. 219]:

> If it should be said, Divide 10 into two parts the product of which
> is 30 or 40, it is clear that this case is impossible. Nevertheless,
> we will work thus: ...

Cardano then applies the usual procedure with the given data, which amounts to solving $X^2 - 10X + 40 = 0$, and comes up with the solution: these parts are $5 + \sqrt{-15}$ and $5 - \sqrt{-15}$. He then justifies his result:

Putting aside the mental tortures involved,[†] multiply $5 + \sqrt{-15}$ by $5 - \sqrt{-15}$, making $25 - (-15)$ which is $+15$. Hence this product is 40. [...] So progresses arithmetic subtlety the end of which, as is said, is as refined as it is useless.

However, with the "casus irreducibilis" of cubic equations, complex numbers were imposed upon mathematicians. The operations on these numbers are clearly taught, in a nearly modern way, by Rafaele Bombelli (c. 1526–1573), in his influential treatise: "Algebra" (1572). In this book, Bombelli boldly applies to cube roots of complex numbers the same simplification procedure as in (b) above, and he obtains, for instance

$$\sqrt[3]{2 + \sqrt{-121}} = 2 + \sqrt{-1} \qquad \text{and} \qquad \sqrt[3]{2 - \sqrt{-121}} = 2 - \sqrt{-1},$$

from which it follows that Cardano's formula gives indeed 4 for a solution of $X^3 = 15X + 4$.

Complex numbers thus appeared, not to solve quadratic equations which lack solutions (and do not need any), but to explain why Cardano's formula, efficient as it may seen, fails in certain cases to provide expected solutions to cubic equations.

[†]In the original text: "dismissis incruciationibus." Perhaps Cardano played on words here, since another translation for this passage is: "the cross-multiples having canceled out," referring to the fact that in the product $(5 + \sqrt{-15})(5 - \sqrt{-15})$, the terms $5\sqrt{-15}$ and $-5\sqrt{-15}$ cancel out.

Chapter 3

Quartic Equations

3.1 The unnaturalness of quartic equations

The solution of quartic equations was found soon after that of cubic equations. It is due to Ludovico Ferrari (1522–1565), a pupil of Cardano, and it first appeared in the "Ars Magna."

Ferrari's method is very ingenious, relying mainly on transformation of equations, but it aroused less interest than the solution of cubic equations. This is clearly shown by its place in the "Ars Magna": while Cardano spends thirteen chapters to discuss the various cases of cubic equations, Ferrari's method is briefly sketched in the penultimate chapter.

The reason for this relative disregard may be found in the introduction of the "Ars Magna" [11, p. 9]:

> Although a long series of rules might be added and a long discourse given about them, we conclude our detailed consideration with the cubic, others being merely mentioned, even if generally, in passing. For as *positio* [the first power] refers to a line, *quadratum* [the square] to a surface, and *cubum* [the cube] to a solid body, it would be very foolish for us to go beyond this point. Nature does not permit it.

This passage shows the equivocal status of algebra in the sixteenth century. Its logical foundations were still geometric, as in the classical Greek period; in this framework, each quantity has a dimension and only quantities of the same dimension can be added or equated. For instance, an equation like $x^2 + b = ax$ makes sense only if x and a are line segments and b is an area, and equations of degree higher than three don't make any sense at all.[*]

However, from an arithmetical point of view, quantities are regarded as dimensionless numbers, which can be raised to any power and equated unconcernedly. This way of thought was clearly prevalent among Babylonians, since the very statement of the problem: "I have subtracted from the area the side of my square: 14.30" is utter nonsense from a geometric point of view. The Arabic algebra also stresses arithmetic, although al-Khowarizmi provides geometric proofs of his rules (see §1.4).

In the "Ars Magna," both the geometric and the arithmetic approaches to equations are present. On one hand, Cardano tries to base his results on Euclid's "Elements," and on the other hand, he gives the solution of equations of degree 4. He also solves some equations of higher degree, such as $X^9 + 3X^6 + 10 = 15X^3$ [11, p. 159], in spite of his initial statement that it would be "foolish" to go beyond degree 3. However, the arithmetic approach, which would eventually predominate, still suffered from its lack of a logical base until the early seventeenth century (see §4.1).

3.2 Ferrari's method

In this section, we use modern notations to discuss Ferrari's solution of quartic equations. Let

$$X^4 + aX^3 + bX^2 + cX + d = 0$$

be an arbitrary quartic equation. By the change of variable $Y = X + \frac{a}{4}$ the cubic term cancels out, and the equation becomes

$$Y^4 + pY^2 + qY + r = 0 \qquad (3.1)$$

---

[*] A way out of this difficulty was eventually found by Descartes. In "La Geometrie" [16, p. 5], published in 1637, he introduces the following convention: if a unit line segment e is chosen, then the square x^2 of a line segment x is the side of the rectangle constructed on e which has the same area as the square with side x. Thus, x^2 is a line segment, and arbitrary powers of x can be interpreted as line segments in a similar way.

with

$$p = b - 6\left(\frac{a}{4}\right)^2$$

$$q = c - \frac{a}{2}b + \left(\frac{a}{2}\right)^3 \qquad (3.2)$$

$$r = d - \frac{a}{4}c + \left(\frac{a}{4}\right)^2 b - 3\left(\frac{a}{4}\right)^4.$$

Moving the linear terms to the right-hand side and completing the square on the left-hand side, we obtain

$$\left(Y^2 + \frac{p}{2}\right)^2 = -qY - r + \left(\frac{p}{2}\right)^2.$$

If we add a quantity u to the expression squared in the left-hand side, we get

$$\left(Y^2 + \frac{p}{2} + u\right)^2 = -qY - r + \left(\frac{p}{2}\right)^2 + 2uY^2 + pu + u^2. \qquad (3.3)$$

The idea is to determine u in such a way that the right-hand side also becomes a square. Looking at the terms in Y^2 and in Y, it is easily seen that if the right-hand side is a square, then it is the square of $\sqrt{2u}Y - \frac{q}{2\sqrt{2u}}$; therefore, we should have

$$-qY - r + \left(\frac{p}{2}\right)^2 + 2uY^2 + pu + u^2 = \left(\sqrt{2u}Y - \frac{q}{2\sqrt{2u}}\right)^2 \qquad (3.4)$$

and, equating the independent terms, we see that this equation holds if and only if

$$-r + \left(\frac{p}{2}\right)^2 + pu + u^2 = \frac{q^2}{8u}$$

or equivalently, after clearing the denominator and rearranging terms,

$$8u^3 + 8pu^2 + (2p^2 - 8r)u - q^2 = 0. \qquad (3.5)$$

Therefore, by solving this cubic equation, we can find a quantity u for which equation (3.4) holds. Returning to equation (3.3), we then have

$$\left(Y^2 + \frac{p}{2} + u\right)^2 = \left(\sqrt{2u}Y - \frac{q}{2\sqrt{2u}}\right)^2,$$

whence

$$Y^2 + \frac{p}{2} + u = \pm\left(\sqrt{2u}Y - \frac{q}{2\sqrt{2u}}\right).$$

The values of Y are then obtained by solving the two quadratic equations above (one corresponding to the sign $+$ for the right-hand side, the other to the sign $-$).

To complete the discussion, it remains to consider the case where $u = 0$ is a root of equation (3.5), since the calculations above implicitly assume $u \neq 0$. But this case occurs only if $q = 0$ and then the initial equation (3.1) is

$$Y^4 + pY^2 + r = 0.$$

This equation is easily solved, since it is a quadratic equation in Y^2.

In summary, the solutions of

$$X^4 + aX^3 + bX^2 + cX + d = 0$$

are obtained as follows: let p, q and r be defined as before (see (3.2)) and let u be a solution of (3.5). If $q \neq 0$, the solutions of the initial quartic equation are

$$X = \varepsilon \sqrt{\frac{u}{2}} + \varepsilon' \sqrt{-\frac{u}{2} - \frac{p}{2} - \frac{\varepsilon q}{2\sqrt{2u}}} - \frac{a}{4}$$

where ε and ε' can be independently $+1$ or -1. If $q = 0$, the solutions are

$$X = \varepsilon \sqrt{-\frac{p}{2} + \varepsilon' \sqrt{\left(\frac{p}{2}\right)^2 - r}} - \frac{a}{4}$$

where ε and ε' can be independently $+1$ or -1.

Equation (3.5), on which the solution of the quartic equation depends, is called the *resolvent cubic* equation (relative to the given quartic equation). Depending on the way equations are set up, one may come up with other resolvent cubic equations. For instance, from equation (3.1) one could pass to

$$(Y^2 + v)^2 = (-pY^2 - qY - r) + 2vY^2 + v^2$$

where v is an arbitrary quantity (which plays the same rôle as $\frac{p}{2} + u$ in the preceding discussion). The condition on v for the right-hand side to be a perfect square is then

$$8v^3 - 4pv^2 - 8rv + 4pr - q^2 = 0. \tag{3.6}$$

After having determined v such that this condition holds, one finishes as before.

This second method is clearly equivalent to the previous one, by the change of variable $v = \frac{p}{2} + u$. Therefore, equation (3.6), which is obtained from (3.5) by this change of variable, is also entitled to be called the resolvent cubic.

Chapter 4

The Creation of Polynomials

4.1 The rise of symbolic algebra

In comparison to the rapid development of the theory of equations around the middle of the sixteenth century, progress during the next two centuries was rather slow. The solution of cubic and quartic equations was a very important breakthrough, and it took some time before the circle of new ideas arising from these solutions was fully explored and understood, and new advances were possible.

First of all, it was necessary to devise appropriate notations for handling equations. In the solution of cubic and quartic equations, Cardano was straining to the utmost the capabilities of the algebraic system available to him. Indeed, his notations were rudimentary: the only symbolism he uses consists in abbreviations such as p : for "plus," m : for "minus" and $\Re$ for "radix" (= root).

For instance, the equation $X^2 + 2X = 48$ is written as

1. *quad. p* : 2 *pos. aeq.* 48

(*quad.* is for "quadratum," *pos.* for "positiones" and *aeq.* for "aequatur"), and

$$(5 + \sqrt{-15})(5 - \sqrt{-15}) = 25 - (-15) = 40$$

is written (see Cajori [10, §140])

$$
\begin{array}{l}
5p : \Re\, m : 15 \\
5m : \Re\, m : 15 \\
\hline
25m : m : 15 \quad \tilde{q}d\ est\ 40.
\end{array}
$$

Using this embryonic notation, transformation of equations was clearly a tour de force, and a more efficient notation had to develop in order to enlighten this new part of algebra.

This development was rather erratic. Advances made by some authors were not immediately taken up by others, and the process of normalization of notations took a long time. For example, the symbols $+$ and $-$ were already used in Germany since the end of the fifteenth century (Cajori [10, §201]), but they were not widely accepted before the early seventeenth century, and the sign $=$ for equality, first proposed by R. Recorde in 1557, had to struggle with Descartes' symbol ∞ for nearly two centuries (Cajori [10, §267]).

These are relatively minor points, since it may be assumed that p :, m : and *aeq.* were as convenient to Cardano as $+$, $-$ and $=$ are to us. There is one point, however, where a new notation was vital. In effect, it helped create a new mathematical object: polynomials. There is indeed a significant step from

> 1. *quad. p* : 2 *pos. aeq.* 48

which is the mere statement of a problem, to the calculation with polynomials like $X^2 + 2X - 48$, and this step was considerably facilitated by a suitable notation.

Significant as it may be, the evolution from equations to polynomials is rather subtle, and leading mathematicians of this period rarely took the time to clarify their views on the subject; the rise of the concept of polynomial was most often overshadowed by its application to the theory of equations, and it can only be gathered from indirect indications.

Two milestones in this evolution are "L'Arithmetique" (1585) of Simon Stevin (1548–1620) and "In Artem Analyticem Isagoge" (= Introduction to the Analytic Art) (1591) of François Viète (1540–1603).

4.1.1   *L'Arithmetique*

This book combines notational advances made by Bombelli and earlier authors (see Cajori [10, §296]) with theoretical advances made by Pedro Nunes (1502–1578) (see Bosmans [5, p. 165]) to present a comprehensive treatment of polynomials. Stevin's notation for polynomials, which he terms "multinomials" [55, p. 521] or "integral algebraic numbers" [55, p. 518] (see also pp. 570 ff) has a surprising touch of modernity: the indeterminate is denoted by ①, its square by ②, its cube by ③, etc., and the independent term is indicated by ⓪ (sometimes omitted), so that a "multinomial" appears as an expression like

$$3③ + 5② - 4① + 6⓪ \qquad (\text{or } 3③ + 5② - 4① + 6).$$

Such an expression could be regarded (from a modern point of view) as a finite sequence of real numbers, or, better, as a sequence of real numbers which are all zero except for a finite number of them, or as a function from $\mathbb{N}$ to $\mathbb{R}$ with finite support (compare §5.1).

This exponential notation (which was not unprecedented) probably helped to abolish the psychological barrier of the third degree (see §3.1), by placing all the powers of the unknown on an equal footing. It is however rather unfortunate for equations with several unknowns.

Most important is Stevin's observation that the operations on "integral algebraic numbers" share many features with those on "integral arithmetic numbers" (= integers). In particular, he shows [55, Probleme 53, p. 577] that Euclid's algorithm for determining the greatest common divisor of two integers applies nearly without change to find the greatest common divisor of two polynomials (see §5.2, and particularly p. 44).

Although the concept of polynomial is quite clear, the way equations are set up is rather awkward in "L'Arithmetique," since equations are replaced by proportions and the solution of equations is called by Stevin "the rule of three of quantities." In modern notation, the idea is to replace the solution of an equation like

$$X^2 - aX - b = 0$$

by the following problem: find the fourth proportional u in

$$\frac{X^2}{aX + b} = \frac{X}{u}$$

or, more generally, find $P(u)$ in

$$\frac{X^2}{aX + b} = \frac{P(X)}{P(u)},$$

where $P(X)$ is an arbitrary polynomial in X, see [55, p. 592]. (Of course, the solutions which are equal to zero must then be rejected.)

In Stevin's words [55, p. 595]:

> Given three terms, of which the first (2), the second (1) (0), the third an arbitrary algebraic number: To find their fourth proportional term.

This fancy approach to equations may have been prompted by Stevin's methodical treatment of polynomials: an equality like

$$X^2 = aX + b$$

would mean that the *polynomials* X^2 and $aX + b$ are equal; but polynomials are
equal if and only if the coefficients of similar powers of the unknown are the same
in both polynomials, and this is clearly not the case here, since X^2 appears on
the left-hand side but not on the right-hand side. Stevin's own explanation [55,
pp. 581–582], while not quite convincing, at least shows that he was fully aware
of this notational difficulty:

> The reason why we call rule of three, or invention of the fourth
> proportional of quantities, that which is commonly called equa-
> tion of quantities: [...] Because this word "equation" let the
> beginners think that it was some singular matter, which how-
> ever is common in the usual arithmetic, since we seek to three
> given terms a fourth proportional. As that, which is called equa-
> tion, does not consist in the equality of absolute quantities, but
> in equality of their values, so this proportion is concerned with
> the value of quantities, as the same is usual in everyday life.

This approach met with little success. Even Albert Girard, the first editor of
Stevin's works, did not follow Stevin's set up in his own work, and it was soon
abandoned.

Stevin's formal treatment of polynomials is rather isolated too; in later works,
polynomials were most often considered as functions, although formal operations
like Euclid's algorithm were performed. For instance, here is the definition of an
equation according to René Descartes (1596–1650) [16, p. 156]:

> An equation consists of several terms, some known and some
> unknown, some of which are together equal to the rest; or rather,
> all of which taken together are equal to nothing; for this is often
> the best form to consider.

A polynomial then appears as "the sum of an equation" [16, p. 159]. On the
whole, the idea of polynomial in the seventeenth century is not very different
from the modern notion, and the need for a more formal definition was not felt
for a long time, but one can get some feeling of the difference between Descartes'
view and ours from the following excerpt of "La Geometrie" (1637) [16, p. 159]:
(the emphasis is mine)

> Multiplying together the two *equations* $x - 2 = 0$ and $x - 3 = 0$,
> we have $x^2 - 5x + 6 = 0$, or $x^2 = 5x - 6$. This is an equation
> in which x has the value 2 and *at the same time* the value 3.

4.1.2 *In Artem Analyticem Isagoge*

A major advance in notation with far-reaching consequences was François Viète's idea, put forward in his "Introduction to the Analytic Art" (1591), of designating by letters all the quantities, known or unknown, occurring in a problem. Although letters were occasionally used for unknowns as early as the third century A.D. (by Diophantus of Alexandria, see Cajori [10, §101]), the use of letters for known quantities was very new. It also proved to be very useful, since for the first time it was possible to replace various numerical examples by a single "generic" example, from which all the others could be deduced by assigning values to the letters. However, it should be observed that this progress did not reach its full extent in Viète's works, since Viète completely disregards negative numbers; therefore, his letters always stand for *positive* numbers only.

This slight limitation notwithstanding, the idea had another important consequence: by using symbols as his primary means of expression and showing how to calculate with those symbols, Viète initiated a completely formal treatment of algebraic expressions, which he called *logistice speciosa* [65, p. 17] (as opposed to the *logistice numerosa*, which deals with numbers). This "symbolic logistic" gave some substance, some legitimacy to algebraic calculations, which allowed Viète to free himself from the geometric diagrams used so far as justifications.

However, Viète's calculations are somewhat hindered by his insistence that each coefficient in an equation be endowed with a dimension, in such a way that all the terms have the same dimension: the "prime and perpetual law of equations" is that "homogeneous terms must be compared with homogeneous terms" [65, p. 15]. Moreover, Viète's notation is not as advanced as it could be, since he does not use numerical exponents. For instance, instead of

Let $A^3 + 3BA = 2Z$,

Viète writes (see Cajori [10, §177])

Proponatur A cubus $+$ B plano 3 in A aequari Z solido 2,

insisting that B and Z have degree 2 and 3 respectively.

These minor flaws were soon corrected. In "La Geometrie" (1637) [16], René Descartes shaped the notation that is still in use today (except for his above-mentioned sign ∞ for equality). Thus, in less than one century, algebraic notation had dramatically improved, reaching the same level of generality and the same versatility as ours. These notational advances fostered a deeper understanding of the nature of equations, and the theory of equations was soon advanced in some important points, such as the number of roots and the relations between roots and coefficients of an equation.

4.2 Relations between roots and coefficients

Cardano's observations on the number of roots of cubic and quartic equations (see §2.3) were substantially generalized during the next century. Progress in plane trigonometry brought rather unexpected insights into this question.

In 1593, at the end of the preface of his book "Ideae Mathematicae" [48] (see also Goldstine [27, §1.6]), Adriaan van Roomen* (1561–1615) issued the following challenge to "all the mathematicians throughout the whole world":

Find the solution of the equation

$$45X - 3795X^3 + 95634X^5 - 1138500X^7 + 7811375X^9 - 34512075X^{11}$$
$$+ 105306075X^{13} - 232676280X^{15} + 384942375X^{17} - 488494125X^{19}$$
$$+ 483841800X^{21} - 378658800X^{23} + 236030652X^{25} - 117679100X^{27}$$
$$+ 46955700X^{29} - 14945040X^{31} + 3764565X^{33} - 740259X^{35}$$
$$+ 111150X^{37} - 12300X^{39} + 945X^{41} - 45X^{43} + X^{45} = A.$$

He gave the following examples, the second of which is erroneous:

(a) if $A = \sqrt{2 + \sqrt{2 + \sqrt{2 + \sqrt{2}}}}$, then $X = \sqrt{2 - \sqrt{2 + \sqrt{2 + \sqrt{2 + \sqrt{3}}}}}$;

(b) if $A = \sqrt{2 + \sqrt{2 - \sqrt{2 - \sqrt{2 - \sqrt{2 - \sqrt{2}}}}}}$

$$\text{[it should be } A = \sqrt{2 - \sqrt{2 - \sqrt{2 + \sqrt{2 + \sqrt{2}}}}}],$$

then $X = \sqrt{2 - \sqrt{2 + \sqrt{2 + \sqrt{2 + \sqrt{2 + \sqrt{2}}}}}}$

$$\text{[it should be } X = \sqrt{2 - \sqrt{2 + \sqrt{2 + \sqrt{2 + \sqrt{2 + \sqrt{3}}}}}}];$$

*then professor at the University of Louvain.

PROBLEMA MATHEMATICVM
omnibus totius orbis Mathematicis ad conſtruendū propoſitum.

SI duorum terminorum prioris ad poſteriorem pro-
portio ſit, ut 1 (1) ad 45 (1) -- 3⁻95 (3) + 9, 5634
(5) -- 113, 8500 (7) + 781, 1375 (9) -- 3451, 2075 (11) +
1, 0530, 6075 (13) -- 2, 3267, 6280 (15) + 3, 8494, 2375
(17) -- 4, 8849, 4125 (19) + 4, 8384, 1800 (21) -- 3, 7865,
8800 (23) + 2, 3603, 0652 (25) -- 1, 1767, 9100 (27) + 4695,
5700 (29) -- 1494, 5040 (31) + 376, 4565 (33) -- 74, 0259
(35) + 11, 1150 (37) -- 1, 2300 (39) + 945 (41) -- 45 (43) +
1 (45), decurque terminus poſterior, invenire priorem.

Exemplum primum datum.

SIt terminus poſterior *r bin.* 2 + *r bin.* 2 + *r bin* 2 + *r* 2. quęritur terminus
prior. SOLVTIO. Dico terminū priorem eſſe *r bin.* 2 - *r bin.* 2 + *r bin,* 2 +
r bin. 2 + *r* 3.

Exemplum ſecundum datum.

Sit terminus poſterior *r bin.* 2 + *r bin.* 2 - *r bin.* 2 - *r bin.* 2 - *r bin.* 2 - *r* 2.
quæritur terminus prior. SOLVTIO. Terminus prior eſt *r bin.* 2 - *r bin.* 2 +
r bin. 2 + *r bin.* 2 + *r bin.* 2 + *r* 2.

Exemplum tertium datum.

Sit terminus poſterior *r bin.* 2 + *r* 2, quæritur terminus prior.
SOL. Terminus prior eſt *r bin.* 2 - *r quadrin.* 2 + $r\frac{3}{16}$ + $r\frac{15}{16}$ + *r bin.* $\frac{5-r\,5}{8}$ $\frac{5}{64}$
Si in numeris abſolutis ſolinomijs id proponere libuerit : Sit poſterior termi-
nus *r* 3 $\frac{4142,1356,2373,0950,4880,1688,7242,0969,8078,5696,7187,5375.}{10000,0000,0000,0000,0000,0000,0000,0000,0000,0000,0000.}$
Quæritur terminus prior. SOLVTIO. Terminus prior erit
r $\frac{27,4093,0490,8522,5243,1015,8831,2112,6838,8180.}{10000,0000,0000,0000,0000,0000,0000,0800,0000,0000.}$

EXEMPLVM QVÆSITVM.

SIt poſterior terminus *r trinomia* 1 $\frac{3}{4}$ -- *r* $\frac{1}{16}$ -- *r bin.* 1 $\frac{7}{8}$ -- *r* $\frac{45}{64}$
quæritur terminus prior. Hoc exemplum omnibus Mathematicis ad con-
ſtruendum ſit propoſitum. Non dubito quin *Ludolf van Cillen* ejus ſolu-
tionem, ſaltem in numeris ſolinomijs ſit inventurus.

M E-

(c) if $A = \sqrt{2 + \sqrt{2}}$

$$= \sqrt{3.4142135623730950488016887242096980785696718753755},$$

then

$$X = \sqrt{2 - \sqrt{2 + \sqrt{\tfrac{3}{16}} + \sqrt{\tfrac{15}{16}} + \sqrt{\tfrac{5}{8} - \sqrt{\tfrac{5}{64}}}}}$$

$$= \sqrt{0.00274093049085225243101588312112683881 80},$$

and he asked for a solution when

$$A = \sqrt{\tfrac{7}{4} - \sqrt{\tfrac{5}{16}} - \sqrt{\tfrac{15}{8}} - \sqrt{\tfrac{45}{64}}}.$$

Of course, this was not just any 45th degree equation; its coefficients had been very carefully chosen.

When this problem was submitted to Viète, he recognized that the left-hand side of the equation is the polynomial by which $2 \sin 45\alpha$ is expressed as a function of $2 \sin \alpha$ (see equation (4.2) below, p. 33). Therefore, it suffices to find an arc α such that $2 \sin 45\alpha = A$, and the solution of Van Roomen's equation is $X = 2 \sin \alpha$. In Van Roomen's examples:

(a) $A = 2 \sin \tfrac{15\pi}{2^5}$, and $X = 2 \sin \tfrac{\pi}{2^5 \cdot 3}$,

(b) A should be $2 \sin \tfrac{15\pi}{2^6}$, and $X = 2 \sin \tfrac{\pi}{2^6 \cdot 3}$,

(c) $A = 2 \sin \tfrac{3\pi}{2^3}$, and $X = 2 \sin \tfrac{\pi}{2^3 \cdot 3 \cdot 5}$,

and in the proposed problem $A = 2 \sin \tfrac{\pi}{3 \cdot 5}$, whence $X = 2 \sin \tfrac{\pi}{3^3 \cdot 5^2}$. That Van Roomen's examples correspond to these arcs can be verified by the formulas

$$2 \sin \tfrac{\alpha}{2} = \pm \sqrt{2 - 2 \cos \alpha} \qquad 2 \cos \tfrac{\alpha}{2} = \pm \sqrt{2 + 2 \cos \alpha}$$

$$2 \cos \tfrac{\pi}{3} = 1 \qquad\qquad\qquad 2 \sin \tfrac{\pi}{3} = \sqrt{3}$$

$$2 \cos \tfrac{2\pi}{5} = \tfrac{\sqrt{5}-1}{2} \qquad\qquad 2 \sin \tfrac{2\pi}{5} = \sqrt{\tfrac{5+\sqrt{5}}{2}}$$

(see also remark 7.6). From these last results, the value of $\sin \tfrac{\pi}{15}$ and of $\cos \tfrac{\pi}{15}$ can be calculated by the addition formulas, since $\tfrac{\pi}{15} = \tfrac{2\pi}{5} - \tfrac{\pi}{3}$.

It turns out that the solution $2 \sin \tfrac{\pi}{3^3 \cdot 5^2}$ of Van Roomen's equation could also be expressed by radicals, but this expression, which does not involve square roots only, is of little use for the determination of its numerical value since it requires the extraction of roots of complex numbers, see Remark 7.6, p. 85. Only the numerical value, suitably approximated to the ninth place, is given by Viète. But Viète does not stop there. While Van Roomen asked for *the* solution of his 45th degree equation, Viète shows that this equation has 23 positive solutions, and, in passing, he points out that it also has 22 negative solutions [64, Cap. 6]. Indeed, if α is an arc such that $2 \sin 45\alpha = A$, then, letting

$$\alpha_k = \alpha + k\frac{2\pi}{45},$$

one also has $2 \sin 45\alpha_k = A$ for all $k = 0, 1, \ldots, 44$, so that $2 \sin \alpha_k$ is a solution of Van Roomen's equation. If $A \geq 0$ (and $A \leq 2$), then one can choose α between 0 and $\frac{\pi}{90}$, whence

$$2 \sin \alpha_k \begin{cases} \geq 0 & \text{for } k = 0, \ldots, 22, \\ < 0 & \text{for } k = 23, \ldots, 44. \end{cases}$$

Another interesting feature of Viète's brilliant solution [64] is that, instead of solving directly Van Roomen's equation, which amounts, as we have seen, to the division of an arc into 45 parts, Viète decomposes the problem: since $45 = 3^2 \cdot 5$, the problem can be solved by the trisection of the arc, followed by the trisection of the resulting arc and the division into 5 parts of the arc thus obtained. As Viète shows, $2 \sin n\alpha$ is given as a function of $2 \sin \alpha$ by an equation of degree n, for n odd (see equation (4.2) below), whence the solutions of Van Roomen's equation of degree 45 can be obtained by solving successively two equations of degree 3 and one equation of degree 5. This idea of solving an equation step by step was to play a central role in Lagrange's and Gauss' investigations, two hundred years later (see Chapters 10 and 12).

In modern language, Viète's results on the division of arcs can be stated as follows: for any integer $n \geq 1$, let $[\frac{n}{2}]$ be the greatest integer which is less than (or equal to) $\frac{n}{2}$, and define

$$f_n(X) = \sum_{i=0}^{[\frac{n}{2}]} (-1)^i \frac{n}{n-i} \binom{n-i}{i} X^{n-2i}$$

where $\binom{n-i}{i} = \frac{(n-i)!}{i!(n-2i)!}$ is the binomial coefficient. Then, for all $n \geq 1$,

$$2 \cos n\alpha = f_n(2 \cos \alpha) \tag{4.1}$$

and for all *odd* $n \geq 1$,

$$2 \sin n\alpha = (-1)^{(n-1)/2} f_n(2 \sin \alpha). \tag{4.2}$$

Formula (4.1) can be proved by induction on n, using

$$2 \cos(n+1)\alpha = (2 \cos \alpha)(2 \cos n\alpha) - 2 \cos(n-1)\alpha$$

and (4.2) is easily deduced from (4.1), by applying (4.1) to $\beta = \frac{\pi}{2} - \alpha$, which is such that $\cos \beta = \sin \alpha$. (The original formulation is not quite so general, but Viète shows how to compute recursively the coefficients of f_n, see [64, Cap. 9], [65, pp. 432 ff] or Goldstine [27, §1.6].)

For each integer $n \geq 1$, the equation

$$f_n(X) = A$$

has degree n, and the same arguments as for Van Roomen's equation show that this equation has n solutions (at least when $|A| \leq 2$). These examples, which are quite explicit for $n = 3, 5, 7$ in Viète's works [64, Cap. 9], [65, pp. 445 ff], may have been influential in the progressive emergence of the idea that equations of degree n have n roots, although this idea was still somewhat obscured by Viète's insistence on considering positive roots only (see also §6.1).

In later works, such as "De Recognitione Aequationum" (On Understanding Equations), published posthumously in 1615, Viète also stressed the importance of understanding the structure of equations, meaning by this the relations between roots and coefficients. However, the theoretical tools at his disposal were not sufficiently developed, and he failed to grasp these relations in their full generality. For example, he shows [65, pp. 210–211] that if an equation[†]

$$B^p A - A^3 = Z^s \tag{4.3}$$

(in the indeterminate A) has two roots A and E, then assuming $A > E$, one has

$$B^p = A^2 + E^2 + AE$$
$$Z^s = A^2 E + E^2 A.$$

The proof is as follows: since

$$B^p A - A^3 = Z^s \qquad \text{and} \qquad B^p E - E^3 = Z^s,$$

one has $B^p A - A^3 = B^p E - E^3$, whence

$$B^p(A - E) = A^3 - E^3$$

and, dividing both sides by $A - E$,

$$B^p = A^2 + E^2 + AE.$$

[†]The superscripts of B and Z indicate the dimensions: p is for *plano* and s for *solido*.

The formula for Z^s is then obtained by substituting for B^p in the initial equation (4.3).

The structure of equations was eventually discovered in its proper generality and its simplest form by Albert Girard (1595–1632), and published in "Invention nouvelle en l'algebre" (1629) [26].

> As the next theorem needs new terms, the definitions will be given first. [26, p. E2 v°]

Girard calls an equation *incomplete* it it lacks at least one term (i.e. if at least one of the coefficients is zero); the various terms are called *minglings* ("meslés") and the last is called the *closure*. The *first faction* of the solutions is their sum, the *second faction* is the sum of their products two by two, the *third* is the sum of their products three by three and the *last* is their product. Finally, an equation is in the *alternative order* when the odd powers of the unknown are on one side of the equality and the even powers on the other side, and when moreover the coefficient of the highest power is 1.

Girard's main theorem is then [26, p. E4]

> All the equations of algebra receive as many solutions as the exponent of the highest quantity demonstrates, except the in-complete ones: & the first faction of the solutions is equal to the number of the first mingling, the second faction of the same is equal to the number of the second mingling; the third to the third, & so on, so that the last faction is equal to the closure, & this according to the signs that can be observed in the alternative order.

The restriction to complete equations is not easy to explain. Half a page later, Girard points out that incomplete equations have not always as many solutions, and that in this case some solutions are imaginary ("impossible" is Girard's own word). However, it is clear that even complete equations may have imaginary solutions (consider for instance $x^2 + x + 1 = 0$), and this fact could not have escaped Girard.

At any rate, Girard claims that the relations between roots and coefficients also hold in this case, provided that the equation be completed by adding powers of the unknown with coefficient 0. Therefore, the theorem asserts that each equation

$$X^n + s_2 X^{n-2} + s_4 X^{n-4} + \cdots = s_1 X^{n-1} + s_3 X^{n-3} + s_5 X^{n-5} + \cdots$$

or

$$X^n - s_1 X^{n-1} + s_2 X^{n-2} - s_3 X^{n-3} + \cdots + (-1)^n s_n = 0$$

has n roots $x_1, \ldots, x_n$ such that

$$s_1 = \sum_{i=1}^{n} x_i, \quad s_2 = \sum_{i<j} x_i x_j, \quad s_3 = \sum_{i<j<k} x_i x_j x_k, \quad \ldots, \tag{4.4}$$

$$s_n = x_1 x_2 \ldots x_n.$$

It should be observed that this "theorem" is not nearly as precise as the modern formulation of the fundamental theorem of algebra, since Girard does not explicitly assert that the roots are of the form $a + b\sqrt{-1}$. It is therefore more a *postulate* than a theorem: it claims the existence of "impossible" roots of polynomials, but it is essentially unprovable[‡] since nothing else is said about these roots, except (implicitly) that one can calculate with them as if they were numbers. Of course, in all the examples, it turns out that the impossible roots are of the form $a + b\sqrt{-1}$, but Girard nowhere explains what he has in mind.

> One could say of what use are these solutions which are impossible, I answer for three things, for the certitude of the general rule, & because there is no other solution, & for its utility. [26, p. F r°]

Girard elaborates as follows on the utility of impossible roots: if one seeks the (positive) values of $(x + 1)^2 + 2$, where x is such that $x^4 = 4x - 3$, then since the solutions of this last equation are $1, 1, -1 + \sqrt{-2}$ and $-1 - \sqrt{-2}$, one gets $(x + 1)^2 + 2 = 6, 6, 0$ or 0, so that 6 is the unique result. One would never have been so sure of that without the impossible roots.

Of course, Girard does not provide the faintest hint of a proof of his theorem. It would have been very interesting to see at least how he found the relations (4.4) between the roots and the coefficients of an equation. These relations readily follow from the identification of coefficients in

$$X^n - s_1 X^{n-1} + s_2 X^{n-2} - \cdots + (-1)^n s_n = (X - x_1)(X - x_2) \cdots (X - x_n)$$

but this equality was probably not known to Girard. Indeed, Girard does not seem to have been aware of the fact that a number a is a root of a polynomial $P(X)$ if

---

[‡]At least, the proof and, for that matter, even a correct statement of the theorem were very far beyond the reach of seventeenth century mathematicians: see §9.2.

and only if $X - a$ divides $P(X)$: this observation is usually credited to Descartes [16, p. 159], although Nunes may have been aware of it as early as 1567, and perhaps earlier (see Bosmans [5, pp. 163 ff]).

On the subject of impossible roots, Descartes first seems more cautious than Girard (the emphasis is mine):

> Every equation *can* have as many distinct roots as the number of dimensions of the unknown quantity in the equation. [16, p. 159]

This at least can be proved by Descartes' preceding observation (see theorem 5.15). However, Descartes further states:

> Neither the true nor the false roots are always real; sometimes they are imaginary; that is, while we can always conceive of as many roots for each equation as I have already assigned, yet there is not always a definite quantity corresponding to each root so conceived of. [16, p. 175]

Around the middle of the seventeenth century, the fact that the number of solutions of an equation is equal to the degree was becoming common knowledge, like a piece of mathematical "folklore," accepted without proof and never questioned. At least, it was a good working hypothesis, and mathematicians started to calculate formally with roots of equations without worrying about their nature, pushing further Girard's results to discover what kind of information can be obtained rationally from the coefficients of an equation.

Girard himself had shown (omitting the details of his calculations) that the sum of the squares of the solutions, the sum of their cubes, and the sum of their fourth powers, can be calculated from the coefficients [26, p. F2 r°]: if $x_1, \ldots, x_n$ are the solutions of

$$X^n - s_1 X^{n-1} + s_2 X^{n-2} - s_3 X^{n-3} + \cdots + (-1)^n s_n = 0,$$

let

$$\sigma_k = \sum_{i=1}^{n} x_i^k$$

for any integer k; then

$$\sigma_1 = s_1,$$
$$\sigma_2 = s_1^2 - 2s_2,$$
$$\sigma_3 = s_1^3 - 3s_1 s_2 + 3s_3,$$
$$\sigma_4 = s_1^4 - 4s_1^2 s_2 + 4s_1 s_3 + 2s_2^2 - 4s_4.$$

Around 1666, general formulas for the sum of any power of the solutions were found by Isaac Newton (1642–1727) [45, v. I, p. 519] (who was probably unaware of Girard's work: see footnote (12) in [45, v. I, p. 518]). Newton's clever observation is that, while the formulas for σ_1, σ_2, ... in terms of $s_1, \ldots, s_n$ do not seem to follow a simple pattern, yet there are simple formulas expressing σ_k in terms of $s_1, \ldots, s_n$ and $\sigma_1, \ldots, \sigma_{k-1}$. These formulas can be used to calculate recursively the various σ_k in terms of $s_1, \ldots, s_n$. Newton's formulas are

$$\sigma_1 = s_1,$$
$$\sigma_2 = s_1 \sigma_1 - 2s_2,$$
$$\sigma_3 = s_1 \sigma_2 - s_2 \sigma_1 + 3s_3,$$
$$\sigma_4 = s_1 \sigma_3 - s_2 \sigma_2 + s_3 \sigma_1 - 4s_4,$$
$$\sigma_5 = s_1 \sigma_4 - s_2 \sigma_3 + s_3 \sigma_2 - s_4 \sigma_1 + 5s_5,$$

and, generally,

$$\sigma_k = \begin{cases} \sum_{i=1}^{k-1} (-1)^{i+1} s_i \sigma_{k-i} + (-1)^{k+1} k\, s_k & \text{for } k \leq n, \\ \sum_{i=1}^{n} (-1)^{i+1} s_i \sigma_{k-i} & \text{for } k > n. \end{cases}$$

These formulas (for $k \leq n$) were published without proof in "Arithmetica Universalis" (1707) [46, p. 107] (see also [45, vol. I, p. 519; vol. V, p. 361]). Various ingenious proofs have since been proposed (see for instance Bourbaki [6, App. 1 n° 3]); the following elementary proof is perhaps not very different from Newton's own calculations.

For any integers a, b with $1 \leq a \leq n$ and $b \geq 1$, let $\tau(a, b)$ be the sum of the various different terms obtained from $x_1^b x_2 \ldots x_a$ by permutation of $x_1, \ldots, x_n$; thus

$$\tau(a, 1) = s_a, \qquad \tau(1, b) = \sigma_b$$

and, for $a, b \geq 2$,

$$\tau(a, b) = \sum_{i_1=1}^{n} \sum_{\substack{i_2 < i_3 < \ldots < i_a \\ i_k \neq i_1}} x_{i_1}^b x_{i_2} \ldots x_{i_a}.$$

Since, for $b \geq 2$, each term $x_{i_1}^b x_{i_2} \ldots x_{i_a}$ can be obtained as a product

$$x_{i_1}^b x_{i_2} \ldots x_{i_a} = \left(x_{i_1} x_{i_2} \ldots x_{i_a}\right)\left(x_{i_1}^{b-1}\right),$$

the product $s_a \sigma_{b-1}$ yields all the terms of $\tau(a, b)$; moreover, this product also yields terms like $x_{i_1}^{b-1} x_{i_2} \ldots x_{i_{a+1}}$, if $a \leq n - 1$. Furthermore, if $b = 2$, then each of these latter terms is obtained $(a + 1)$ times, while they are obtained only once if $b > 2$. Therefore, we have the following results:

$$\tau(a, b) = s_a \sigma_{b-1} - \tau(a + 1, b - 1) \qquad \text{for } a < n \text{ and } b > 2, \qquad (4.5)$$

$$\tau(a, 2) = s_a \sigma_1 - (a + 1)s_{a+1} \qquad \text{for } a < n, \qquad (4.6)$$

$$\tau(n, b) = s_n \sigma_{b-1} \qquad \text{for } b \geq 2. \qquad (4.7)$$

Since $\tau(1, k) = \sigma_k$, equation (4.5) with $a = 1$ and $b = k$ yields

$$\sigma_k = s_1 \sigma_{k-1} - \tau(2, k - 1).$$

Equation (4.5) with $a = 2$ and $b = k - 1$ can then be used to eliminate $\tau(2, k-1)$, yielding

$$\sigma_k = s_1 \sigma_{k-1} - s_2 \sigma_{k-2} + \tau(3, k - 2).$$

Next, we use (4.5) with $a = 3$ and $b = k - 2$ to eliminate $\tau(3, k - 2)$, and so on. After a certain number of steps, we obtain

$$\sigma_k = s_1 \sigma_{k-1} - s_2 \sigma_{k-2} + \cdots + (-1)^k \tau(k - 1, 2) \qquad \text{if } k \leq n,$$

whence, using (4.6),

$$\sigma_k = s_1 \sigma_{k-1} - s_2 \sigma_{k-2} + \cdots + (-1)^k s_{k-1} \sigma_1 + (-1)^{k+1} k\, s_k.$$

If $k > n$, we obtain

$$\sigma_k = s_1 \sigma_{k-1} - s_2 \sigma_{k-2} + \cdots + (-1)^{n+1} \tau(n, k + 1 - n)$$

whence, by (4.7),

$$\sigma_k = s_1 \sigma_{k-1} - s_2 \sigma_{k-2} + \cdots + (-1)^{n+1} s_n \sigma_{k-n},$$

completing the proof.

This, of course, was not Newton's most prominent achievement, even if we consider only his contributions to the theory of equations. Indeed, Newton was much more interested in numerical aspects (see for instance Goldstine [27, chapter 2]).

Numerical methods to find the roots of polynomial equations were at first one of the several aims of the theory of equations, developed by some of the same authors who developed other aspects: see for instance Cardano's "golden rule" [11, ch. 30], Stevin's "Appendice algebraique" [55, pp. 740–745] or Viète's "De numerosa Potestatum ad Exegesim Resolutione" (On the numerical resolution of powers by exegetics) [65, pp. 311–370]. These numerical methods were much more successful for the solution of explicit numerical equations than the algebraic formulas "by radicals"; indeed, algebraic formulas are available only for degrees up to 4 and they are by no means more accurate than the numerical methods (see also §2.3). Therefore, the numerical solution of equations soon developed into a new branch of mathematics, growing more accurate and powerful while the algebraic theory of equations was progressively stalled.

Since the discussion of numerical methods falls beyond the scope of this book, we take the occasion of this period of relatively low activity in algebra to justify a pause in the historical exposition. We turn in the next chapter to a modern exposition of the above-mentioned results on polynomials in one indeterminate, in order to show on which mathematical base later results were grounded.

Chapter 5

A Modern Approach to Polynomials

5.1 Definitions

In modern terminology, a polynomial in one indeterminate with coefficients in a ring A can be defined as a map

$$P \colon \mathbb{N} \to A$$

such that the set supp $P = \{n \in \mathbb{N} \mid P_n \neq 0\}$, called the *support* of P, is finite. The addition of polynomials is the usual addition of maps,

$$(P + Q)_n = P_n + Q_n$$

and the product is the convolution product

$$(PQ)_n = \sum_{i+j=n} P_i \cdot Q_j.$$

Every element $a \in A$ is identified with the polynomial $a \colon \mathbb{N} \to A$ which maps 0 to a and n to 0 for $n \neq 0$. Denoting by

$$X \colon \mathbb{N} \to A$$

the polynomial which maps 1 onto the unit element $1 \in A$ and the other integers to 0, it is then easily seen that every polynomial P can be uniquely written as

$$P = \sum_{i \in \mathbb{N}} P_i \cdot X^i.$$

Therefore, we shall henceforth denote by

$$a_0 + a_1 X + \cdots + a_n X^n \tag{5.1}$$

(as is usual!) the polynomial which maps $i \in \mathbb{N}$ to a_i for $i = 0, \ldots, n$ and to 0 for $i > n$. Accordingly, the set of all polynomials with coefficients in A (or polynomials *over* A) is denoted $A[X]$. Straightforward calculations show that $A[X]$ is a ring, which is commutative if and only if A is commutative.

The ring of polynomials in any number m of indeterminates over A can be similarly defined as the ring of maps from $\mathbb{N}^m$ to A with finite support, with the convolution product.

Of course, the definition above is not quite natural. The naive approach to polynomials is to consider expressions like (5.1), where X is an undefined object, called an *indeterminate*, or a *variable*. While this terminology will be retained in the sequel, it should be observed that, without any other proper definition, to say something is an indeterminate or a variable is hardly a definition. Moreover, it fosters confusion between the polynomial

$$P(X) = a_0 + a_1 X + \cdots + a_n X^n$$

and the associated polynomial function

$$P(\cdot)\colon A \to A$$

which maps $x \in A$ onto $P(x) = a_0 + a_1 x + \cdots + a_n x^n$. This same confusion has prompted the use of the term *constant* polynomials for the elements of A, considered as polynomials. While this confusion is not so serious when A is a field with infinitely many elements (see Corollary 5.16 below, p. 52), it could be harmful when A is finite. For instance, if $A = \{a_1, \ldots, a_n\}$ (with $n \geq 2$), then the polynomial $(X - a_1) \cdots (X - a_n)$ is not the zero polynomial since the coefficient of X^n is 1, but its associated polynomial function maps every element of A to 0.

The *degree* of a non-zero polynomial P is the greatest integer n for which the coefficient of X^n in the expression of P is not zero; this coefficient is called the *leading coefficient* of P, and P is said to be *monic* if its leading coefficient is 1.

The degree of P is denoted by $\deg P$. One also sets $\deg 0 = -\infty$, so that the following relations hold without restriction if A is a domain (i.e. a ring in which $ab = 0$ implies $a = 0$ or $b = 0$):

$$\deg(P + Q) \leq \max(\deg P, \deg Q),$$
$$\deg(PQ) = \deg P + \deg Q.$$

When A is a (commutative) field, the ring $A[X]$ has a field of fractions $A(X)$, constructed as follows: the elements of $A(X)$ are the equivalence classes of couples f/g, where $f, g \in A[X]$ and $g \neq 0$, under the equivalence relation

$$f/g = f'/g' \quad \text{if} \quad g'f = f'g.$$

The addition of equivalence classes is defined by

$$(f_1/g_1) + (f_2/g_2) = (f_1 g_2 + f_2 g_1)/g_1 g_2$$

and the multiplication by

$$(f_1/g_1) \cdot (f_2/g_2) = (f_1 f_2)/(g_1 g_2).$$

It is then easily verified that $A(X)$ is a field, called the field of *rational fractions* in one indeterminate X over the field A. The same construction can be applied to the ring of polynomials in m indeterminates and yields the field of rational fractions in m indeterminates over the field A.

However, the ring $A[X]$ of polynomials in one indeterminate over a field A has particularly nice properties, which follow from the Euclidean division algorithm. These properties are reviewed in the next section.

5.2 Euclidean division

From now, we only consider polynomials over a field F.

5.1. THEOREM (EUCLIDEAN DIVISION PROPERTY). *Let P_1, $P_2 \in F[X]$. If $P_2 \neq 0$, then there exist polynomials Q, $R \in F[X]$ such that*

$$P_1 = P_2 Q + R \quad \text{and} \quad \deg R < \deg P_2.$$

Moreover, the polynomials Q and R are uniquely determined by these properties.

The polynomials Q and R are called respectively the *quotient* and the *remainder* of the division of P_1 by P_2.

Proof. The existence of Q and R is proved by induction on $\deg P_1$. If $\deg P_1 < \deg P_2$, we set $Q = 0$ and $R = P_1$. If $\deg P_1 - \deg P_2 = d \geq 0$, then letting $c \in F$ be the quotient of the leading coefficient of P_1 by that of P_2, we have

$$\deg(P_1 - cX^d P_2) < \deg P_1.$$

Therefore, by the induction hypothesis, there exist Q and $R \in F[X]$ such that

$$P_1 - cX^d P_2 = P_2 Q + R \quad \text{and} \quad \deg R < \deg P_2.$$

This equation yields

$$P_1 = P_2(Q + cX^d) + R$$

whence $Q + cX^d$ and R satisfy the required properties.

To prove the uniqueness of Q and R, assume

$$P_1 = P_2 Q_1 + R_1 = P_2 Q_2 + R_2$$

with $\deg R_1 < \deg P_2$ and $\deg R_2 < \deg P_2$. Then $R_1 - R_2 = P_2(Q_2 - Q_1)$ and this equality is impossible if both sides are non-zero, since the degree of the right-hand side is then at least equal to $\deg P_2$, while the degree of the left-hand side is strictly less than $\deg P_2$. $\qquad\qquad\qquad\qquad\qquad\qquad\square$

5.2. DEFINITIONS. Let $P_1, P_2 \in F[X]$. We say P_2 *divides* P_1 if

$$P_1 = P_2 Q \qquad \text{for some } Q \in F[X]$$

or, equivalently when $P_2 \neq 0$, if the remainder of the division of P_1 by P_2 is 0.

A *greatest common divisor* (GCD) of P_1 and P_2 is a polynomial $D \in F[X]$ which has the following two properties:

(a) D divides P_1 and P_2,
(b) if S is a polynomial which divides P_1 and P_2 then S divides D.

If 1 is a GCD of P_1 and P_2, then P_1 and P_2 are said to be *relatively prime*.

Since it is by no means obvious that any two polynomials P_1, P_2 have a GCD, our first objective is to devise a method of finding such a GCD, thereby proving its existence. We shall closely follow Euclid's algorithm for finding the GCD of two integers or the greatest common measure of two line segments, assuming (without loss of generality) that $\deg P_1 \geq \deg P_2$.

If $P_2 = 0$, then P_1 is a GCD of P_1 and P_2. Otherwise, we divide P_1 by P_2:

$$P_1 = P_2 Q_1 + R_1. \qquad\qquad (E.1)$$

Next, we divide P_2 by the remainder R_1, provided that it is not zero,

$$P_2 = R_1 Q_2 + R_2 \qquad\qquad (E.2)$$

then we divide the first remainder by the second, and so on, as long as the remainders are not zero:

$$R_1 = R_2 Q_3 + R_3, \qquad (E.3)$$

$$R_2 = R_3 Q_4 + R_4, \qquad (E.4)$$

$$\cdots \qquad \cdots$$

$$R_{n-2} = R_{n-1} Q_n + R_n, \qquad (E.n)$$

$$R_{n-1} = R_n Q_{n+1} + R_{n+1}. \qquad (E.n+1)$$

Since $\deg P_2 > \deg R_1 > \deg R_2 > \ldots$, this sequence of integers cannot extend indefinitely. Therefore, $R_{n+1} = 0$ for some n.

Claim: If $R_{n+1} = 0$, then R_n is a GCD of P_1 and P_2. (If $n = 0$, then set $R_n = P_2$.)

To see that R_n divides P_1 and P_2, observe that equation $(E.n+1)$, together with $R_{n+1} = 0$, implies that R_n divides R_{n-1}. It then follows from equation $(E.n)$ that R_n also divides R_{n-2}. Going up in the sequence of equations $(E.n)$, $(E.n-1), (E.n-2), \ldots, (E.2), (E.1)$, we conclude recursively that R_n divides $R_{n-3}, \ldots, R_2, R_1, P_2$ and P_1.

Assume next that P_1 and P_2 are both divisible by some polynomial S. Then equation $(E.1)$ shows that S also divides R_1. Since it divides P_2 and R_1, S also divides R_2, by equation $(E.2)$. Going down in the above sequence of equations $(E.2), (E.3), \ldots, (E.n)$, we finally see that S divides R_n. This completes the proof that R_n is a GCD of P_1 and P_2.

We next observe that the GCD of two polynomials is not unique (except over the field with two elements), as the following theorem shows.

5.3. THEOREM. *Any two polynomials $P_1, P_2 \in F[X]$ which are not both zero have a unique monic greatest common divisor D_1, and a polynomial $D \in F[X]$ is a greatest common divisor of P_1 and P_2 if and only if $D = cD_1$ for some $c \in F^\times (= F - \{0\})$. Moreover, if D is a greatest common divisor of P_1 and P_2, then*

$$D = P_1 U_1 + P_2 U_2 \qquad \text{for some } U_1, U_2 \in F[X].$$

Proof. Euclid's algorithm already yields a greatest common divisor R_n of P_1 and P_2. Dividing R_n by its leading coefficient, we get a monic GCD of P_1 and P_2. Now, assume D and D' are GCDs of P_1 and P_2. Then D divides D' since D satisfies condition (a) and D' satisfies condition (b). The same argument, with D

and D' interchanged, shows that D' divides D. Let $D' = DQ$ and $D = D'Q'$ for some $Q, Q' \in F[X]$. It follows that $QQ' = 1$, so that Q and Q' are constants, which are inverse of each other. This proves at once the second statement and the uniqueness of the monic GCD of P_1 and P_2, since D and D' cannot be both monic unless $Q = Q' = 1$.

Suppose D is any GCD of P_1 and P_2; then D and the greatest common divisor R_n found by Euclid's algorithm are related by $D = cR_n$ for some $c \in F^\times$. Therefore, it suffices to prove the last statement for R_n. To this end, we consider again the sequence of equations $(E.1), \ldots, (E.n)$. From equation $(E.n)$, we get

$$R_n = R_{n-2} - R_{n-1}Q_n.$$

We then use equation $(E.n - 1)$ to eliminate R_{n-1} in this expression of R_n, and we obtain

$$R_n = -R_{n-3}Q_n + R_{n-2}(1 + Q_{n-1}Q_n).$$

This shows that R_n is a sum of multiples of R_{n-2} and R_{n-3}. Now, R_{n-2} can be eliminated using $(E.n - 2)$. We thus obtain an expression of R_n as a sum of multiples of R_{n-3} and R_{n-4}. Going up in the sequence of equations $(E.n - 3)$, $(E.n - 4), \ldots, (E.1)$, we end up with an expression of R_n as a sum of multiples of P_1 and P_2,

$$R_n = P_1U_1 + P_2U_2$$

for some $U_1, U_2 \in F[X]$.

The argument above can be made more transparent with only a sprinkle of matrix algebra. We may rewrite equation $(E.1)$ in the form

$$\begin{pmatrix} P_1 \\ P_2 \end{pmatrix} = \begin{pmatrix} Q_1 & 1 \\ 1 & 0 \end{pmatrix} \begin{pmatrix} P_2 \\ R_1 \end{pmatrix},$$

equation $(E.2)$ in the form

$$\begin{pmatrix} P_2 \\ R_1 \end{pmatrix} = \begin{pmatrix} Q_2 & 1 \\ 1 & 0 \end{pmatrix} \begin{pmatrix} R_1 \\ R_2 \end{pmatrix},$$

and so on, finishing with equation $(E.n + 1)$ (with $R_{n+1} = 0$) in the form

$$\begin{pmatrix} R_{n-1} \\ R_n \end{pmatrix} = \begin{pmatrix} Q_{n+1} & 1 \\ 1 & 0 \end{pmatrix} \begin{pmatrix} R_n \\ 0 \end{pmatrix}.$$

Combining the matrix equations above, we get

$$\begin{pmatrix} P_1 \\ P_2 \end{pmatrix} = \begin{pmatrix} Q_1 & 1 \\ 1 & 0 \end{pmatrix} \begin{pmatrix} Q_2 & 1 \\ 1 & 0 \end{pmatrix} \cdots \begin{pmatrix} Q_{n+1} & 1 \\ 1 & 0 \end{pmatrix} \begin{pmatrix} R_n \\ 0 \end{pmatrix}.$$

Each matrix $\begin{pmatrix} Q_i & 1 \\ 1 & 0 \end{pmatrix}$ is invertible, with inverse $\begin{pmatrix} 0 & 1 \\ 1 & -Q_i \end{pmatrix}$, hence the preceding equation yields, after multiplying each side on the left successively by $\begin{pmatrix} 0 & 1 \\ 1 & -Q_1 \end{pmatrix}$, $\begin{pmatrix} 0 & 1 \\ 1 & -Q_2 \end{pmatrix}, \ldots,$

$$\begin{pmatrix} 0 & 1 \\ 1 & -Q_{n+1} \end{pmatrix} \cdots \begin{pmatrix} 0 & 1 \\ 1 & -Q_2 \end{pmatrix} \begin{pmatrix} 0 & 1 \\ 1 & -Q_1 \end{pmatrix} \begin{pmatrix} P_1 \\ P_2 \end{pmatrix} = \begin{pmatrix} R_n \\ 0 \end{pmatrix}.$$

If $U_1, \ldots, U_4 \in F[X]$ are such that

$$\begin{pmatrix} 0 & 1 \\ 1 & -Q_{n+1} \end{pmatrix} \cdots \begin{pmatrix} 0 & 1 \\ 1 & -Q_2 \end{pmatrix} \begin{pmatrix} 0 & 1 \\ 1 & -Q_1 \end{pmatrix} = \begin{pmatrix} U_1 & U_2 \\ U_3 & U_4 \end{pmatrix},$$

it follows that

$$U_1 P_1 + U_2 P_2 = R_n$$

(and $U_3 P_1 + U_4 P_2 = 0$). $\qquad\qquad \square$

Because of its repeated use in the sequel, the following special case seems to be worth pointing out explicitly:

5.4. COROLLARY. *If P_1, P_2 are relatively prime polynomials in $F[X]$, then there exist polynomials U_1, $U_2 \in F[X]$ such that*

$$P_1 U_1 + P_2 U_2 = 1.$$

5.5. Remarks. (a) The proof given above is effective: Euclid's algorithm yields a procedure for constructing the polynomials U_1, U_2 in Theorem 5.3 and Corollary 5.4.

(b) Since the GCD of two polynomials P_1, $P_2 \in F[X]$ can be found by rational calculations (i.e. calculations involving only the four basic operations of arithmetic), it does not depend on particular properties of the field F. The point of this observation is that if the field F is embedded in a larger field K, then the polynomials P_1, $P_2 \in F[X]$ can be regarded as polynomials over K, but their monic GCD in $K[X]$ is the same as their monic GCD in $F[X]$. This is noteworthy in view of the fact that the *irreducible* factors of P_1 and P_2 depend on the base field F, as is clear from Example 5.7(b) below.

5.3 Irreducible polynomials

5.6. DEFINITION. A polynomial $P \in F[X]$ is said to be *irreducible in $F[X]$* (or *over F*) if $\deg P > 0$ and P is not divisible by any polynomial $Q \in F[X]$ such that $0 < \deg Q < \deg P$.

From this definition, it follows that if a polynomial D divides an irreducible polynomial P, then either D is a constant or $\deg D = \deg P$. In the latter case, the quotient of P by D is a constant, whence D is the product of P by a non-zero constant. In particular, for any polynomial $S \in F[X]$, either 1 or P is a GCD of P and S. Consequently, either P divides S or P is relatively prime to S.

5.7. *Examples.* (a) By definition, it is clear that every polynomial of degree 1 is irreducible. It will be proved later that over the field of complex numbers, only these polynomials are irreducible.

(b) Theorem 5.12 below (p. 51) will show that if a polynomial of degree at least 2 has a root in the base field, then it is not irreducible. The converse is true for polynomials of degree 2 or 3; namely, if a polynomial of degree 2 or 3 has no root in the base field, then it is irreducible over this field, see Corollary 5.13 (p. 51). It follows that, for instance, the polynomial $X^2 - 2$ is irreducible over $\mathbb{Q}$, but not over $\mathbb{R}$. Thus, the irreducibility of polynomials of degree at least 2 depends on the base field (compare Remark 5.5(b)).

(c) A polynomial (of degree at least 4) may be reducible over a field without having any root in this field. For instance, in $\mathbb{Q}[X]$, the polynomial $X^4 + 4$ is reducible since

$$X^4 + 4 = (X^2 + 2X + 2)(X^2 - 2X + 2).$$

Remark. To determine whether a given polynomial with rational coefficients is irreducible or not over $\mathbb{Q}$ may be difficult, although a systematic procedure has been devised by Kronecker, see Van der Waerden [61, §32]. This procedure is not unlike that which is used to find the rational roots of polynomials with rational coefficients, see §6.3.

5.8. THEOREM. *Every non constant polynomial $P \in F[X]$ is a finite product*

$$P = c \cdot P_1 \cdot \ldots \cdot P_n$$

where $c \in F^\times$ and $P_1, \ldots, P_n$ are monic irreducible polynomials (not necessarily distinct). Moreover, this factorization is unique, except for the order of the factors.

Proof. The existence of the above factorization is easily proved by induction on $\deg P$. If $\deg P = 1$ or, more generally, if P is irreducible, then $P = c \cdot P_1$ where c is the leading coefficient of P and $P_1 = c^{-1}P$ is irreducible and monic. If P is reducible, then it can be written as a product of two polynomials of degree strictly less than $\deg P$. By induction, each of these two polynomials has a finite factorization as above, and these factorizations multiply up to a factorization of P.

To prove that the factorization is unique, we shall use the following lemma:

5.9. LEMMA. *If a polynomial divides a product of r factors and is relatively prime to the first $r - 1$ factors, then it divides the last one.*

Proof. It suffices to consider the case $r = 2$, since the general case then easily follows by induction.

Assume that a polynomial S divides a product $T \cdot U$ and is relatively prime to T. By Corollary 5.4, p. 47, we can find polynomials V, W such that $SV + TW = 1$. Multiplying each side of this equality by U, we obtain

$$S(UV) + (TU)W = U.$$

Now, S divides the left-hand side, since it divides TU by hypothesis; it then follows that S divides U. $\quad\square$

End of the proof of Theorem 5.8. It remains to prove the uniqueness (up to the order of factors) of factorizations. Assume

$$P = cP_1 \ldots P_n = dQ_1 \ldots Q_m \tag{5.2}$$

where $c, d \in F^\times$ and $P_1, \ldots, P_n, Q_1, \ldots, Q_m$ are monic irreducible polynomials.

First, $c = d$ since c and d are both equal to the leading coefficient of P. Therefore, (5.2) yields

$$P_1 \ldots P_n = Q_1 \ldots Q_m. \tag{5.3}$$

Next, since P_1 divides the product $Q_1 \ldots Q_m$, it follows from Lemma 5.9 that it cannot be relatively prime to all the factors. Changing the numbering of $Q_1, \ldots, Q_m$ if necessary, we may assume that P_1 is not relatively prime to Q_1, so that their monic GCD, which we denote by D, is not 1. Since D divides P_1, which is irreducible, it is equal to P_1 up to a constant factor. Since moreover D and P_1 are both monic, the constant factor is 1, whence $D = P_1$.

We may argue similarly with Q_1, since Q_1 is also monic and irreducible, and we thus obtain $D = Q_1$, whence

$$P_1 = Q_1.$$

Canceling P_1 ($= Q_1$) in equation (5.3), we get a similar equality, with one factor less on each side:

$$P_2 \ldots P_n = Q_2 \ldots Q_m.$$

Using inductively the same argument, we get (changing the numbering of Q_2, $\ldots$, Q_m if necessary)

$$P_2 = Q_2, \quad P_3 = Q_3, \quad \ldots \quad P_n = Q_n,$$

and it follows that $n \leq m$. If $n < m$, then comparing the degrees of both sides of (5.3) we get

$$\deg Q_{n+1} = \cdots = \deg Q_m = 0$$

and this is absurd since Q_i is irreducible for $i = 1, \ldots, m$. Therefore, $n = m$ and the proof is complete.   □

Lemma 5.9 has another consequence which is worth noting in view of its repeated use in the sequel:

5.10. PROPOSITION. *If a polynomial is divisible by pairwise relatively prime polynomials, then it is divisible by their product.*

Proof. Let $P_1, \ldots, P_r$ be pairwise relatively prime polynomials which divide a polynomial P. We argue by induction on r, the case $r = 1$ being trivial. By the induction hypothesis,

$$P = P_1 \ldots P_{r-1} Q$$

for some polynomial Q. Since P_r divides P, Lemma 5.9 shows that P_r divides Q, so that $P_1 \ldots P_r$ divides P.   □

5.4 Roots

As in the preceding sections, F denotes a field. For any polynomial

$$P = a_0 + a_1 X + \cdots + a_n X^n \in F[X]$$

we denote by $P(\cdot)$ the associated polynomial function

$$P(\cdot): F \to F$$

which maps any $x \in F$ to $P(x) = a_0 + a_1 x + \ldots + a_n x^n$. It is readily verified that for any two polynomials $P, Q \in F[X]$ and any $x \in F$,

$$(P + Q)(x) = P(x) + Q(x) \qquad \text{and} \qquad (P \cdot Q)(x) = P(x) \cdot Q(x).$$

Therefore, the map $P \mapsto P(\cdot)$ is a homomorphism from the ring $F[X]$ to the ring of functions from F to F.

5.11. DEFINITION. An element $a \in F$ is a *root* of a polynomial $P \in F[X]$ if $P(a) = 0$.

5.12. THEOREM. *An element $a \in F$ is a root of a polynomial $P \in F[X]$ if and only if $(X - a)$ divides P.*

Proof. Since $\deg(X - a) = 1$, the remainder R of the division of P by $(X - a)$ is a constant polynomial. Evaluating at a the polynomial functions associated with each side of the equation

$$P = (X - a)Q + R$$

we get

$$P(a) = (a - a)Q(a) + R,$$

whence $P(a) = R$. This shows that $P(a) = 0$ if and only if the remainder of the division of P by $X - a$ is 0. The theorem follows, since the last condition means that $(X - a)$ divides P. $\qquad \square$

5.13. COROLLARY. *Let $P \in F[X]$ be a polynomial of degree 2 or 3. Then P is irreducible over F if and only if it has no root in F.*

Proof. This readily follows from the theorem, since the hypothesis on $\deg P$ implies that if P is not irreducible, then it has a factor of degree 1, whence a factor of the form $X - a$. $\qquad \square$

5.14. DEFINITIONS. The *multiplicity* of a root a of a nonzero polynomial P is the exponent of the highest power of $X - a$ which divides P. Thus, the multiplicity is m if $(X - a)^m$ divides P but $(X - a)^{m+1}$ does not divide P. A root is called *simple* when its multiplicity is 1; otherwise it is called *multiple*.

When the multiplicity of a as a root of P is considered as a function of P, it is also called the *valuation* of P at a, and denoted $v_a(P)$; we then set $v_a(P) = 0$ if a is not a root of P. By convention, we also set $v_a(0) = \infty$, so that the following relations hold for any $P, Q \in F[X]$ and any $a \in F$

$$v_a(P + Q) \geq \min(v_a(P), v_a(Q)), \qquad v_a(PQ) = v_a(P) + v_a(Q).$$

These properties are exactly the same as those of the function

$$- \deg \colon F[X] \to \mathbb{Z} \cup \{\infty\}$$

which maps any polynomial to the *opposite* of its degree, see p. 42. Accordingly, $(-\deg)$ is sometimes considered as the valuation "at infinity."

5.15. THEOREM. *Every non-zero polynomial $P \in F[X]$ has a finite number of roots. If $a_1, \ldots, a_r$ are the various roots of P in F, with respective multiplicities $m_1, \ldots, m_r$, then $\deg P \geq m_1 + \ldots + m_r$ and*

$$P = (X - a_1)^{m_1} \cdots (X - a_r)^{m_r} Q$$

for some polynomial $Q \in F[X]$ which has no root in F.

Proof. If $a_1, \ldots, a_r$ are distinct roots of P in F, with respective multiplicities $m_1, \ldots, m_r$, then the polynomials $(X - a_1)^{m_1}, \ldots, (X - a_r)^{m_r}$ are relatively prime and divide P, whence Proposition 5.10 shows that

$$P = (X - a_1)^{m_1} \cdots (X - a_r)^{m_r} Q,$$

for some polynomial Q.

It readily follows that

$$\deg P \geq m_1 + \cdots + m_r;$$

hence P cannot have infinitely many roots. The rest follows from the observation that if $a_1, \ldots, a_r$ are *all* the roots of P in F, then Q has no root. $\qquad\square$

5.16. COROLLARY. *Let $P, Q \in F[X]$. If F is infinite, then $P = Q$ if and only if the associated polynomial functions $P(\cdot)$ and $Q(\cdot)$ are equal.*

Proof. If $P(\cdot) = Q(\cdot)$, then all the elements in F are roots of $P - Q$, whence $P - Q = 0$, since F is infinite. The converse is clear. $\qquad\square$

5.5 Multiple roots and derivatives

The aim of this section is to derive a method of determining whether a polynomial has multiple roots without actually finding the roots, and to reduce to 1 the multiplicity of the roots. (More precisely, the method yields, for any given polynomial P, a polynomial P_s which has the same roots as P, each with multiplicity 1).

This method is due to Johann Hudde (1633–1704). It uses the derivative of polynomials, which was introduced purely algebraically by Hudde in his letter "De Reductione Aequationum" (On the reduction of equations) (1657) [31] and subsequently applied to find maxima and minima of polynomials and rational fractions in his letter "De Maximis et Minimis" (1658) [32].

5.17. DEFINITION. The *derivative* ∂P of a polynomial $P = a_0 + a_1 X + \cdots + a_n X^n$ with coefficients in a field F is the polynomial

$$\partial P = a_1 + 2a_2 X + 3a_3 X^2 + \cdots + na_n X^{n-1}.$$

Straightforward calculations show that the following (familiar) relations hold:

$$\partial(P + Q) = \partial P + \partial Q, \qquad \partial(PQ) = (\partial P)Q + P(\partial Q).$$

Remark. The integers which appear in the coefficients of ∂P are regarded as elements in F; thus, n stands for $1 + 1 + \cdots + 1$ (n terms), where 1 is the unit element of F. This requires some caution, since it could happen that $n = 0$ in F even if $n \neq 0$ (as an integer). For instance, if F is the field with two elements $\{0, 1\}$, then $2 = 1 + 1 = 0$ in F, whence non-constant polynomials like $X^2 + 1$ have derivative zero over F.

5.18. LEMMA. *Let $a \in F$ be a root of a polynomial $P \in F[X]$. Then a is a multiple root of P (i.e. $v_a(P) \geq 2$) if and only if a is a root of ∂P.*

Proof. Since a is a root of P, one has $P = (X - a)Q$ for some $Q \in F[X]$, whence

$$\partial P = Q + (X - a)\partial Q.$$

This equality shows that $X - a$ divides ∂P if and only if it divides Q. Since this last condition amounts to $(X - a)^2$ divides P, i.e. $v_a(P) \geq 2$, the lemma follows. $\qquad\square$

5.19. PROPOSITION. *Let $P \in F[X]$ be a polynomial which splits into a product of linear factors* over some field K containing F. A necessary and sufficient condition for the roots of P in K to be all simple is that P and ∂P be relatively prime.*

Proof. If some root a of P in K is not simple, then, by the preceding lemma, P and ∂P have the common factor $X - a$ in $K[X]$; therefore, they are not relatively prime in $K[X]$, whence also not relatively prime in $F[X]$ (see Remark 5.5(b)).

Conversely, if P and ∂P are not relatively prime, they have in $K[X]$ a common irreducible factor, which has degree 1 since all the irreducible factors of P in $K[X]$ are linear. We can thus find a polynomial $X - a \in K[X]$ which divides both P and ∂P, and it follows from the preceding lemma that a is a multiple root of P in K.   □

To improve Lemma 5.18, we now assume that the characteristic of F is 0, which means that every non-zero integer is non-zero in F. (The characteristic of a field F is either 0 or the smallest integer $n > 0$ such that $n = 0$ in F).

5.20. PROPOSITION. *Assume that* char $F = 0$ *and let* $a \in F$ *be a root of a non-zero polynomial* $P \in F[X]$. *Then*

$$v_a(\partial P) = v_a(P) - 1.$$

Proof. Let $m = v_a(P)$ and let $P = (X - a)^m Q$ where $Q \in F[X]$ is not divisible by $X - a$. Then

$$\partial P = (X - a)^{m-1}\big(mQ + (X - a)\partial Q\big),$$

and the hypothesis on the characteristic of F ensures that $mQ \neq 0$. Then, since $X - a$ does not divide Q, it does not divide $mQ + (X - a)\partial Q$ either, whence $v_a(\partial P) = m - 1$.   □

As an application of this result, we now derive Hudde's method to reduce to 1 the multiplicity of the roots of a non-zero polynomial P over a field F of characteristic zero. Let D be a GCD of P and ∂P, and let $P_s = P/D$.

5.21. THEOREM. *Let K be an arbitrary field containing F. The roots of P_s in K are the same as those of P, and every root of P_s in K is simple, i.e. has multiplicity 1.*

*In §9.2, it is proved that this condition holds for every (non constant) polynomial.

Proof. Since P_s is a quotient of P, it is clear that every root of P_s is a root of P. Conversely, if $a \in K$ is a root of P, of multiplicity m, then the preceding proposition shows that $v_a(\partial P) = m - 1$, and it follows that $(X - a)^{m-1}$ is the highest power of $X - a$ which divides both P and ∂P. It is therefore the highest power of $X - a$ which divides D. Consequently, P_s is divisible by $X - a$ but not by $(X - a)^2$, which means that a is a simple root of P_s. $\qquad\square$

5.22. COROLLARY. *If $P \in F[X]$ is irreducible, then its roots in every field K containing F are simple.*

Proof. Since P is irreducible and does not divide ∂P, since $\deg \partial P = \deg P - 1$, the constant polynomial 1 is a GCD of P and ∂P. Therefore, $P_s = P$ and the preceding theorem shows that every root of P in any field K containing F is simple. $\qquad\square$

This corollary does not hold if char $F \neq 0$. For instance, if char $F = 2$ and $a \in F$ is not a square in F, then $X^2 - a$ is irreducible in $F[X]$, but it has $\sqrt{a}$ as a double root in $F(\sqrt{a})$. (Observe that $\sqrt{a} = -\sqrt{a}$ since the characteristic is 2.)

5.23. Remarks. **(a)** Since a GCD of P and ∂P can be calculated by Euclid's algorithm (see p. 44), it is not necessary to find the roots of P in order to construct the polynomial P_s. Thus, there is no serious restriction if we henceforth assume, when trying to solve an equation $P = 0$, that all the roots of P in F or in any field containing F are simple.

(b) In his work, Hudde does not explicitly introduce the derivative polynomial ∂P, but indirectly he uses it. His formulation [31, Reg. 10, pp. 433 ff] is as follows : to reduce to 1 the multiplicity of the roots of a polynomial

$$P = a_0 + a_1 X + a_2 X^2 + \cdots + a_n X^n,$$

form a new polynomial P_1 by multiplying the coefficients of P by the terms of an arbitrary arithmetical progression $m, m + r, m + 2r, \ldots, m + nr$, so

$$P_1 = a_0 m + a_1(m + r)X + a_2(m + 2r)X^2 + \cdots + a_n(m + nr)X^n.$$

Then, the quotient of P by a greatest common divisor D_1 of P and P_1 is the required polynomial.

The relation between this rule and its modern translation is easy to see, since $P_1 = mP + rX\partial P$. Therefore, $D_1 = D$ up to a non-zero constant factor, and possibly to a factor X in D_1, if 0 is a root of P. So, Hudde's method yields the same equation with simple roots as its modern equivalent, except that Hudde's equation lacks the root 0 whenever 0 is a root of the initial equation.

5.6   Common roots of two polynomials

As the preceding discussion shows, it is sometimes useful to determine whether two polynomials P, Q are relatively prime or not. The most straightforward method is of course to calculate a GCD of P and Q, but there is another construction, due to L. Euler (1707–1783) (with different notations): the *resultant* of P and Q. This construction is also basic for elimination theory; it will be used in §§6.4 and 10.1.

Let

$$P = a_n X^n + a_{n-1} X^{n-1} + \cdots + a_1 X + a_0 \qquad (a_n \neq 0)$$

and

$$Q = b_m X^m + b_{m-1} X^{m-1} + \cdots + b_1 X + b_0 \qquad (b_m \neq 0)$$

be polynomials over a field F. The *resultant* of P and Q is the following $(m+n) \times (m+n)$ determinant:

$$R = \det \begin{pmatrix}
a_n & 0 & \cdots & 0 & b_m & 0 & \cdots & \cdots & 0 \\
a_{n-1} & a_n & & \vdots & b_{m-1} & b_m & & & \vdots \\
\vdots & a_{n-1} & \ddots & 0 & \vdots & b_{m-1} & \ddots & & \vdots \\
\vdots & \vdots & \ddots & a_n & b_1 & \vdots & \ddots & \ddots & 0 \\
a_1 & \vdots & & a_{n-1} & b_0 & b_1 & & \ddots & b_m \\
a_0 & a_1 & & \vdots & 0 & b_0 & \ddots & & b_{m-1} \\
0 & a_0 & \ddots & \vdots & \vdots & \vdots & \ddots & \ddots & \vdots \\
\vdots & & \ddots & a_1 & \vdots & & & \ddots & b_1 \\
0 & \cdots & 0 & a_0 & 0 & \cdots & \cdots & 0 & b_0
\end{pmatrix}$$

$$\underbrace{\hspace{4cm}}_{m \text{ columns}} \underbrace{\hspace{4cm}}_{n \text{ columns}}$$

5.24. THEOREM. *Assume[†] P and Q split into products of linear factors over some field K containing F, and let R denote the resultant of P and Q. The following conditions are then equivalent:*

    (a) P and Q are not relatively prime;

    (b) P and Q have a common root in K;

    (c) R = 0.

Proof. *(a) ⇒ (b)* Let D be a GCD of P and Q. By hypothesis, D is not constant. Moreover, since D divides P (and Q) which splits into linear factors over K, the irreducible factors of D in $K[X]$ have degree 1. Therefore, D has at least one root in K, which is also a common root of P and Q since D divides P and Q.

(b) ⇒ (c) Let $u \in K$ be a common root of P and Q. Let

$$P = (X - u)P_1 \quad \text{and} \quad Q = (X - u)Q_1$$

where $P_1, Q_1 \in K[X]$. Then

$$PQ_1 = QP_1 \quad (= \frac{PQ}{X - u}). \tag{5.4}$$

From this equality, we obtain a system of $m + n$ equations by equating the coefficients of like terms. More precisely, if

$$P = a_n X^n + a_{n-1} X^{n-1} + \cdots + a_1 X + a_0,$$
$$Q = b_m X^m + b_{m-1} X^{m-1} + \cdots + b_1 X + b_0,$$

as above, and if

$$P_1 = -(z_1 X^{n-1} + z_2 X^{n-2} + \cdots + z_{n-1} X + z_n)$$

and

$$Q_1 = y_1 X^{m-1} + y_2 X^{m-2} + \cdots + y_{m-1} X + y_m,$$

then the coefficient of X^k in $PQ_1 - Q_1 P$ is

$$\sum_{i+j=k} (a_i y_{m-j} + b_i z_{n-j}) = \sum_{r-s=k-m} a_r y_s + \sum_{t-u=k-n} b_t z_u$$

(where we set $a_r = 0$ (resp. $b_t = 0$) if $r > n$ or $r < 0$ (resp. if $t > m$ or $t < 0$) and $y_s = 0$ (resp. $z_u = 0$) if $s > m$ or $s < 1$ (resp. $u > n$ or $u <$

[†]Theorem 9.3, p. 116, will show that this hypothesis always holds.

1)). Therefore, equation (5.4) is equivalent to the following system of equations, whose left sides are the coefficients of X^{n+m-1}, $X^{n+m-2}, \ldots, X^n, X^{n-1}, \ldots,$ X and the constant term in $PQ_1 - QP_1$:

$$
\begin{cases}
a_n y_1 & +b_m z_1 & = 0 \\
a_{n-1} y_1 + a_n y_2 & +b_{m-1} z_1 + b_m z_2 & = 0 \\
\vdots \quad \ddots & \vdots & \vdots \\
\cdots +a_n y_m \quad +b_1 z_1 \quad +\cdots & & = 0 \\
\cdots +a_{n-1} y_m \quad +b_0 z_1 \quad +\cdots & & = 0 \\
\vdots & \ddots \quad \vdots \quad \vdots \\
a_0 y_m & +b_0 z_n = 0
\end{cases}
\tag{5.5}
$$

This system can be regarded as a system of $m+n$ homogeneous linear equations in the indeterminates $y_1, \ldots, y_m, z_1, \ldots, z_n$. It is easily verified that the coefficient matrix of this system is the matrix which appears in the definition of R. It follows that $R = 0$, since $y_1, \ldots, y_m, z_1, \ldots, z_n$ is a non-trivial solution of this system.

(c) $\Rightarrow$ *(a)* Assume now that $R = 0$; then, reversing the steps in the proof of *(b)* $\Rightarrow$ *(c)*, we observe that the system (5.5) has a non-trivial solution and conclude that there exist non-zero polynomials P_1 and Q_1 such that

$$
PQ_1 = QP_1, \qquad \deg P_1 \leq n - 1, \qquad \deg Q_1 \leq m - 1.
$$

(The inequalities $\deg P_1 < n - 1$ and $\deg Q_1 < m - 1$ occur when $y_1 = z_1 = 0$.) The preceding equality shows that P divides QP_1. If P is relatively prime to Q, then it divides P_1, by Lemma 5.9, p. 49. This is impossible since $\deg P_1 < \deg P$; therefore, P and Q are not relatively prime. $\qquad\square$

Appendix: Decomposition of rational fractions in sums of partial fractions

To find a primitive of a rational fraction $\frac{P}{Q}$ (where $P, Q \in \mathbb{R}[X]$ and $Q \neq 0$), it is customary to decompose it as a sum of partial fractions in the following way: factor Q into irreducible factors

$$
Q = Q_1^{m_1} \cdots Q_r^{m_r}
$$

where $Q_1, \ldots, Q_r$ are distinct irreducible polynomials. Then there exist polynomials $P_0, P_1, \ldots, P_r$ such that

$$\frac{P}{Q} = P_0 + \frac{P_1}{Q_1^{m_1}} + \cdots + \frac{P_r}{Q_r^{m_r}}$$

and $\deg P_i < \deg Q_i^{m_i}$ for $i = 1, \ldots, r$.

The existence of the polynomials $P_0, \ldots, P_r$ follows by induction on r from the following proposition:

5.25. PROPOSITION. *If $Q = S_1 S_2$ where S_1 and S_2 are relatively prime polynomials, then there exist polynomials P_0, P_1, P_2 such that*

$$\frac{P}{Q} = P_0 + \frac{P_1}{S_1} + \frac{P_2}{S_2}$$

and $\deg P_i < \deg S_i$ for $i = 1, 2$.

Proof. Since S_1 and S_2 are relatively prime, Corollary 5.4 (p. 47) yields

$$1 = S_1 T_1 + S_2 T_2$$

for some polynomials T_1, T_2. Multiplying each side by $\frac{P}{Q}$, we get

$$\frac{P}{Q} = \frac{P T_1}{S_2} + \frac{P T_2}{S_1}.$$

By Euclidean division of $P T_1$ by S_2 and of $P T_2$ by S_1, we have

$$P T_1 = S_2 U_2 + P_2 \quad \text{and} \quad P T_2 = S_1 U_1 + P_1$$

for some polynomials U_1, U_2, P_1, P_2 with $\deg P_i < \deg S_i$ for $i = 1, 2$. Substituting for $P T_1$ and $P T_2$ in the preceding equation, we obtain

$$\frac{P}{Q} = (U_1 + U_2) + \frac{P_2}{S_2} + \frac{P_1}{S_1}.$$

$\square$

To facilitate integration, each partial fraction $\frac{P}{Q^m}$ (where Q is irreducible and $\deg P < \deg Q^m$) can be decomposed further as

$$\frac{P}{Q^m} = \frac{P_1}{Q} + \frac{P_2}{Q^2} + \cdots + \frac{P_m}{Q^m}$$

with $\deg P_i < \deg Q$ for $i = 1, \ldots, m$.

To obtain this decomposition, let P_1 be the quotient of the Euclidean division of P by Q^{m-1},

$$P = P_1 Q^{m-1} + R_1 \qquad \text{with} \qquad \deg R_1 < \deg Q^{m-1}.$$

Since $\deg P < \deg Q^m$ it follows that $\deg P_1 < \deg Q$. Let then P_2 be the quotient of the Euclidean division of R_1 by Q^{m-2}, and so on. Then

$$P = P_1 Q^{m-1} + P_2 Q^{m-2} + \cdots + P_{m-1} Q + P_m, \qquad (5.6)$$

and the required decomposition follows after the division of both sides by Q^m.

Remark. The right-hand side of (5.6) is the "Q-adic" expansion of P. When P and Q are replaced by integers, equation (5.6) shows that the integer P is written as $P_1 P_2 \ldots P_m$ in the base Q.

Chapter 6

Alternative Methods for Cubic and Quartic Equations

With their improved notations, mathematicians of the seventeenth century devised new methods for solving cubic and quartic equations. The aim of this chapter is to review some of these advances, in particular the important method proposed by Ehrenfried Walter Tschirnhaus in 1683.

6.1 Viète on cubic equations

Viète's contribution to the theory of cubic equations is twofold: in "De Recognitione Aequationum" he gave a trigonometric solution for the irreducible case and in "De Emendatione Aequationum" a solution for the general case which requires the extraction of only one cube root. These methods were both posthumously published in "De Aequationum Recognitione et Emendatione Tractatus Duo" (1615) ("Two Treatises on the Understanding and Amendment of Equations," see [65]).

6.1.1 *Trigonometric solution for the irreducible case*

The irreducible case of cubic equations

$$X^3 + pX + q = 0$$

occurs when $\left(\frac{p}{3}\right)^3 + \left(\frac{q}{2}\right)^2 < 0$ (see § 2.3(c)). This inequality of course implies $p < 0$, hence there is no loss of generality if the above equation is written as

$$X^3 - 3a^2 X = a^2 b, \tag{6.1}$$

and the condition $\left(\frac{p}{3}\right)^3 + \left(\frac{q}{2}\right)^2 < 0$ becomes $a > |\frac{b}{2}|$. (Note that one can obviously assume $a > 0$, since only a^2 occurs in the given equation.)

From the formula for the cosine of a sum of arcs (or from the general formula (4.1) in Chapter 4, p. 33), it follows that for all $\alpha \in \mathbb{R}$

$$(2a \cos \alpha)^3 - 3a^2(2a \cos \alpha) = 2a^3 \cos 3\alpha.$$

Comparing with equation (6.1), we see that if α is an arc such that

$$\cos 3\alpha = \frac{b}{2a},$$

then $2a \cos \alpha$ is a solution of equation (6.1). The two other solutions are easily derived from this one; if $\alpha_k = \alpha + k\frac{2\pi}{3}$, with $k = 1, 2$, then we also have

$$\cos 3\alpha_k = \frac{b}{2a},$$

whence the solutions of equation (6.1) are

$$2a \cos \alpha,$$
$$2a \cos\left(\alpha + \tfrac{2\pi}{3}\right) = -a \cos \alpha - a\sqrt{3} \sin \alpha,$$
$$\text{and} \qquad 2a \cos\left(\alpha + \tfrac{4\pi}{3}\right) = -a \cos \alpha + a\sqrt{3} \sin \alpha.$$

Since Viète systematically avoids negative numbers, he gives only the first solution, which is positive if $b > 0$ (see [65, p. 174]). However, he points out immediately afterwards that $a \cos \alpha + a\sqrt{3} \sin \alpha$ and $a \cos \alpha - a\sqrt{3} \sin \alpha$ are solutions of the equation

$$3a^2 Y - Y^3 = a^2 b.$$

This shows how clear was Viète's notion of the number of roots of cubic equations.

6.1.2   *Algebraic solution for the general case*

Viète suggested (in [65, p. 287]) an ingenious change of variable to solve the equation

$$X^3 + pX + q = 0. \qquad (6.2)$$

Setting $X = \frac{p}{3Y} - Y$ and substituting in the equation above, he gets for Y the equation

$$Y^6 - qY^3 - \left(\frac{p}{3}\right)^3 = 0,$$

whence Y^3 can be found by solving a quadratic equation. The solutions are

$$Y^3 = \frac{q}{2} \pm \sqrt{\left(\frac{p}{3}\right)^3 + \left(\frac{q}{2}\right)^2}.$$

Therefore, a solution of the cubic equation (6.2) is given by

$$X = \frac{p}{3Y} - Y,$$

where

$$Y^3 = \frac{q}{2} + \sqrt{\left(\frac{p}{3}\right)^3 + \left(\frac{q}{2}\right)^2}.$$

Remarks. (a) If the other determination of Y^3 is chosen, namely

$$Y'^3 = \frac{q}{2} - \sqrt{\left(\frac{p}{3}\right)^3 + \left(\frac{q}{2}\right)^2},$$

then the value of X does not change. Indeed, since $(YY')^3 = -\left(\frac{p}{3}\right)^3$, we have

$$\frac{p}{3Y} = -Y' \quad \text{and} \quad \frac{p}{3Y'} = -Y,$$

whence

$$\frac{p}{3Y} - Y = \frac{p}{3Y'} - Y'.$$

Incidentally, this remark also shows that Viète's method yields the same result as Cardano's formula, since after substituting for Y and Y' in the formula $X = -Y - Y'$, we obtain

$$X = \sqrt[3]{-\frac{q}{2} + \sqrt{\left(\frac{p}{3}\right)^3 + \left(\frac{q}{2}\right)^2}} + \sqrt[3]{-\frac{q}{2} - \sqrt{\left(\frac{p}{3}\right)^3 + \left(\frac{q}{2}\right)^2}}.$$

(b) The case where $Y = 0$ occurs only if $p = 0$; this case is therefore readily solved.

(c) Viète gives only one root, because in the original formulation the equation is*

$$A^3 + 3B^p A = 2Z^s,$$

which has only one real root if B^p is positive, since the function $A^3 + 3B^p A$ is then monotonically increasing and therefore takes the value $2Z^s$ only once.

---

*The exponents p, s are for *plano* and *solido*; unknowns are always designated by vowels.

6.2 Descartes on quartic equations

New insights into the solution of equations arose from the arithmetic of polynomials. In "La Geometrie," Descartes recommends the following way of attacking
equations of any degree: "First, try to put the given equation into the form of an
equation of the same degree obtained by multiplying together two others, each of a
lower degree" [16, p. 192]. He himself shows how this method can be successfully
applied to quartic equations [16, pp. 180ff].

After canceling out the cubic term, as in Ferrari's method (Chapter 3), the
general quartic equation is set in the form

$$X^4 + pX^2 + qX + r = 0, \tag{6.3}$$

and we may assume $q \neq 0$, otherwise the equation is quadratic in X^2 and is
therefore easily solved. We then determine a, b, c, d in such a way that

$$X^4 + pX^2 + qX + r = (X^2 + aX + b)(X^2 + cX + d).$$

Equating the coefficients of similar powers of X, we obtain from this equation

$$0 = a + c, \tag{6.4}$$

$$p = b + d + ac, \tag{6.5}$$

$$q = ad + bc, \tag{6.6}$$

$$r = bd. \tag{6.7}$$

From equations (6.4), (6.5), (6.6), the values of b, c and d are easily derived in
terms of a:

$$c = -a,$$

$$b = \frac{a^2}{2} + \frac{p}{2} - \frac{q}{2a},$$

$$d = \frac{a^2}{2} + \frac{p}{2} + \frac{q}{2a}.$$

(Observe that $a \neq 0$ since $q \neq 0$.) Substituting for b and d in equation (6.7), we
get the following equation for a:

$$a^6 + 2pa^4 + (p^2 - 4r)a^2 - q^2 = 0. \tag{6.8}$$

This is a cubic equation in a^2, which can therefore be solved. If a is a solution of this equation, then the given equation (6.3) factors into two quadratic equations

$$X^2 + aX + \frac{a^2}{2} + \frac{p}{2} - \frac{q}{2a} = 0 \quad \text{and} \quad X^2 - aX + \frac{a^2}{2} + \frac{p}{2} + \frac{q}{2a} = 0,$$

whence the solutions are easily found.

6.3 Rational solutions for equations with rational coefficients

The rational solutions of equations with rational coefficients of arbitrary degree can be found by a finite trial and error process. This seems to have been first observed by Albert Girard [26, D.4 v°]; it also appears in "La Geometrie" [16, p. 176].

Let

$$a_n X^n + a_{n-1} X^{n-1} + \cdots + a_1 X + a_0 = 0 \tag{6.9}$$

be an equation with rational coefficients $a_i \in \mathbb{Q}$ for $i = 0, \ldots, n$. Multiplying each side by a common multiple of the denominators of the coefficients if necessary, we may assume $a_i \in \mathbb{Z}$ for $i = 0, \ldots, n$. Multiplying then each side by a_n^{n-1}, the equation becomes

$$(a_n X)^n + a_{n-1}(a_n X)^{n-1} + a_{n-2} a_n (a_n X)^{n-2} + \cdots$$
$$+ a_1 a_n^{n-2}(a_n X) + a_0 a_n^{n-1} = 0.$$

Letting $Y = a_n X$, we are then reduced to a *monic* equation with integral coefficients

$$Y^n + b_{n-1} Y^{n-1} + b_{n-2} Y^{n-2} + \cdots + b_1 Y + b_0 = 0 \quad (b_i \in \mathbb{Z}). \tag{6.10}$$

6.1. THEOREM. *All the rational roots of a monic equation with integral coefficients are integers which divide the independent term.*

Proof. Discarding the null roots and dividing the left-hand side of (6.10) by a suitable power of Y, we may assume $b_0 \neq 0$. Let then $y \in \mathbb{Q}$ be a rational root of (6.10). Write $y = \frac{y_1}{y_2}$ where y_1, y_2 are relatively prime integers. From

$$\left(\frac{y_1}{y_2}\right)^n + b_{n-1}\left(\frac{y_1}{y_2}\right)^{n-1} + \cdots + b_1\left(\frac{y_1}{y_2}\right) + b_0 = 0$$

it follows, after multiplication by y_2^n and rearrangement of the terms,

$$y_1^n = -y_2\left(b_{n-1}y_1^{n-1} + \cdots + b_1 y_1 y_2^{n-2} + b_0 y_2^{n-1}\right).$$

This equation shows that each prime factor of y_2 divides y_1^n, whence also y_1; since y_1 and y_2 are relatively prime, this is impossible unless y_2 has no prime factor. Therefore $y_2 = \pm 1$ and $y \in \mathbb{Z}$.

To prove that y divides b_0, consider again equation (6.10) and separate off on one side the constant term; we obtain

$$b_0 = -y\left(y^{n-1} + b_{n-1}y^{n-2} + \cdots + b_1\right).$$

This equation shows that y divides b_0, since the factor between brackets is an integer.   □

Tracing back through the transformations from equation (6.9) to equation (6.10), we get the following result:

6.2. COROLLARY. *Each rational solution of the equation with integral coefficients*

$$a_n X^n + a_{n-1} X^{n-1} + \cdots + a_1 X + a_0 = 0 \qquad (a_i \in \mathbb{Z})$$

has the form $y a_n^{-n}$, *where* $y \in \mathbb{Z}$ *is a divisor of* $a_0 a_n^{n-1}$.

This last condition if very useful, in that it gives a bound on the number of trials which are necessary to find a rational root of the proposed equation, provided that $a_0 \neq 0$. Of course, this can always be assumed, after dividing by a suitable power of X.

For example, the theorem (or its corollary) shows that an equation like

$$X^n + a_{n-1} X^{n-1} + \cdots + a_1 X \pm 1 = 0$$

with $a_i \in \mathbb{Z}$ for $i = 1, \ldots, n-1$, has no rational root, except possibly $+1$ or -1.

"Other example, once very difficult" [26, E r°]: the rational solutions of

$$X^3 = 7X - 6$$

are among ± 1, ± 2, ± 3, ± 6. Trying successively the various possibilities, one finds 1, 2 and -3 as solutions.

6.4 Tschirnhaus' method

Although research on the theory of equations was not quite as active at the end of the seventeenth century, substantial progress arose from a 4-page note by Tschirnhaus [58], in 1683. This note proposes a uniform method to solve equations of any degree.

The basic idea is very simple; it starts from the observation that it is always possible to remove the second term of any equation

$$X^n + a_{n-1}X^{n-1} + \cdots + a_1 X + a_0 = 0$$

by the simple change of variable $Y = X + \frac{a_{n-1}}{n}$. (See e.g. §2.2 and §3.2). By allowing more general changes of variable, such as

$$Y = X^m + b_{m-1}X^{m-1} + \cdots + b_1 X + b_0, \qquad (6.11)$$

Tschirnhaus aims to cancel out several terms of the proposed equation. More precisely, by a suitable choice of the m parameters $b_0, b_1, \ldots, b_{m-1}$, the above change of variable yields an equation in Y of the form

$$Y^n + c_{n-1}Y^{n-1} + \cdots + c_1 Y + c_0 = 0$$

in which any m coefficients c_i can be chosen to vanish. Roughly speaking, this is because the m parameters $b_0, \ldots, b_{m-1}$ provide m degrees of freedom, which can be used to fulfill m conditions. In particular, taking $m = n - 1$, all the terms except the first and the last could be removed, hence the equation in Y takes the form

$$Y^n + c_0 = 0,$$

and is thus readily solved by radicals. Plugging in the solution $Y = \sqrt[n]{-c_0}$ in equation (6.11), we then obtain a solution of the proposed equation of degree n by solving an equation of degree $m = n - 1$, namely

$$X^{n-1} + b_{n-2}X^{n-2} + \cdots + b_1 X + b_0 = \sqrt[n]{-c_0}.$$

Arguing by induction on the degree, it thus follows that equations of any degree can be solved by radicals.

There is however a major obstacle, which was soon noticed by Leibniz [43, p. 449, p. 403]: the conditions which ensure that all the coefficients $c_1, \ldots, c_{n-1}$ vanish yield a system of equations of various degrees in the parameters b_i, and this system is very difficult to solve. Indeed, solving this system actually amounts to solving a single equation of degree $1 \cdot 2 \cdot \ldots \cdot (n - 1) = (n - 1)!$. It thus appears that this method does not work for $n > 3$, unless the resulting equation of degree $(n - 1)!$ has some particular features which make it reducible to equations of degree less than n. This turns out to be the case for $n = 4$: the resulting sextic can be seen to factor into a product of factors of degree 2 whose coefficients are solutions of cubic equations (see Lagrange [40, Art. 41–45]), but for $n \geq 5$ no such simplification is apparent. (Note that for composite n, Tschirnhaus' method can be applied differently, and possibly more easily. For instance if $n = 4$, then canceling out the coefficients of Y and Y^3 reduces the equation in Y to a quadratic equation in Y^2.)

To discuss Tschirnhaus' method in some detail, we start by explaining how the equation in Y can be found. This is a special instance of a general type of problem which is dealt with by elimination theory. The problem is to eliminate the indeterminate X between the two equations

$$X^n + a_{n-1}X^{n-1} + a_{n-2}X^{n-2} + \cdots + a_1 X + a_0 = 0, \qquad (6.12)$$
$$X^m + b_{m-1}X^{m-1} + b_{m-2}X^{m-2} + \cdots + b_1 X + b_0 = Y, \qquad (6.13)$$

(with $m < n$), i.e. to find an equation $R(Y) = 0$, called a *resulting equation*, which has the following properties:

(a) whenever x and y are such that equations (6.12) and (6.13) hold, then $R(y) = 0$,

(b) whenever y is such that $R(y) = 0$, then equations (6.12) and (6.13) have a common root x.

This last property shows that if $R(Y) = 0$ can be solved, then one (at least) of the roots of equation (6.12) is among the roots of equation (6.13).

The properties of $R(Y)$ can be rephrased as follows, considering (6.12) and (6.13) as equations in X with coefficients in the field of rational fractions in Y: $R(Y) = 0$ if and only if the polynomials

$$P(X) = X^n + a_{n-1}X^{n-1} + \cdots + a_1 X + a_0$$

and

$$Q(X) = X^m + b_{m-1}X^{m-1} + \cdots + b_1 X + (b_0 - Y)$$

have a common root. As Theorem 5.24 (p. 56) shows, a solution of this problem is the resultant[†] $R(Y)$ of P and Q, defined as the determinant of the following matrix:

$$
\begin{pmatrix}
1 & 0 & \cdots & 0 & 1 & 0 & \cdots & \cdots & 0 \\
a_{n-1} & 1 & & \vdots & b_{m-1} & 1 & & & \vdots \\
\vdots & a_{n-1} & \ddots & 0 & \vdots & b_{m-1} & \ddots & & \vdots \\
\vdots & \vdots & \ddots & 1 & b_1 & \vdots & \ddots & \ddots & 0 \\
a_i & \vdots & & a_{n-1} & b_0 - Y & b_1 & & \ddots & 1 \\
a_0 & a_1 & & \vdots & 0 & b_0 - Y & \ddots & & b_{m-1} \\
0 & a_0 & \ddots & \vdots & \vdots & & \ddots & \ddots & \vdots \\
\vdots & & \ddots & a_1 & \vdots & & & \ddots & b_1 \\
0 & \cdots & 0 & a_0 & 0 & \cdots & & 0 & b_0 - Y
\end{pmatrix}
$$

$\underbrace{}_{m \text{ columns}}$ $\underbrace{}_{n \text{ columns}}$

Since the indeterminate Y appears only in the last n columns, it is easily verified that R is a polynomial of degree n in Y. Moreover, since the determinant is an alternating sum of products of entries from different rows and columns, it follows that products of only k factors b_i occur in the coefficient of Y^{n-k}. Therefore,

$$R(Y) = c_n Y^n + c_{n-1} Y^{n-1} + \cdots + c_1 Y + c_0$$

where c_{n-k} is a polynomial of degree k in $b_0, \ldots, b_{m-1}$. (Actually, $c_n = (-1)^n$.)

In order to cancel out $c_{n-1}, c_{n-2}, \ldots, c_1$, consider now $m = n - 1$. The preceding discussion shows that $c_{n-1} = c_{n-2} = \ldots = c_1 = 0$ is a system of $n - 1$ equations of degrees $1, 2, 3, \ldots, n - 1$ in the variables $b_0, \ldots, b_{n-2}$. Between these equations, $n - 2$ variables can be eliminated, and the resulting equation in a single variable has degree $1 \cdot 2 \cdot 3 \cdot \ldots \cdot (n - 1) = (n - 1)!$ (see for instance Weber [67, §53]). This was proved only much later by Bezout, but by considering some examples one soon realizes that the solution of the above system of equations is far from easy.

[†]It is slightly anachronistic to resort to determinants in this context, since they came into use somewhat later, but the actual calculations in elimination theory were equivalent and they in fact motivated the development of determinants.

Let us consider for instance the cubic equation

$$X^3 + pX + q = 0 \qquad (p \neq 0) \tag{6.14}$$

and let

$$Y = X^2 + b_1 X + b_0. \tag{6.15}$$

Elimination of X between these two equations according to the method explained above yields the following resulting equation in Y:

$$c_3 Y^3 + c_2 Y^2 + c_1 Y + c_0 = 0 \tag{6.16}$$

where

$$c_3 = -1,$$
$$c_2 = 3b_0 - 2p,$$
$$c_1 = 4pb_0 - 3qb_1 - 3b_0^2 - pb_1^2 - p^2,$$
$$c_0 = q^2 + p^2 b_0 - pqb_1 + 3qb_0 b_1 - 2pb_0^2 + b_0^3 - qb_1^3 + pb_0 b_1^2.$$

Thus, in order to cancel out c_2 and c_1, it suffices to let $b_0 = \frac{2p}{3}$ and to choose for b_1 a root of the quadratic equation

$$pb_1^2 + 3qb_1 - \frac{p^2}{3} = 0,$$

for instance

$$b_1 = \frac{3}{p} \left(\sqrt{\left(\frac{p}{3}\right)^3 + \left(\frac{q}{2}\right)^2} - \frac{q}{2} \right).$$

With the above choice of b_0 and b_1, and letting $A = \sqrt{\left(\frac{p}{3}\right)^3 + \left(\frac{q}{2}\right)^2}$, we have

$$c_0 = 2^3 A^3 \left(\frac{3}{p}\right)^3 \left(A - \frac{q}{2}\right).$$

Therefore, a root of the resulting equation (6.16) in Y is

$$Y = \frac{6A}{p} \sqrt[3]{A - \frac{q}{2}}.$$

A root of the proposed cubic equation (6.14) is then found by solving the quadratic equation (6.15), which is now

$$X^2 + \frac{3}{p}\left(A - \frac{q}{2}\right)X + \frac{2p}{3} = \frac{6A}{p}\sqrt[3]{A - \frac{q}{2}}. \tag{6.17}$$

However, in general only one of the roots of this quadratic equation is a root of the proposed cubic equation (6.14). A better way to solve (6.14) is to find the common roots of (6.14) and (6.17), which are the roots of their greatest common divisor.

Letting $B = \sqrt[3]{A - \frac{q}{2}}$, one gets by Euclid's algorithm (p. 44) the following greatest common divisor, if $A \neq 0$:

$$2A\left(\frac{3}{p}\right)^2\left(B^2 + \frac{p}{3}\right)\left(BX + \frac{p}{3} - B^2\right).$$

(It is easy to see that $B^2 + \frac{p}{3} \neq 0$ if $A \neq 0$ and $p \neq 0$.) There is thus only one common root of (6.14) and (6.17); namely

$$X = \frac{B^2 - \frac{p}{3}}{B}.$$

Since $B = \sqrt[3]{-\frac{q}{2} + \sqrt{\left(\frac{p}{3}\right)^3 + \left(\frac{q}{2}\right)^2}}$, it is easily verified that

$$-\frac{p}{3B} = \sqrt[3]{-\frac{q}{2} - \sqrt{\left(\frac{p}{3}\right)^3 + \left(\frac{q}{2}\right)^2}};$$

thus the above formula for X is identical to Cardano's formula. (Compare also Viète's method in § 6.1.)

If $A = 0$, then the left-hand side of (6.17) divides the given cubic polynomial, so both roots of (6.17) are roots of the proposed equation.

Chapter 7

Roots of Unity

7.1 Introduction

New branches of mathematics, such as analytic geometry and differential calculus, came into being during the seventeenth century, and it is therefore not surprising that investigations in the algebraic theory of equations came to a near standstill at the end of this century, being pursued only occasionally by leading mathematicians such as Tschirnhaus. However, progress in other branches indirectly brought some new advances in algebra. A case in point is the well-known "de Moivre's formula": for every integer n and every $\alpha \in \mathbb{R}$,

$$(\cos \alpha + i \sin \alpha)^n = \cos(n\alpha) + i \sin(n\alpha), \qquad (7.1)$$

which is easily proved by induction on n, since from the addition formulas for sines and cosines it readily follows that

$$(\cos \alpha + i \sin \alpha)(\cos \beta + i \sin \beta) = \cos(\alpha + \beta) + i \sin(\alpha + \beta). \qquad (7.2)$$

This formula (7.1), and its proof through (7.2), were first given by Euler in 1748 (see Smith [54, vol. 2, p. 450]) but it was already implicit in earlier works by Cotes and by de Moivre. Actually, the above proof, simple as it is, is deceitful, since it does not keep any record of the slow evolution which led to de Moivre's formula. It is the purpose of this chapter to sketch this evolution and to discuss the significance of de Moivre's formula for the algebraic theory of equations.

7.2   The origin of de Moivre's formula

While the differential calculus was being shaped by Leibniz and Newton, the integration (or primitivation) of rational fractions was unavoidable. Very soon, formulas equivalent to

$$\int x^n dx = \frac{x^{n+1}}{n+1} \qquad \text{for } n \neq -1$$

and

$$\int \frac{dx}{x} = \log x$$

became familiar, and the integration of any rational fraction in which the denominator is a power of a linear polynomial easily follows by a change of variable. Moreover, around 1675, Leibniz had also obtained

$$\int \frac{dx}{x^2 + 1} = \tan^{-1} x,$$

from which the integration of other rational fractions can be derived.

The integration of rational fractions is the main theme of a 1702 paper by Leibniz in the Acta Eruditorum of Leipzig : "Specimen novum Analyseos pro Scientia infiniti circa Summas et Quadraturas" ("New specimen of the Analysis for the Science of the infinite about Sums and Quadratures" [42, n° 24]). In this paper, Leibniz points out the usefulness of the decomposition of rational fractions into sums of partial fractions (see the appendix to Chapter 5) to reduce the integration of rational fractions to the integration of $\frac{dx}{x}$ and $\frac{dx}{x^2+1}$ or, in his words, to the quadrature of the hyperbola or the circle. Since this decomposition requires that the denominator be factored in a product of irreducible polynomials, he is thus led to investigate the factorization of real polynomials, coming close to the "fundamental theorem of algebra" according to which every real polynomial of positive degree is a product of factors of degree 1 or 2 (see Chapter 9).

> Now, this leads us to a question of utmost importance: whether all the rational quadratures may be reduced to the quadrature of the hyperbola and of the circle, which by our analysis above amounts to the following: whether every algebraic equation or real integral formula in which the indeterminate is rational can be decomposed into simple or plane real factors [= real factors of degree 1 or 2]. [42, p. 359]

Leibniz then proposes the following counterexample: since

$$x^4 + a^4 = (x^2 + a^2\sqrt{-1})(x^2 - a^2\sqrt{-1}),$$

it follows that

$$x^4 + a^4 =$$
$$\left(x + a\sqrt{\sqrt{-1}}\right)\left(x - a\sqrt{\sqrt{-1}}\right)\left(x + a\sqrt{-\sqrt{-1}}\right)\left(x - a\sqrt{-\sqrt{-1}}\right).$$

Failing to observe that

$$\sqrt{\sqrt{-1}} = \frac{1 + \sqrt{-1}}{\sqrt{2}} \quad \text{and} \quad \sqrt{-\sqrt{-1}} = \frac{1 - \sqrt{-1}}{\sqrt{2}},$$

he draws the erroneous conclusion that no non-trivial combination of the four factors above yields a real divisor of $x^4 + a^4$.

> Therefore, $\int \frac{dx}{x^4+a^4}$ cannot be reduced to the squaring of the circle or the hyperbola by our analysis above, but founds a new kind of its own. [42, p. 360]

Even without deeper considerations about complex numbers, Leibniz could have avoided this mistake if he had observed that, by adding and subtracting $2a^2x^2$, one gets*

$$x^4 + a^4 = (x^2 + a^2)^2 - 2a^2x^2$$
$$= (x^2 + a^2 + \sqrt{2}ax)(x^2 + a^2 - \sqrt{2}ax).$$

As it appears from [45, v. IV, pp. 205 ff], Newton had also tried his hand at the same questions as early as 1676, and he had obtained this factorization of $x^4 + a^4$, as well as factorizations of $1 \pm x^n$ for various values of the integer n, (see the appendix), but in 1702 he presumably did not care enough about mathematics any more to point out the mistake in Leibniz's paper, had he been aware of it.

Leibniz's argument was definitively refuted by Roger Cotes (1682–1716), who thoroughly investigated the factorization of the binomials $a^n \pm x^n$, obtaining the

---

*This was pointed out by N. Bernoulli in the Acta Eruditorum of 1719.

following formulas:

$$a^{2m} + x^{2m} = \prod_{k=0}^{m-1}\left(a^2 - 2a\cos\left(\frac{(2k+1)\pi}{2m}\right)x + x^2\right), \tag{7.3}$$

$$a^{2m+1} + x^{2m+1} = (a + x)\prod_{k=0}^{m-1}\left(a^2 - 2a\cos\left(\frac{(2k+1)\pi}{2m+1}\right)x + x^2\right), \tag{7.4}$$

$$a^{2m} - x^{2m} = (a - x)(a + x)\prod_{k=1}^{m-1}\left(a^2 - 2a\cos\left(\frac{2k\pi}{2m}\right)x + x^2\right), \tag{7.5}$$

$$a^{2m+1} - x^{2m+1} = (a - x)\prod_{k=1}^{m}\left(a^2 - 2a\cos\left(\frac{2k\pi}{2m+1}\right)x + x^2\right). \tag{7.6}$$

These formulas appear in a compilation of Cotes' papers entitled "Theoremata tum Logometrica tum Trigonometrica Datarum Fluxionum Fluentes exhibentia, per Methodum Mensurarum Ulterius extensam" (1722) ("Theorems, some logometric, some trigonometric, which yield the fluents of given fluxions by the method of measures further developed" [15, pp. 113–114]), in a very elegant form: to find the factors of $a^\lambda \pm x^\lambda$, it is prescribed to divide a circle of radius a into 2λ equal parts AB, BC, CD, DE, EF, etc. Let O be the center of the circle and let P be a point on the radius OA, at a distance $OP = x \ (< a)$ from O. Then

$$a^\lambda - x^\lambda = OA^\lambda - OP^\lambda = AP \cdot CP \cdot EP \cdot \text{etc.,} \qquad \text{and}$$
$$a^\lambda + x^\lambda = OA^\lambda + OP^\lambda = BP \cdot DP \cdot FP \cdot \text{etc.}$$

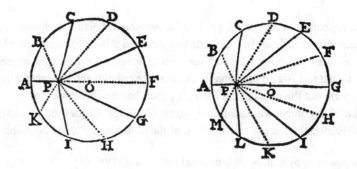

Exempli gratia fi λ fit *5*, dividatur circumferentia in 10 partes æquales, eritque $A\mathcal{P} \times C\mathcal{P} \times E\mathcal{P} \times G\mathcal{P} \times I\mathcal{P} = OA^s - O\mathcal{P}^s$ existente $\mathcal{P}$ intra circulum: & $B\mathcal{P} \times D\mathcal{P} \times F\mathcal{P} \times H\mathcal{P} \times K\mathcal{P} = OA^s$ $+O\mathcal{P}^s$. Similiter fi λ fit *6*, divisa circumferentia in 12 partes æquales: erit $A\mathcal{P} \times C\mathcal{P} \times E\mathcal{P} \times G\mathcal{P} \times I\mathcal{P} \times L\mathcal{P} = OA^6 - O\mathcal{P}^6$, existente $\mathcal{P}$ intra circulum; & $B\mathcal{P} \times D\mathcal{P} \times F\mathcal{P} \times H\mathcal{P} \times K\mathcal{P}$ $\times M\mathcal{P} = OA^6 + O\mathcal{P}^6$.

[15, p. 114] (Univ. Cath. Louvain, Centre général de Documentation)

To check that this formulation is equivalent to the previous one, it suffices to observe that, on the figure below,

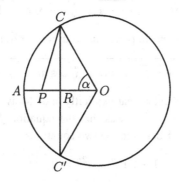

we have, by Pythagoras' theorem,

$$CP^2 = PR^2 + RC^2.$$

Since

$$PR = OP - OR = x - a\cos\alpha \qquad \text{and} \qquad RC = a\sin\alpha,$$

it follows that

$$CP = \sqrt{x^2 - 2ax\cos\alpha + a^2}.$$

Therefore,

$$CP \cdot C'P = x^2 - 2ax\cos\alpha + a^2.$$

Cotes' formulas were given without justification, but a proof was eventually supplied in 1730 by Abraham de Moivre (1667–1754), who had already obtained some interesting results on the division of the circle. In a 1707 paper entitled "Aequationum quaerundam Potestatis tertiae, quintae, septimae, novae, &c superiorum, ad infinitum usque pergendo, in terminis finitis, ad imstar Regularum pro Cubicis quae vocantur, *Cardani*, Resolutio Analytica" ("The analytic solution in finite terms of certain equations of the third, fifth, seventh, ninth and other higher powers, by rules similar to those called *Cardano's* for the cubics", see Smith [54, vol. 2, pp. 441 ff]), he had observed that the equation

$$f_n(X) = 2a \qquad (n \text{ odd})$$

where f_n is the polynomial by which $2 \cos n\alpha$ is expressed as a function of $2 \cos \alpha$ (see equation (4.1) in Chapter 4, p. 33), has the solution

$$X = \sqrt[n]{a + \sqrt{a^2 - 1}} + \sqrt[n]{a - \sqrt{a^2 - 1}},$$

for any value of a, whatsoever. In particular, if $a = \cos n\alpha$, it follows that

$$2 \cos \alpha = \sqrt[n]{\cos n\alpha + \sqrt{-1} \sin n\alpha} + \sqrt[n]{\cos n\alpha - \sqrt{-1} \sin n\alpha}, \qquad (7.7)$$

although this formula does not appear explicitly in de Moivre's paper of 1707.

De Moivre's basic observation was that the equation $f_n(X) = 2a$ can be obtained by elimination of z between the two equations

$$1 - 2az^n + z^{2n} = 0, \qquad (7.8)$$

$$1 - Xz + z^2 = 0. \qquad (7.9)$$

Indeed, equation (7.9) yields

$$1 + z^2 = Xz$$

and, squaring both sides of this equation, we get

$$1 + z^4 = (X^2 - 2)z^2.$$

These last equations show that, for $n = 1, 2$,

$$1 + z^{2n} = f_n(X)z^n. \qquad (7.10)$$

From these initial steps, it is easily verified by induction that equation (7.10) holds for every integer n, using the recurrence formula

$$f_{n+1}(X) = X f_n(X) - f_{n-1}(X) \qquad (7.11)$$

(see §4.2). Comparing (7.10) with (7.8), we obtain $f_n(X) = 2a$. Now, dividing by z both sides of (7.9), it follows that $X = z + z^{-1}$, while (7.8) yields

$$z^n = a \pm \sqrt{a^2 - 1}.$$

We thus obtain several equivalent expressions for X:

$$X = \sqrt[n]{a + \sqrt{a^2 - 1}} + \left(\sqrt[n]{a + \sqrt{a^2 - 1}}\right)^{-1},$$

$$\text{or} \quad X = \sqrt[n]{a + \sqrt{a^2 - 1}} + \sqrt[n]{a - \sqrt{a^2 - 1}},$$

$$\text{or} \quad X = \left(\sqrt[n]{a - \sqrt{a^2 - 1}}\right)^{-1} + \sqrt[n]{a - \sqrt{a^2 - 1}},$$

$$\text{or} \quad X = \left(\sqrt[n]{a - \sqrt{a^2 - 1}}\right)^{-1} + \left(\sqrt[n]{a + \sqrt{a^2 - 1}}\right)^{-1}.$$

The equivalence of these expressions is easily derived from the equation

$$(a + \sqrt{a^2 - 1})(a - \sqrt{a^2 - 1}) = 1.$$

De Moivre repeatedly returned to these questions in the sequel, displaying formula (7.7) quite explicitly on p. 1 of his book "Miscellanea Analytica" (1730) (see Smith [54, vol. 2, p. 446]). It is noteworthy that for $X = 2\cos\alpha$ and $a = \cos n\alpha$, the values of z obtained by solving equations (7.8) and (7.9) are

$$\sqrt[n]{a \pm \sqrt{a^2 - 1}} = \sqrt[n]{\cos n\alpha \pm \sqrt{-1}\sin n\alpha}$$

and

$$\frac{X}{2} \pm \sqrt{\left(\frac{X}{2}\right)^2 - 1} = \cos\alpha \pm \sqrt{-1}\sin\alpha,$$

so that

$$\sqrt[n]{\cos n\alpha \pm \sqrt{-1}\sin n\alpha} = \cos\alpha \pm \sqrt{-1}\sin\alpha, \tag{7.12}$$

but this was never written out explicitly by de Moivre.

Nevertheless, de Moivre's approach turned out to be quite fruitful, since Cotes' formulas can be easily proved by pushing the preceding calculations a little further (see Exercise 1).

In 1739, de Moivre used the trigonometric representation of complex numbers and presumably also his formula, which certainly was thoroughly familiar to him by then, to extract the n-th root of the "impossible binomial $a + \sqrt{-b}$" (see Smith

[54, vol. 2, p. 449]). He states the procedure as follows: let φ be an angle such that

$$\cos \varphi = \frac{a}{\sqrt{a^2 + b}};$$

then the n-th roots of $a + \sqrt{-b}$ are

$$\sqrt[2n]{a^2 + b}\left(\cos \psi + \sqrt{\cos^2 \psi - 1}\right),$$

where ψ ranges over $\frac{\varphi}{n}$, $\frac{2\pi-\varphi}{n}$, $\frac{2\pi+\varphi}{n}$, $\frac{4\pi-\varphi}{n}$, $\frac{4\pi+\varphi}{n}$, etc. until the number of them is equal to n. (This result is correct up to the sign of the imaginary part: see Proposition 7.1 below, p. 81.)

As a result of this work, the credibility of the "fundamental theorem of algebra" was significantly enhanced, since the objection that Leibniz had raised was definitely answered: it was clear that extraction of roots of complex numbers does not produce imaginary numbers of a new kind. Moreover, since equations of degree at most 4 can be solved by radicals, it follows from de Moivre's result that polynomials of degree at most 4 split into products of linear factors over the field of complex numbers. It was not long afterwards that the first attempts to prove the fundamental theorem were made (without formulas for the solution of higher degree equations by radicals), and we shall come back to this topic in Chapter 9.

Another consequence with far-reaching implications is that the n-th root of any (non-zero) number is ambiguous: it has n different determinations. Therefore, every formula which involves the extraction of a root needs some clarification as to *which* root should be chosen. This observation, which was conspicuously used as a starting-point in Vandermonde's subsequent investigations, sheds a completely new light on the problem of solving equations by radicals, and even on known solutions. Indeed, Cardano's formula, as it appears in §2.2 (see equation (2.3), p. 16), involves the extraction of two cube roots; if we consider various determinations of these cube roots, we obtain the three solutions of the cubic equation: this solves the puzzle of §2.3(a).[†]

Moreover, even de Moivre's formula as it appears in (7.12) above is ambiguous. To express it properly, one has to raise $\cos \alpha + \sqrt{-1} \sin \alpha$ to the n-th power instead of extracting the n-th root of $\cos n\alpha + \sqrt{-1} \sin n\alpha$. This viewpoint was adopted by Euler in his "Introductio in Analysin Infinitorum" (1748) (see Smith [54, vol. 2, p. 450]), in which he proves de Moivre's formula (7.1) as in §7.1. Later

---

[†]With the notations of §2.2, the cube roots should be determined in such a way that their product be $-\frac{p}{3}$, as it appears from the proof of Cardano's formula in §2.2 or alternatively from Viète's method in §6.1.2.

in the same book, comparing the power series expansions of the exponential and of the sine and cosine functions, Euler also states

$$e^{\alpha\sqrt{-1}} = \cos\alpha + \sqrt{-1}\sin\alpha,$$

a relation from which de Moivre's formula readily follows. Of course, once de Moivre's formula is established, the other major results of this section, and in particular Cotes' formulas, can be seen as easy applications. We devote the next section to a streamlined exposition of de Moivre's results on the roots of complex numbers along these lines.

7.3 The roots of unity

Let a and b be real numbers, not both zero, and denote by $\sqrt{a^2 + b^2}$ the (real, positive) square root of $a^2 + b^2$. Since

$$\left(\frac{a}{\sqrt{a^2 + b^2}}\right)^2 + \left(\frac{b}{\sqrt{a^2 + b^2}}\right)^2 = 1,$$

there is a unique angle φ such that $0 \leq \varphi < 2\pi$,

$$\cos\varphi = \frac{a}{\sqrt{a^2 + b^2}} \qquad \text{and} \qquad \sin\varphi = \frac{b}{\sqrt{a^2 + b^2}}.$$

We thus obtain the trigonometric expression of the complex number $a + bi \neq 0$, namely

$$a + bi = \sqrt{a^2 + b^2}(\cos\varphi + i\sin\varphi).$$

7.1. PROPOSITION. *For any positive integer n, the n distinct n-th roots of $a + bi$ are*

$$\sqrt[2n]{a^2 + b^2}\left(\cos\frac{\varphi + 2k\pi}{n} + i\sin\frac{\varphi + 2k\pi}{n}\right) \tag{7.13}$$

for $k = 0, \ldots, n - 1$.

In this formula, $\sqrt[2n]{a^2 + b^2}$ is the unique real positive $2n$-th root of $a^2 + b^2$.

Proof. De Moivre's formula (7.1) yields

$$\left(\sqrt[2n]{a^2 + b^2}\left(\cos\frac{\varphi + 2k\pi}{n} + i\sin\frac{\varphi + 2k\pi}{n}\right)\right)^n = \sqrt{a^2 + b^2}(\cos\varphi + i\sin\varphi),$$

so that each of the expressions (7.13) is an n-th root of $a + bi$. Moreover, these expressions are easily seen to be pairwise distinct for $k = 0, \ldots, n - 1$, since for these values of k it is impossible that two among the angles $\frac{\varphi + 2k\pi}{n}$ differ by a multiple of 2π.   □

7.2. DEFINITION. A complex number ζ is called an n-th root of unity, for some integer n, if $\zeta^n = 1$.

The set of all n-th roots of unity is denoted by μ_n. Thus,

$$X^n - 1 = \prod_{\zeta \in \mu_n} (X - \zeta).$$

By the preceding proposition we have

$$\mu_n = \left\{ e^{2k\pi i/n} = \cos \frac{2k\pi}{n} + i \sin \frac{2k\pi}{n} \mid k = 0, \ldots, n - 1 \right\},$$

hence

$$X^n - 1 = \prod_{k=0}^{n-1} \left(X - \cos \frac{2k\pi}{n} - i \sin \frac{2k\pi}{n} \right). \tag{7.14}$$

This formula can be used to produce a factorization of $X^n - 1$ into real factors; indeed, if $k + \ell = n$, then

$$\cos \frac{2k\pi}{n} = \cos \frac{2\ell\pi}{n} \quad \text{and} \quad \sin \frac{2k\pi}{n} = -\sin \frac{2\ell\pi}{n},$$

whence

$$\left(X - \cos \frac{2k\pi}{n} - i \sin \frac{2k\pi}{n} \right)\left(X - \cos \frac{2\ell\pi}{n} - i \sin \frac{2\ell\pi}{n} \right) =$$
$$X^2 - 2\cos\left(\frac{2k\pi}{n}\right)X + 1.$$

Therefore, multiplying the corresponding pairs of factors in the right-hand side of (7.14), we obtain

$$X^n - 1 = (X - 1) \prod_{k=1}^{\frac{n-1}{2}} \left(X^2 - 2\cos\left(\frac{2k\pi}{n}\right)X + 1 \right) \qquad \text{if } n \text{ is odd,}$$

$$X^n - 1 = (X - 1)(X + 1) \prod_{k=1}^{\frac{n}{2}-1} \left(X^2 - 2\cos\left(\frac{2k\pi}{n}\right)X + 1 \right) \quad \text{if } n \text{ is even.}$$

Substituting a/x for X in these formulas and multiplying each side by x^n to clear denominators, we recover Cotes' formulas (7.5) and (7.6). Formulas (7.3) and (7.4) can be similarly derived by considering n-th roots of -1 instead of n-th roots of 1.

It should be observed that, in a rectangular coordinate system, the points $\left(\cos\frac{2k\pi}{n}, \sin\frac{2k\pi}{n}\right)$ for $k = 0, 1, \ldots, n-1$, which represent the n-th roots unity in the planar representation of $\mathbb{C}$, are the vertices of a regular polygon with n sides: they divide the unit circle into n equal parts. For this reason, the theory which is concerned with n-th roots of unity or with the values of the cosine and sine functions at $\frac{2k\pi}{n}$ for integers k, n, is called *cyclotomy*, meaning literally "division of the circle" [into equal parts].

Likewise, the n-th roots of any non-zero complex number are represented in the plane of complex numbers by the vertices of a regular n-gon, as Proposition 7.1 shows.

That the roots of 1 deserve special interest comes from the fact that, if an n-th root u of some complex number v has been found, then the various determinations of $\sqrt[n]{v}$ are the products ωu, where ω runs over the set of n-th roots of unity. This is easily seen from

$$(\omega u)^n = \omega^n u^n = 1 \cdot u^n = v$$

or, equivalently, from Proposition 7.1.

While the n-th roots of unity have been determined above by a trigonometric expression, yet the problem of deciding whether these roots of unity have an expression by radicals has been untouched. We now turn to this problem, and prove, after some ideas of de Moivre:

7.3. THEOREM. *Let n be a positive integer. If, for each prime factor p of n, the p-th roots of unity can be expressed by radicals, then the n-th roots of unity can be expressed by radicals.*

This theorem follows by induction from the following result:

7.4. LEMMA. *Let r and s be positive integers. If $\xi_1, \ldots, \xi_r$ (resp. $\eta_1, \ldots, \eta_s$) are the r-th roots of unity (resp. the s-th roots of unity), then the rs-th roots of unity are of the form $\xi_i \sqrt[r]{\eta_j}$ for $i = 1, \ldots, r$ and $j = 1, \ldots, s$.*

Proof. From the factorization of $Y^s - 1$, it follows by letting $Y = X^r$ that

$$X^{rs} - 1 = \prod_{j=1}^{s} (X^r - \eta_j).$$

Therefore, the rs-th roots of unity are the r-th roots of the various η_j, for $j = 1$, $\ldots$, s. □

Proof of Theorem 7.3. We argue by induction on the number of factors of n. If n is a prime number, then there is nothing to prove. Assume then that $n = rs$ for some positive integers r, $s \neq 1$. Then the number of factors of r (resp. s) is strictly less than the number of factors of n, whence, by induction, the r-th roots of unity $\xi_1, \ldots, \xi_r$ and the s-th roots of unity $\eta_1, \ldots, \eta_s$ can be expressed by radicals. Since the n-th roots of unity are of the form $\xi_i \sqrt[r]{\eta_j}$, these can also be expressed by radicals. □

7.5. Remark. Of course, the expressions thus obtained are not necessarily the simplest ones or the most suitable for the actual calculation of these roots. For instance, since the 4-th roots of unity are ± 1 and $\pm\sqrt{-1}$, the 8-th roots of unity are obtained as

$$\pm 1, \ \pm\sqrt{-1}, \ \pm\sqrt{\sqrt{-1}} \text{ and } \pm\sqrt{-\sqrt{-1}},$$

while they can also be expressed as

$$\pm 1, \ \pm\sqrt{-1}, \ \frac{1 \pm \sqrt{-1}}{\sqrt{2}} \text{ and } -\frac{1 \pm \sqrt{-1}}{\sqrt{2}},$$

since $\cos\frac{\pi}{4} = \sin\frac{\pi}{4} = \frac{1}{\sqrt{2}}$.

Moreover, the result of Lemma 7.4 can be improved when r and s are relatively prime: in this case, one of the determinations of $\sqrt[r]{\eta_j}$ is an s-th root of unity η_k, so that the rs-th roots of unity are the products of the form $\xi_i \eta_k$ ($i = 1, \ldots, r$ and $k = 1, \ldots, s$), see Remark 7.13 (p. 90) and Exercise 3.

Theorem 7.3 shows that, in order to find expressions by radicals for the n-th roots of unity, for any integer n, it suffices to consider the case where n is prime. Since the equation $X^n - 1 = 0$ has the obvious root $X = 1$, we may divide $X^n - 1$ by $X - 1$, and the question reduces to the following problem: solve by radicals the equation

$$X^{n-1} + X^{n-2} + \cdots + X + 1 = 0, \tag{7.15}$$

for n prime. This equation is readily solved for $n = 2$ and 3: the roots are -1 for $n = 2$, and $\frac{-1 \pm \sqrt{-3}}{2}$ for $n = 3$.

For $n \geq 5$, the following trick (due to de Moivre) is useful: after division by $X^{\frac{n-1}{2}}$, the change of variable $Y = X + X^{-1}$ transforms equation (7.15) into an equation of degree $\frac{n-1}{2}$ in Y. (This trick succeeds because in the polynomial

(7.15) the coefficients of the terms which are symmetric with respect to the middle term are equal.) Thus, for $n = 5$, we first divide each side of

$$X^4 + X^3 + X^2 + X + 1 = 0$$

by X^2, and the change of variable $Y = X + X^{-1}$ transforms the resulting equation $X^2 + X + 1 + X^{-1} + X^{-2} = 0$ into

$$Y^2 + Y - 1 = 0.$$

We thus find

$$Y = \frac{-1 \pm \sqrt{5}}{2},$$

and the values of X are obtained by solving $X + X^{-1} = Y$ for the various values of Y. Thus, the 5-th roots of unity (other than 1) are the roots of the equations

$$X^2 - \frac{-1 + \sqrt{5}}{2}X + 1 = 0 \quad \text{and} \quad X^2 - \frac{-1 - \sqrt{5}}{2}X + 1 = 0,$$

which are

$$\frac{\sqrt{5} - 1 \pm \sqrt{-10 - 2\sqrt{5}}}{4} \quad \text{and} \quad \frac{-\sqrt{5} - 1 \pm \sqrt{-10 + 2\sqrt{5}}}{4}.$$

Similarly, for $n = 7$, de Moivre's trick yields for Y $(= X + X^{-1})$ the cubic equation

$$Y^3 + Y^2 - 2Y - 1 = 0$$

which can be solved by radicals; the 7-th roots of unity can therefore be expressed by radicals.

However, for the next prime number, which is 11, de Moivre's trick yields an equation of degree 5, for which no general formula by radicals is known. Solving this equation was one of the greatest achievements of Vandermonde (see Chapter 11).

7.6. Remarks. (a) Since the roots of equation (7.15) are $e^{2k\pi i/n}$ for $k = 1, \dots, n - 1$, the roots of the equation in Y $(= X + X^{-1})$ are

$$e^{2k\pi i/n} + e^{-2k\pi i/n} = 2\cos\frac{2k\pi}{n} \quad \text{for } k = 1, \dots, \frac{n - 1}{2}.$$

Therefore, the calculations above yield expressions by radicals for $2\cos\frac{2\pi}{5}$ and $2\cos\frac{4\pi}{5}$ as the positive and the negative root of $Y^2 + Y - 1 = 0$,

$$2\cos\frac{2\pi}{5} = \frac{\sqrt{5}-1}{2} \qquad \text{and} \qquad 2\cos\frac{4\pi}{5} = -\frac{\sqrt{5}+1}{2}.$$

(b) From Theorem 7.3 and the results above, it follows that the $2 \cdot 3^3 \cdot 5^2$-th roots of unity can be expressed by radicals. Hence, $\sin\frac{\pi}{3^3 \cdot 5^2}$ can also be expressed by radicals, since

$$\sin\frac{\pi}{3^3 \cdot 5^2} = \frac{1}{2i}\left(e^{2\pi i/2\cdot 3^3 \cdot 5^2} - e^{-2\pi i/2\cdot 3^3 \cdot 5^2}\right).$$

This explains why Van Roomen was able to give a solution by radicals in his third example, see §4.2.

7.4   Primitive roots and cyclotomic polynomials

In this section, we complete our discussion of the elementary aspects of the theory of roots of unity by listing several results which are more or less straightforward consequences of de Moivre's formula.[†] They are a natural outgrowth of the theory developed so far, and became known during the second half of the eighteenth century. The central notion is the following:

7.7. DEFINITIONS. The *exponent* of a root of unity ζ is the smallest integer $e > 0$ such that $\zeta^e = 1$. For instance, the exponent of 1 is 1 and the exponent of -1 is 2, although 1 is an n-th root of unity for every n and -1 is an n-th root of unity for every even n.

The n-th roots of unity of exponent n are also called *primitive n-th roots of unity.*

We aim to give a complete description of the primitive n-th roots of unity and to show that these roots are indeed primitive, in the sense that the other n-th roots of unity can be obtained as powers of any such root. A basic ingredient in the proofs is the following number-theoretic proposition:

[†]More precisely, these results follow from the fact that the set μ_n of n-th roots of unity is a finite subgroup of the multiplicative group of complex numbers.

7.8. THEOREM. *Let d be the (positive) greatest common divisor of two integers* n_1, n_2. *Then there exist integers* m_1, m_2 *such that*

$$d = n_1 m_1 + n_2 m_2.$$

In particular, if n_1 and n_2 are relatively prime, then there exist integers m_1, m_2 such that

$$n_1 m_1 + n_2 m_2 = 1.$$

Proof. Duplicate the arguments in the proof of Theorem 5.3 (p. 45) with integers instead of polynomials. □

By the way, it is useful to note that most of the arguments in §§5.2 and 5.3 can be carried out with integers instead of polynomials, using the absolute value of integers instead of the degree of polynomials; the point is that we also have a Euclidean division property for integers: if m and n are integers and if $m \neq 0$, then there are integers q, r such that

$$n = mq + r \qquad \text{and} \qquad 0 \leq r < |m|.$$

Moreover, the integers q and r are uniquely determined by these properties. Thus, mimicking the proofs in §§5.2 and 5.3, we get a proof of the unique factorization of integers into prime factors and obtain along the way other useful results like: "If an integer divides a product of r factors and is relatively prime to the $(r - 1)$ first factors, then it divides the last one" (compare Lemma 5.9) or "If an integer is divisible by pairwise relatively prime integers, then it is divisible by their product" (compare Proposition 5.10).

Theorem 7.8 is sometimes known as "Bezout's theorem." This is clearly a misnomer, since it can be traced back at least to Bachet de Méziriac, in his "Problèmes plaisans et délectables qui se font par les nombres" (1624). However, it could be fair to associate the name of Bezout to a similar statement for polynomials with coefficients in a field of rational fractions in one or several indeterminates (i.e. Theorem 5.3 with $F = K(X_1, \ldots, X_n)$ for some field K). This last result implies the existence, for any two polynomials $P_1(X_1, \ldots, X_{n+1})$ and $P_2(X_1, \ldots, X_{n+1})$ in $n+1$ indeterminates, of polynomials $Q_1(X_1, \ldots, X_{n+1})$, $Q_2(X_1, \ldots, X_{n+1})$ and $D(X_1, \ldots, X_n)$ such that

$$P_1 Q_1 + P_2 Q_2 = D.$$

Since the indeterminate X_{n+1} does not appear in D, one says that D is obtained by elimination of X_{n+1} between P_1 and P_2. (The relation with the resultant of P_1 and P_2 (considered as polynomials in X_{n+1}) is pointed out in Exercise 7.)

We now come back to roots of unity with the following result which characterizes the exponent of roots of unity:

7.9. LEMMA. *Let e be the exponent of a root of unity ζ and let m be an integer. Then $\zeta^m = 1$ if and only if e divides m. In particular, the exponent of an n-th root of unity divides n.*

Proof. If e divides m, then $m = ef$ for some integer f, and since $\zeta^e = 1$, it readily follows from $\zeta^m = (\zeta^e)^f$ that $\zeta^m = 1$.

Conversely, assume $\zeta^m = 1$ and let d be the greatest common divisor of m and e. By Theorem 7.8, there are integers r and s such that

$$mr + es = d.$$

Since $\zeta^d = (\zeta^m)^r(\zeta^e)^s$, we have $\zeta^d = 1$; but since e is the smallest exponent for which the power of ζ is 1, it follows that $d \geq e$, whence $d = e$ since d divides e. Therefore, e divides m. $\qquad\square$

7.10. PROPOSITION. *Let ζ and η be roots of unity of exponents e and f respectively. If e and f are relatively prime, then $\zeta\eta$ is a root of unity of exponent ef.*

Proof. Since $\zeta^e = 1$ and $\eta^f = 1$, we have

$$(\zeta\eta)^{ef} = 1,$$

hence it follows from the preceding lemma that the exponent of $\zeta\eta$, which we denote by k, divides ef.

On the other hand, from $(\zeta\eta)^k = 1$, it follows that

$$\zeta^k = \eta^{-k},$$

and raising each side to the power f yields

$$\zeta^{kf} = 1.$$

The preceding lemma then shows that e divides kf. Since e is relatively prime to f, by hypothesis, it follows that e divides k. Likewise, interchanging ζ and η, we see that f divides k. Since e and f are relatively prime and divide k, their product ef divides k; but we have already observed that k divides ef, hence $k = ef$. $\qquad\square$

We now show that the primitive n-th roots of unity generate the other n-th roots of unity.

7.11. PROPOSITION. *If ζ is a primitive n-th root of unity, then the n-th roots of unity are of the form*

$$\zeta^0 = 1, \, \zeta, \, \zeta^2, \, \ldots, \, \zeta^{n-1}.$$

Conversely, if ζ is an n-th root of unity such that every n-th root of unity is a power of ζ, then ζ is primitive.

Proof. Since ζ is an n-th root of unity, all the powers of ζ are n-th roots of unity, whence the set

$$S = \{\zeta^i \mid i = 0, 1, \ldots, n - 1\}$$

consists of n-th roots of unity, i.e. $S \subseteq \mu_n$. To prove that all the n-th roots of unity are contained in S, i.e. $S = \mu_n$, it then suffices to prove that S has n distinct elements. This amounts to proving the following claim: *the various powers ζ^i for $i = 0, \ldots, n - 1$ are pairwise distinct.*

Assume on the contrary $\zeta^i = \zeta^j$ for some i, j with $0 \le i < j \le n - 1$. Then $\zeta^{j-i} = 1$, whence, by Lemma 7.9, the exponent n of ζ divides $j - i$. This is impossible, since $0 < j - i < n$, and this contradiction proves the claim. Conversely, assume that every n-th root of unity is a power of ζ. If ζ is not primitive, then $\zeta^m = 1$ for some positive integer $m < n$. Then every power of ζ, hence every n-th root of unity, is an m-th root of unity, i.e. $\mu_n \subseteq \mu_m$. This is clearly impossible, since there are n distinct roots of unity while the number of m-th roots of unity is only m. □

Remark. The proposition above shows that the (multiplicative) group μ_n is generated by a single element: one says that μ_n is a *cyclic* group. The proposition also shows that the generators of μ_n are the primitive n-th roots of unity.

A complete description of the primitive n-th roots of unity will be obtained as a consequence of the following result:

7.12. PROPOSITION. *Let ζ be a primitive n-th root of unity and let k be an integer. Then ζ^k is a primitive n-th root of unity if and only if k is relatively prime to n.*

Proof. Assume first that k and n have a common factor $d \neq 1$. Then

$$(\zeta^k)^{n/d} = (\zeta^n)^{k/d},$$

whence $(\zeta^k)^{n/d} = 1$ since $\zeta^n = 1$. Therefore, the exponent of ζ^k divides n/d, hence ζ^k is not a primitive n-th root of unity.

Conversely, assume that ζ^k is not primitive; then there is a positive integer $m < n$ such that $(\zeta^k)^m = 1$. Since the exponent of ζ is n, it follows from Lemma 7.9 that n divides km. If n is relatively prime to k, then it divides m; but this is impossible since $m < n$. Therefore, n and k are not relatively prime. $\square$

7.13. Remark. Let r and s be relatively prime integers and let η be a primitive s-th root of unity. Proposition 7.12 shows that η^r is a primitive s-th root of unity, hence by Proposition 7.11 the s-th roots of unity are

$$\mu_s = \{\eta^{ri} \mid i = 0, \dots, s - 1\}.$$

Therefore, every s-th root of unity can be written as η^{ri} for some (unique) integer i between 0 and $s - 1$, hence every s-th root of unity has an r-th root in μ_s.

7.14. COROLLARY. The primitive n-th roots of unity are

$$e^{2k\pi i/n} = \cos \frac{2k\pi}{n} + i \sin \frac{2k\pi}{n}$$

where k runs over the positive integers which are less than n and relatively prime to n. In particular, if n is prime, then every n-th root of unity except 1 is primitive.

Proof. Let $\zeta = \cos \frac{2\pi}{n} + i \sin \frac{2\pi}{n}$. Since

$$\mu_n = \left\{ \cos \frac{2k\pi}{n} + i \sin \frac{2k\pi}{n} \mid k = 0, \dots, n - 1 \right\},$$

we have by de Moivre's formula

$$\mu_n = \{\zeta^k \mid k = 0, \dots, n - 1\}.$$

Therefore, Proposition 7.11 shows that ζ is a primitive n-th root of unity and it follows from Proposition 7.12 that every primitive n-th root of unity is of the form ζ^k where k is a positive integer relatively prime to n between 0 and $n - 1$. $\square$

We now introduce the polynomials which have as roots the primitive n-th roots of unity. Because of their relation with the division of the circle, these polynomials are called *cyclotomic polynomials*.

7.15. DEFINITION. The cyclotomic polynomials Φ_n ($n = 1, 2, 3, \ldots$) are defined inductively by $\Phi_1(X) = X - 1$ and, for $n \geq 2$,

$$\Phi_n(X) = \frac{X^n - 1}{\prod_{\substack{d|n \\ d \neq n}} \Phi_d(X)}$$

where d runs over the set of divisors of n, with $d \neq n$.

In particular, if p is a prime number, we have

$$\Phi_p(X) = \frac{X^p - 1}{X - 1} = X^{p-1} + X^{p-2} + \cdots + X + 1.$$

However, if n is not prime, it is not clear *a priori* that Φ_n is a polynomial. We prove this and the fact that the roots of Φ_n are the primitive n-th roots of unity simultaneously:

7.16. PROPOSITION. *For every integer $n \geq 1$, the rational fraction Φ_n is a monic polynomial with integral coefficients, and*

$$\Phi_n(X) = \prod_{\zeta}(X - \zeta),$$

where ζ runs over the set of primitive n-th roots of unity.

Proof. We argue by induction on n. Since the proposition is trivial for $n = 1$, we assume that it holds for all integers up to $n - 1$. Then, for every divisor d of n, with $d \neq n$, we have

$$\Phi_d(X) = \prod_{\zeta}(X - \zeta) \in \mathbb{Z}[X],$$

where ζ runs over the roots of unity of exponent d. Since the roots of exponent d are n-th roots of unity, it follows that Φ_d divides $X^n - 1$. Moreover, if d_1 and d_2 are distinct divisors of n, then Φ_{d_1} and Φ_{d_2} are relatively prime since their roots are pairwise distinct. Therefore, by Proposition 5.10, the product $\prod_d \Phi_d(X) \in \mathbb{Z}[X]$ divides $X^n - 1$, and it follows that $\Phi_n(X)$ is a polynomial in $\mathbb{Q}[X]$, which is monic since $X^n - 1$ and Φ_d (for every proper divisor d of n) are monic. Moreover, the proof of the Euclidean division property (Theorem 5.1, p. 43) shows that the quotient of a polynomial in $\mathbb{Z}[X]$ by a monic polynomial in $\mathbb{Z}[X]$ is in $\mathbb{Z}[X]$. Therefore, $\Phi_n(X) \in \mathbb{Z}[X]$.

Now, the effect of dividing $X^n - 1$ by $\prod_{\substack{d|n \\ d \neq n}} \Phi_d(X)$ is to remove from

$$X^n - 1 = \prod_{\zeta \in \mu_n} (X - \zeta)$$

all the factors $X - \zeta$ where ζ has exponent $d \neq n$. Therefore, in $\Phi_n(X)$ remain all the factors $X - \zeta$ where ζ has exponent n, and only those factors. Thus

$$\Phi_n(X) = \Pi_\zeta (X - \zeta)$$

where ζ runs over the set of primitive n-th roots of unity.   □

Remark. It will be shown in Theorem 12.31 (p. 198) that Φ_n is irreducible in $\mathbb{Q}[X]$, for every n. The proof is easier when n is prime, see Theorem 12.10 (p. 175).

Appendix: Leibniz and Newton on the summation of series

Around 1675, Leibniz obtained the following result, of which he was justifiably proud:

$$1 - \frac{1}{3} + \frac{1}{5} - \frac{1}{7} + \cdots = \frac{\pi}{4}.$$

His method was to use the formula

$$\int \frac{dx}{x^2 + 1} = \tan^{-1} x, \tag{7.16}$$

which yields

$$\int_0^1 \frac{dx}{x^2 + 1} = \frac{\pi}{4},$$

together with the power series expansion of $(x^2 + 1)^{-1}$, namely

$$\frac{1}{x^2 + 1} = 1 - x^2 + x^4 - x^6 + x^8 - \cdots$$

from which it follows that

$$\int_0^1 \frac{dx}{x^2 + 1} = \int_0^1 dx - \int_0^1 x^2 dx + \int_0^1 x^4 dx - \cdots$$

$$= 1 - \frac{1}{3} + \frac{1}{5} - \cdots$$

Of course, the fact that the integral of the power series expansion of $(x^2+1)^{-1}$ is equal to the series of the integrals of the terms needs some justification, which was supplied much later.

In 1676, Newton sent to Leibniz the following variant:

$$1 + \frac{1}{3} - \frac{1}{5} - \frac{1}{7} + \frac{1}{9} + \frac{1}{11} - \frac{1}{13} - \frac{1}{15} + \cdots = \frac{\pi}{2\sqrt{2}},$$

with the terse hint that it depended on the reduction of an integrand to partial fractions (see [45, v. IV, p. 212]). It is very likely that Newton's result was obtained from the evaluation of $\int_0^1 \frac{(x^2+1)dx}{x^4+1}$. Indeed, the rational fraction $\frac{x^2+1}{x^4+1}$ has the following decomposition into partial fractions:

$$\frac{x^2+1}{x^4+1} = \frac{1}{2}\frac{1}{X^2+\sqrt{2}X+1} + \frac{1}{2}\frac{1}{X^2-\sqrt{2}X+1},$$

and, by the change of variable $Y = \sqrt{2}X + 1$ (resp. $Y = \sqrt{2}X - 1$), it easily follows from (7.16) that

$$\int \frac{dx}{x^2+\sqrt{2}x+1} = \sqrt{2}\tan^{-1}(\sqrt{2}x+1),$$

resp. $$\int \frac{dx}{x^2-\sqrt{2}x+1} = \sqrt{2}\tan^{-1}(\sqrt{2}x-1).$$

Therefore,

$$\int_0^1 \frac{(x^2+1)dx}{x^4+1} = \frac{1}{\sqrt{2}}\left(\tan^{-1}(\sqrt{2}+1) + \tan^{-1}(\sqrt{2}-1)\right),$$

and since $(\sqrt{2}+1)(\sqrt{2}-1) = 1$, we have

$$\tan^{-1}(\sqrt{2}+1) + \tan^{-1}(\sqrt{2}-1) = \frac{\pi}{2},$$

whence

$$\int_0^1 \frac{(x^2+1)dx}{x^4+1} = \frac{\pi}{2\sqrt{2}}.$$

On the other hand, from the power series expansion

$$\frac{1}{x^4+1} = 1 - x^4 + x^8 - x^{12} + x^{16} - \cdots$$

it follows that

$$\frac{x^2 + 1}{x^4 + 1} = 1 + x^2 - x^4 - x^6 + x^8 + x^{10} - \cdots$$

whence

$$\int_0^1 \frac{(x^2 + 1)dx}{x^4 + 1} = \int_0^1 dx + \int_0^1 x^2 dx - \int_0^1 x^4 dx - \int_0^1 x^6 dx + \cdots$$

$$= 1 + \frac{1}{3} - \frac{1}{5} - \frac{1}{7} + \cdots$$

This proves Newton's result.

Exercises

1. *The aim of this first exercise is to prove Cotes' formulas along the lines of de Moivre's calculations with polynomials (§7.2).* Denote by $f_n(X) \in \mathbb{Q}[X]$ the monic polynomial of degree n such that $f_n(2 \cos \alpha) = 2 \cos n\alpha$ (see equation (4.1) in Chapter 4, p. 33) and let, for $n = 1, 2, 3, \ldots$,

$$P_n(X, z) = 1 - f_n(X)z^n + z^{2n} \in \mathbb{Q}[X, z].$$

(a) Prove: $P_1(X, z)$ divides $P_n(X, z)$ in $\mathbb{Q}[X, z]$ for every n. [Hint: Use induction on n and the recurrence relation (7.11), p. 78.]

(b) Prove: $P_n(2 \cos \alpha, z) = \prod_{k=0}^{n-1} P_1\left(2 \cos\left(\alpha + \frac{2k\pi}{n}\right), z\right)$ for $\alpha \in \mathbb{R}$. [Hint: Prove it first for $\alpha \notin \frac{\pi}{n}\mathbb{Z}$, using (a) and observing that for $\alpha \notin \frac{\pi}{n}\mathbb{Z}$ the polynomials in the right hand side are pairwise relatively prime. Since the coefficients of each side are continuous functions of α which are equal outside a discrete subset of $\mathbb{R}$, they are equal for every $\alpha \in \mathbb{R}$.]

(c) Show that for $\alpha = 0$ both sides of the relation in (b) are squares. Extracting the square root of each side, show that

$$1 - z^{2m} = (1 - z)(1 + z) \prod_{k=1}^{m-1} \left(1 - 2 \cos\left(\frac{2k\pi}{2m}\right)z + z^2\right)$$

and

$$1 - z^{2m+1} = (1 - z) \prod_{k=1}^{m} \left(1 - 2 \cos\left(\frac{2k\pi}{2m+1}\right)z + z^2\right).$$

For $\alpha = \frac{\pi}{n}$, derive similarly the formulas

$$1 + z^{2m} = \prod_{k=0}^{m-1}\left(1 - 2\cos\left(\frac{(2k+1)\pi}{2m}\right)z + z^2\right)$$

and

$$1 + z^{2m+1} = (1 + z)\prod_{k=0}^{m-1}\left(1 - 2\cos\left(\frac{(2k+1)\pi}{2m+1}\right)z + z^2\right).$$

Cotes' formulas (7.3)–(7.6) (p. 76) follow by substituting $\frac{x}{a}$ for z and multiplying each side by an appropriate power of a to clear denominators.

2. *The following exercise aims to show some of the remarkable properties of the polynomials f_n of Exercise 1.*

(a) Prove: $f_n(X) - 2\cos n\alpha = \prod_{k=0}^{n-1}(X - 2\cos(\alpha + \frac{2k\pi}{n}))$ for $\alpha \in \mathbb{R}$.
  [Hint: Argue as in Exercise 1(b).]
(b) Prove: $f_m(f_n(X)) = f_{mn}(X)$ for all integers m, n. [Hint: Use (a).]
(c) Prove: $f_m(X) \cdot f_n(X) = f_{m+n}(X) + f_{|m-n|}(X)$ for all integers m, n.
  (To allow $m = n$, set $f_0(X) = 2$.) [Hint: Use induction on n. For $n = 1$, use the recurrence relation (7.11).]

3. Let $\mu_r = \{\xi_1, \ldots, \xi_r\}$ and $\mu_s = \{\eta_1, \ldots, \eta_s\}$. Assuming that r and s are relatively prime, show that

$$\mu_{rs} = \{\xi_i \eta_j \mid i = 1, \ldots, r \text{ and } j = 1, \ldots, s\}.$$

[Hint: It suffices to prove that the products $\xi_i \eta_j$ are pairwise distinct. Use Theorem 7.8.]

4. Let ζ be a root of unity of exponent e and let k be an integer. Find the exponent of ζ^k. (Compare Proposition 7.12.)

5. Show that an n-th root of unity ζ is primitive if and only if $\zeta^d \neq 1$ for every proper factor d of n.

6. Let p be a prime number. Prove that

$$\sum_{\omega \in \mu_p} \omega^i = \begin{cases} 0 & \text{if } i \in \mathbb{Z} \text{ is not divisible by } p, \\ p & \text{if } i \in \mathbb{Z} \text{ is divisible by } p. \end{cases}$$

[Hint: Use Newton's formulas of Chapter 4.]

7. Let $P = a_n X^n + a_{n-1} X^{n-1} + \cdots + a_1 X + a_0$ and $Q = b_m X^m + b_{m-1} X_{m-1} + \cdots + b_1 X + b_0$ (with a_n, $b_m \neq 0$) be polynomials over a field F, and let $\mathcal{A}$ be the square matrix of order $m + n$ which appears in the definition of the resultant R of P and Q (so that $R = \det \mathcal{A}$, see §5.6, p. 56). Verify the following matrix multiplication:

$$\begin{pmatrix} X^{m+n-1} & X^{m+n-2} & \cdots & X & 1 \end{pmatrix} \cdot \mathcal{A} =$$
$$\begin{pmatrix} X^{m-1}P & \cdots & XP & P & X^{n-1}Q & \cdots & XQ & Q \end{pmatrix}.$$

Multiply each side of this matrix equation by the transpose of the matrix of co-factors of $\mathcal{A}$. Considering the last entry of the resulting line matrices, conclude that R is a linear combination of $X^{m-1}P, \ldots, XP, P, X^{n-1}Q, \ldots, XQ$ $\wp$, hence that there exist polynomials $U, V \in F[X]$ such that $\deg U \leq m - 1$, $\deg V \leq n - 1$ and

$$PU + QV = R.$$

Using this result, give an alternative proof of Theorem 5.24.

8. *This last exercise is an introduction to Euler's "totient function"* φ. For any integer $n \geq 2$, let $\varphi(n)$ be the number of integers which are relatively prime to n between 0 and $n - 1$.

    (a) Show that $\varphi(n) = \deg \Phi_n$ and that $\varphi(n)$ is equal to the number of prim-itive n-th roots of unity.

    (b) Show that $\varphi(mn) = \varphi(m)\varphi(n)$ if m and n are relatively prime. [Hint: Compare Exercise 3.]

    (c) Show that $\varphi(p^k) = p^{k-1}(p - 1)$ for any prime number p.

    (d) Derive from (b) and (c) the following formula: if $n = p_1^{k_1} \cdots p_r^{k_r}$ ($p_1, \ldots, p_r$ distinct prime numbers), then

$$\varphi(n) = \prod_{i=1}^{r} p_i^{k_i-1}(p_i - 1) = n \cdot \prod_{i=1}^{r}(1 - p_i^{-1}).$$

    (e) Show that $n = \sum_{d|n} \varphi(d)$ where d runs over the factors of n (including n). [Hint: Compare Definition 7.15.]

Chapter 8

Symmetric Functions

8.1 Introduction

During the first half of the eighteenth century, the structure of equations, as formerly investigated by Viète (§4.2), became clearer and clearer. Calculating formally with roots of equations, mathematicians became aware of the kind of information that can be gathered from the coefficients without solving equations. As Girard had shown (§4.2), for any polynomial

$$X^n - s_1 X^{n-1} + s_2 X^{n-2} - \cdots + (-1)^n s_n =$$
$$(X - x_1)(X - x_2) \cdots (X - x_n) \quad (8.1)$$

we have

$$s_1 = x_1 + \cdots + x_n,$$
$$s_2 = x_1 x_2 + \cdots + x_{n-1} x_n,$$
$$s_3 = x_1 x_2 x_3 + \cdots + x_{n-2} x_{n-1} x_n, \quad (8.2)$$
$$\cdots$$
$$s_n = x_1 x_2 \cdots x_n.$$

The following question naturally arises: what kind of function of the roots x_1, ..., x_n can be calculated from $s_1, \ldots, s_n$?

To translate properly the results of this period, laying on firm ground the formal calculations with roots of polynomials, the roots $x_1, \ldots, x_n$ should be considered as independent indeterminates over some base field F (usually, $F = \mathbb{Q}$, the field of rational numbers). Indeed, every calculation with indeterminates which does not involve divisions by non-constant polynomials can be done as well with

arbitrary elements in a field K containing the base field F. This is a loose translation of the fact that any map from $\{x_1, \ldots, x_n\}$ to K can be (uniquely) extended
to a ring homomorphism from the ring of polynomials $F[x_1, \ldots, x_n]$ to K, mapping a polynomial $P(x_1, \ldots, x_n)$ to the element of K obtained by substituting
for $x_1, \ldots, x_n$ their assigned values in K. (If we wish to allow divisions by non-
constant polynomials, caution is necessary since some denominators may vanish
in K.) Therefore, we introduce the following definition:

8.1 DEFINITION. If $x_1, \ldots, x_n$ are considered as independent indeterminates
over some base field F, the polynomial (8.1) above is called the *general* (or
generic) monic polynomial of degree n over F. Thus, this general polynomial
is a polynomial in one indeterminate X with coefficients in $F[x_1, \ldots, x_n]$; it may
thus be viewed as an element in $F[x_1, \ldots, x_n, X]$.

This polynomial is general (generic) in the following sense: if

$$P = X^n + a_{n-1}X^{n-1} + a_{n-2}X^{n-2} + \cdots + a_1 X + a_0$$

is an arbitrary monic polynomial of degree n over some field K containing F,
which splits* into a product of linear factors in some field L containing K, so that

$$P = (X - u_1) \cdots (X - u_n)$$

with $u_i \in L$ for $i = 1, \ldots, n$, then there is a homomorphism $F[x_1, \ldots, x_n] \to L$
mapping x_i to u_i for $i = 1, \ldots, n$. This ring homomorphism translates any
calculation with $x_1, \ldots, x_n$ into a calculation with $u_1, \ldots, u_n$. For instance, it is
easily seen that this ring homomorphism maps $s_1, s_2, \ldots, s_n$ onto $-a_{n-1}, a_{n-2},$
$\ldots, (-1)^n a_0 \in K$, which means that

$$s_1(u_1, \ldots, u_n) = -a_{n-1}, \quad \text{i.e.} \quad u_1 + \cdots + u_n = -a_{n-1},$$

$$s_2(u_1, \ldots, u_n) = a_{n-2}, \quad \text{i.e.} \quad u_1 u_2 + \cdots + u_{n-1} u_n = a_{n-2},$$

$$\cdots \qquad\qquad\qquad\qquad \cdots$$

$$s_n(u_1, \ldots, u_n) = (-1)^n a_0, \quad \text{i.e.} \quad u_1 \cdots u_n = (-1)^n a_0.$$

After this slight change of viewpoint, from arbitrary elements to indeterminates, the question becomes: which are the rational fractions in n indetermi-

---

*We shall see later that the condition that P splits into a product of linear factors in some field con-
taining K is always fulfilled, see §9.2. At this point this provision cannot be disposed of, however.
(Compare Remark 8.8(a) below, p. 105).

$x_1, \ldots, x_n$ that can be expressed as rational fractions in $s_1, \ldots, s_n$ (where s_1, $\ldots, s_n$ are defined by the equalities (8.2) above)?

The crucial condition turns out to be the following:

8.2. DEFINITION. A polynomial $P(x_1, \ldots, x_n)$ in n indeterminates is *symmetric* if it is not altered when the indeterminates are arbitrarily permuted among themselves; i.e., for every permutation σ of $1, \ldots, n$,

$$P(x_{\sigma(1)}, \ldots, x_{\sigma(n)}) = P(x_1, \ldots, x_n).$$

Similarly, a rational fraction $\frac{P}{Q}$ in n indeterminates is *symmetric* if it is not altered when the indeterminates are permuted; i.e., for every permutation σ of $1, \ldots, n$,

$$\frac{P(x_{\sigma(1)}, \ldots, x_{\sigma(n)})}{Q(x_{\sigma(1)}, \ldots, x_{\sigma(n)})} = \frac{P(x_1, \ldots, x_n)}{Q(x_1, \ldots, x_n)}.$$

Note that this does *not* imply that P and Q are both symmetric, since P and Q can be both multiplied by an arbitrary non-zero polynomial without changing the fraction $\frac{P}{Q}$, but we shall see below (p. 104) that every symmetric rational fraction can be represented as a quotient of symmetric polynomials.

Since the polynomials $s_1, \ldots, s_n$ are symmetric, it is clear that every rational fraction in $s_1, \ldots, s_n$ is a symmetric rational fraction in $x_1, \ldots, x_n$. The converse turns out to be also true, so that the following result holds:

8.3. THEOREM. *A rational fraction in n indeterminates $x_1, \ldots, x_n$ over a field F can be expressed as a rational fraction in $s_1, \ldots, s_n$ if and only if it is symmetric.*

This theorem is in fact a consequence of the analogous result for polynomials:

8.4. THEOREM. *A polynomial in n indeterminates $x_1, \ldots, x_n$ over a field F can be expressed as a polynomial in $s_1, \ldots, s_n$ if and only if it is symmetric.*

These theorems are known as the *fundamental theorems* of symmetric fractions or symmetric polynomials respectively. The polynomials $s_1, \ldots, s_n$ are sometimes called the *elementary* symmetric polynomials, since the others can be expressed in terms of these ones.

Because of their progressive emergence through the calculations of eighteenth century mathematicians, these theorems can hardly be credited to any specific author. (There is not much credit to give anyway, since the proofs are not difficult.)

It seems that they first appeared in print around 1770, in "Meditationes Algebraicae" of Edward Waring (1736–1798) and in "Mémoire sur la résolution des
équations" of A.T. Vandermonde, and in presumably other works. It is noteworthy
that Lagrange in 1770 qualifies the fundamental theorem of symmetric fractions
as "self-evident" [40, Art. 98, p. 372].

Therefore, the most interesting feature that one may expect of a proof is its
effectiveness: it has to provide a method to express any symmetric polynomial as
a polynomial in $s_1, \ldots, s_n$. In the next section, we explain the particularly simple
method suggested by Waring [66, Chapter I, Problem III, Case 3], and thereafter
we shall discuss some applications, but first we point out a convenient notation
which allows to denote symmetric polynomials without writing out all the terms.

NOTATION. We let $\sum x_1^{i_1} x_2^{i_2} \cdots x_n^{i_n}$ be the symmetric polynomial whose terms
are the various *distinct* monomials obtained from $x_1^{i_1} x_2^{i_2} \cdots x_n^{i_n}$ by permutation
of the indeterminates. Observe that this notation is slightly ambiguous, since the
total number of variables is not clear from the notation, as some of the exponents
$i_1, \ldots, i_n$ may be zero. Therefore, the total number of variables should always
be indicated, unless it is clear from the context. For example, as a symmetric
polynomial in two variables,

$$\sum x_1^2 x_2 = x_1^2 x_2 + x_1 x_2^2$$

whereas, as a symmetric polynomial in three variables,

$$\sum x_1^2 x_2 = x_1^2 x_2 + x_1 x_2^2 + x_1^2 x_3 + x_1 x_3^2 + x_2^2 x_3 + x_2 x_3^2.$$

With this notation, the elementary symmetric polynomials can be written simply as

$$s_1 = \sum x_1, \quad s_2 = \sum x_1 x_2, \quad \ldots, \quad s_{n-1} = \sum x_1 \cdots x_{n-1},$$
$$\text{and} \quad s_n = \sum x_1 \cdots x_n \quad (= x_1 \cdots x_n).$$

8.2 Waring's method

In order to define the degree of a polynomial in n indeterminates, we endow the set
$\mathbb{N}^n$ of n-tuples of integers with the lexicographic ordering. Thus $(i_1, \ldots, i_n) \geq$
$(j_1, \ldots, j_n)$ if the first non-zero difference (if any) in the sequence $i_1 - j_1, \ldots,$
$i_n - j_n$ is positive. For any non-zero polynomial $P = P(x_1, \ldots, x_n)$ in n indeterminates $x_1, \ldots, x_n$ over a field, the *degree* of P is then defined as the largest

n-tuple $(i_1, \ldots, i_n) \in \mathbb{N}^n$ for which the coefficient of $x_1^{i_1} \cdots x_n^{i_n}$ in P is non-zero. The degree of P is denoted by $\deg P$. For instance, we have

$$\deg s_1 = (1, 0, 0, \ldots, 0),$$
$$\deg s_2 = (1, 1, 0, \ldots, 0),$$
$$\cdots \tag{8.3}$$
$$\deg s_{n-1} = (1, 1, \ldots, 1, 0),$$
$$\deg s_n = (1, 1, \ldots, 1, 1).$$

By convention, we also set $\deg 0 = -\infty$, and the same relations as for polynomials in one indeterminate hold, namely

$$\deg(P + Q) \leq \max(\deg P, \deg Q), \tag{8.4}$$
$$\deg(PQ) = \deg P + \deg Q. \tag{8.5}$$

Proof of Theorem 8.4. Waring's method to express any symmetric polynomial $P(x_1, \ldots, x_n)$ as a polynomial in $s_1, \ldots, s_n$ is quite similar to Euclid's division algorithm (Theorem 5.1, p. 43). The idea is to match P with a polynomial in $s_1, \ldots, s_n$ which has the same degree as P. Adjusting the leading coefficient, we can arrange that the degree of the difference be less than $\deg P$, and we are finished by induction on the degree.

Let $P \in F[x_1, \ldots, x_n]$ be a non-zero symmetric polynomial, and let

$$\deg P = (i_1, i_2, \ldots, i_n) \in \mathbb{N}^n.$$

We first observe that $i_1 \geq i_2 \geq \cdots \geq i_n$. Indeed, if we can find among the terms of P a term like $a x_1^{i_1} \cdots x_n^{i_n}$ (with $a \neq 0$), then we can also find all the terms obtained from this one by permutation of $x_1, \ldots, x_n$, since P is symmetric. The degrees of these terms are the various n-tuples obtained from $(i_1, \ldots, i_n)$ by permutation of the entries, and the greatest among these n-tuples is the one in which the entries are in not-increasing order.

We can therefore set

$$f = s_1^{i_1 - i_2} s_2^{i_2 - i_3} \cdots s_{n-1}^{i_{n-1} - i_n} s_n^{i_n}.$$

By (8.3) and (8.5), we have

$$\deg f = (i_1 - i_2) \deg(s_1) + (i_2 - i_3) \deg(s_2) + \cdots + i_n \deg(s_n)$$
$$= (i_1 - i_2, 0, \ldots, 0) + (i_2 - i_3, i_2 - i_3, 0, \ldots, 0) + \cdots + (i_n, \ldots, i_n)$$
$$= (i_1, i_2, \ldots, i_n).$$

Moreover, it is readily verified that the leading coefficient of f, which is the coefficient of $x_1^{i_1} \cdots x_n^{i_n}$, is 1, so that

$$f = x_1^{i_1} \cdots x_n^{i_n} + \text{(terms of lower degree)}.$$

Therefore, if $a \in F^\times$ is the leading coefficient of P, so that

$$P = ax_1^{i_1} \cdots x_n^{i_n} + \text{(terms of lower degree)},$$

then, letting $P_1 = P - af$, we see that $\deg P_1 < \deg P$. Moreover, P_1 is symmetric (possibly zero), since P and f are symmetric. We can therefore apply the same arguments to P_1, which has lower degree than P.

In Waring's words:

> The first step of the solution is to find $S^a \times R^{b-a} \times Q^{c-b} \times P^{d-c} \cdots$; among these terms one particular product will be the required sum, while the remaining terms must be identified by the same method and then discarded. [66, p. 13]

To complete the proof, it remains to prove that the process above, by which the degree of the initial symmetric polynomial has been reduced, terminates in a finite number of steps. This readily follows from the following observation:

8.5. LEMMA. $\mathbb{N}^n$ *satisfies the descending chain condition, i.e., it does not contain any infinite strictly decreasing sequence of elements.*

Proof. As the lemma is obvious if $n = 1$, we argue by induction on n. If

$$(i_{11}, i_{12}, \ldots, i_{1n}) > (i_{21}, i_{22}, \ldots, i_{2n}) > \ldots > (i_{m1}, i_{m2}, \ldots, i_{mn}) > \cdots \tag{8.6}$$

is an infinite strictly decreasing sequence in $\mathbb{N}^n$, then the sequence of first entries is not increasing, so that

$$i_{11} \geq i_{21} \geq \cdots \geq i_{m1} \geq \cdots$$

Therefore, this sequence is eventually constant: there is an index M such that

$$i_{m1} = i_{M1} \quad \text{for all } m \geq M.$$

We then delete the first $M - 1$ terms in sequence (8.6), and consider the last $n - 1$ entries of the elements of the remaining (infinite) sequence:

$$(i_{M2}, i_{M3}, \ldots, i_{Mn}) > (i_{(M+1)2}, i_{(M+1)3}, \ldots, i_{(M+1)n}) > \cdots$$

We thus obtain a strictly decreasing sequence of elements in $\mathbb{N}^{n-1}$. The existence of such a sequence contradicts the induction hypothesis. $\qquad\square$

8.6. Example. Let us express the symmetric polynomial in three variables

$$S = \sum x_1^4 x_2 x_3 + \sum x_1^3 x_2^3$$

(that is $S = x_1^4 x_2 x_3 + x_1 x_2^4 x_3 + x_1 x_2 x_3^4 + x_1^3 x_2^3 + x_1^3 x_3^3 + x_2^3 x_3^3$, see the notation set up at the end of §8.1, p. 100) as a polynomial in s_1, s_2, s_3. Since $(4, 1, 1) >$ $(3, 3, 0)$, the degree of S is $(4, 1, 1)$, so we first calculate $s_1^{4-1} s_2^{1-1} s_3^1$,

$$s_1^3 s_3 = \left(\sum x_1\right)^3 (x_1 x_2 x_3) = \sum x_1^4 x_2 x_3 + 3 \sum x_1^3 x_2^2 x_3 + 6 x_1^2 x_2^2 x_3^2,$$

whence

$$S = \sum x_1^3 x_2^3 - 3 \sum x_1^3 x_2^2 x_3 - 6 x_1^2 x_2^2 x_3^2 + s_1^3 s_3.$$

It remains to express in terms of s_1, s_2, s_3 a polynomial of degree $(3, 3, 0)$. Therefore, we calculate the cube of s_2,

$$s_2^3 = \left(\sum x_1 x_2\right)^3 = \sum x_1^3 x_2^3 + 3 \sum x_1^3 x_2^2 x_3 + 6 x_1^2 x_2^2 x_3^2.$$

Substituting for $\sum x_1^3 x_2^3$ in the preceding expression of S, we obtain

$$S = -6 \sum x_1^3 x_2^2 x_3 - 12 x_1^2 x_2^2 x_3^2 + s_1^3 s_3 + s_2^3.$$

Next, in order to eliminate $\sum x_1^3 x_2^2 x_3$, we calculate $s_1 s_2 s_3$,

$$s_1 s_2 s_3 = \left(\sum x_1\right)\left(\sum x_1 x_2\right)(x_1 x_2 x_3) = \sum x_1^3 x_2^2 x_3 + 3 x_1^2 x_2^2 x_3^2,$$

whence

$$S = 6 x_1^2 x_2^2 x_3^2 + s_1^3 s_3 + s_2^3 - 6 s_1 s_2 s_3.$$

Since $x_1^2 x_2^2 x_3^2 = s_3^2$, we finally obtain the required result:

$$S = s_1^3 s_3 + s_2^3 - 6 s_1 s_2 s_3 + 6 s_3^2.$$

From this brief example, it is already clear that the only difficulty in carrying out Waring's method is to write out the various monomials in products like $s_1^{i_1} \cdots s_n^{i_n}$ with their proper coefficient.

Rational fractions: proof of Theorem 8.3. Let P, Q be polynomials in n indeterminates $x_1, \ldots, x_n$ such that the rational fraction $\frac{P}{Q}$ is symmetric. In order to prove that $\frac{P}{Q}$ is a rational fraction in $s_1, \ldots, s_n$, we represent $\frac{P}{Q}$ as a quotient of symmetric polynomials in $x_1, \ldots, x_n$, as follows: if Q is symmetric then P is symmetric too, since $\frac{P}{Q}$ is symmetric, and there is nothing to do. Otherwise, let $Q_1, \ldots, Q_r$ be the various distinct polynomials (other than Q) obtained from Q by permutation of the indeterminates. The product $QQ_1 \cdots Q_r$ is then symmetric since any permutation of the indeterminates merely permutes the factors. Since $\frac{P}{Q}$ is symmetric and $\frac{P}{Q} = \frac{PQ_1 \cdots Q_r}{QQ_1 \cdots Q_r}$, it follows that the polynomial $PQ_1 \cdots Q_r$ is symmetric too. We have thus obtained the required representation of $\frac{P}{Q}$. Now, by the fundamental theorem of symmetric polynomials (Theorem 8.4), there are polynomials f, g such that

$$PQ_1 \cdots Q_r = f(s_1, \ldots, s_n) \qquad \text{and} \qquad QQ_1 \cdots Q_r = g(s_1, \ldots, s_n),$$

whence $\frac{P}{Q} = \frac{f(s_1, \ldots, s_n)}{g(s_1, \ldots, s_n)}$.   □

The fundamental theorem of symmetric polynomials asserts that every symmetric polynomial $P(x_1, \ldots, x_n)$ is of the form

$$P(x_1, \ldots, x_n) = f(s_1, \ldots, s_n)$$

for some polynomial f in n indeterminates. In other words, there is a polynomial $f(y_1, \ldots, y_n)$ in n indeterminates which yields $P(x_1, \ldots, x_n)$ when $s_1, \ldots, s_n$ are substituted for the indeterminates $y_1, \ldots, y_n$. However, it is not clear *a priori* that the expression of P as a polynomial in $s_1, \ldots, s_n$ is unique, or in other words, that there is only one polynomial f for which the above equality holds. Admittedly, the contrary would be surprising, but no one seems to have cared to prove the uniqueness of f before Gauss, who needed it for his second proof (1815) of the fundamental theorem of algebra [25, §5] (see also Smith [54, vol. I, pp. 292–306]).

8.7. THEOREM. *Let f and g be polynomials in n indeterminates $y_1, \ldots, y_n$ over a field F. If f and g yield the same polynomial in $x_1, \ldots, x_n$ when $s_1, \ldots, s_n$ are substituted for $y_1, \ldots, y_n$, i.e. if*

$$f(s_1, \ldots, s_n) = g(s_1, \ldots, s_n) \qquad \text{in } F[x_1, \ldots, x_n],$$

then

$$f(y_1, \ldots, y_n) = g(y_1, \ldots, y_n) \qquad \text{in } F[y_1, \ldots, y_n].$$

Proof. We compare the degree $(i_1, \ldots, i_n)$ of a non-zero monomial

$$m(y_1, \ldots, y_n) = a y_1^{i_1} \cdots y_n^{i_n}$$

to the degree of the monomial

$$m(s_1, \ldots, s_n) = a s_1^{i_1} \cdots s_n^{i_n} \in F[x_1, \ldots, x_n].$$

By (8.3) and (8.5), we have

$$\deg m(s_1, \ldots, s_n) = (i_1 + \cdots + i_n, i_2 + \cdots + i_n, \ldots, i_{n-1} + i_n, i_n).$$

Since the map $(i_1, \ldots, i_n) \mapsto (i_1 + \cdots + i_n, i_2 + \cdots + i_n, \ldots, i_n)$ from $\mathbb{N}^n$ to $\mathbb{N}^n$ is injective, it follows that monomials of different degrees in $F[y_1, \ldots, y_n]$ cannot cancel out in $F[x_1, \ldots, x_n]$ when $s_1, \ldots, s_n$ are substituted for $y_1, \ldots, y_n$. Therefore, every non-zero polynomial $h \in F[y_1, \ldots, y_n]$ yields a non-zero polynomial $h(s_1, \ldots, s_n)$ in $F[x_1, \ldots, x_n]$. Applying this result to $h = f - g$, the theorem follows. $\square$

8.8. Remarks. (a) Let $\varphi \colon F[y_1, \ldots, y_n] \to F[x_1, \ldots, x_n]$ be the ring homomorphism which maps every polynomial $h(y_1, \ldots, y_n)$ to $h(s_1, \ldots, s_n)$. The preceding theorem asserts that φ is injective. Therefore, the image of φ, which is the subring $F[s_1, \ldots, s_n]$ of $F[x_1, \ldots, x_n]$ generated by $s_1, \ldots, s_n$, is isomorphic under φ to a ring of polynomials in n indeterminates. In other words, the polynomials $s_1, \ldots, s_n$ in $F[x_1, \ldots, x_n]$ can be considered as independent indeterminates. This fact is expressed by saying that $s_1, \ldots, s_n$ are *algebraically independent*.

The point of this remark is that the generic monic polynomial of degree n over F

$$X^n - s_1 X^{n-1} + s_2 X^{n-2} - \cdots + (-1)^n s_n$$

is really generic for all monic polynomials of degree n over a field K containing F. Indeed, if

$$X^n + a_{n-1} X^{n-1} + a_{n-2} X^{n-2} + \cdots + a_0$$

is such a polynomial, then, since $s_1, \ldots, s_n$ can be considered as independent indeterminates over F, there is a (unique) ring homomorphism from $F[s_1, \ldots, s_n]$ to K which maps $s_1, s_2, \ldots, s_n$ to $-a_{n-1}, a_{n-2}, \ldots, (-1)^n a_0$. This homomorphism translates calculations with the coefficients of the generic polynomial into calculations with the coefficients of arbitrary polynomials. By contrast with the discussion following Definition 8.1, we do not need to restrict here to polynomials which split into products of linear factors over some extension of the base field; this is precisely why Theorem 8.7 is significant in Gauss' paper [25], since the purpose of this work was to prove the fundamental theorem of algebra, asserting that polynomials split into products of linear factors over the field of complex numbers.

(b) Inspection shows that the hypothesis that the base ring F is a field has not been used in our exposition of Waring's method nor in the proof of Theorem 8.7. Therefore, Theorem 8.4 and Theorem 8.7 are valid over any base ring. Theorem 8.3 also holds over any (commutative) domain, but to generalize it further, some caution is necessary in the very definition of a rational fraction.

8.3   The discriminant

Let

$$\Delta(x_1, \ldots, x_n) = \prod_{1 \leq i < j \leq n} (x_i - x_j) \in \mathbb{Z}[x_1, \ldots, x_n].$$

Every permutation of $x_1, \ldots, x_n$ permutes the factors $x_i - x_j$ among themselves, changing the sign of some of them. So, Δ is either left unchanged or changed into its opposite by a permutation, and Δ^2 is a symmetric polynomial. It follows from Theorem 8.4 (and Theorem 8.7) that

$$\Delta(x_1, \ldots, x_n)^2 = D(s_1, \ldots, s_n)$$

for some (well-defined) polynomial D with integral coefficients, called the *discriminant* of the generic polynomial of degree n.

The discriminant of an arbitrary polynomial over a field F

$$X^n + a_{n-1}X^{n-1} + a_{n-2}X^{n-2} + \cdots + a_0$$

is then defined as $D(-a_{n-1}, a_{n-2}, \ldots, (-1)^n a_0)$, the element in F obtained from $D(s_1, \ldots, s_n)$ by substituting the coefficients of the polynomial for $s_1, \ldots, s_n$.

In degree 2, we readily have

$$\Delta(x_1, x_2)^2 = (x_1 - x_2)^2 = x_1^2 + x_2^2 - 2x_1x_2,$$

and this symmetric polynomial can be expressed in terms of elementary symmetric polynomials as

$$\Delta(x_1, x_2)^2 = (x_1 + x_2)^2 - 4x_1x_2 = s_1^2 - 4s_2.$$

Therefore, the discriminant of the generic polynomial of degree 2 is

$$D(s_1, s_2) = s_1^2 - 4s_2.$$

For degree 3, we shall use some artifices to simplify the calculations. Written out as a sum of monomials, $\Delta(x_1, x_2, x_3) = (x_1 - x_2)(x_1 - x_3)(x_2 - x_3)$ appears as

$$\Delta(x_1, x_2, x_3) = A - B$$

where

$$A = x_1^2 x_2 + x_2^2 x_3 + x_3^2 x_1 \qquad \text{and} \qquad B = x_1 x_2^2 + x_2 x_3^2 + x_3 x_1^2.$$

Therefore,

$$\Delta(x_1, x_2, x_3)^2 = A^2 + B^2 - 2AB = (A + B)^2 - 4AB. \tag{8.7}$$

Now, $A + B$ and AB are symmetric polynomials which are not difficult to express in terms of s_1, s_2, s_3. Straightforward calculations by Waring's method yield the following results (with the notation set up at the end of §8.1 (p. 100), see Example 8.6):

$$A + B = \sum x_1^2 x_2 = s_1 s_2 - 3 s_3,$$
$$AB = \sum x_1^4 x_2 x_3 + \sum x_1^3 x_2^3 + 3 x_1^2 x_2^2 x_3^2 = s_1^3 s_3 + s_2^3 - 6 s_1 s_2 s_3 + 9 s_3^2.$$

The discriminant $D(s_1, s_2, s_3)$, which is equal to $\Delta(x_1, x_2, x_3)^2$, is easily calculated from equation (8.7) above. One finds

$$D(s_1, s_2, s_3) = s_1^2 s_2^2 + 18 s_1 s_2 s_3 - 27 s_3^2 - 4 s_1^3 s_3 - 4 s_2^3.$$

In particular, it follows that the discriminant of $X^3 + pX + q$ is $-27q^2 - 4p^3$. Denoting this discriminant by d, we thus have

$$d = -2^2 3^3 \left(\left(\frac{p}{3} \right)^3 + \left(\frac{q}{2} \right)^2 \right). \tag{8.8}$$

We will now show what kind of information on the roots of polynomials with real coefficients can be obtained from the discriminant. We shall need the following easy result:

8.9. LEMMA. *If $a \in \mathbb{C}$ is a root of a polynomial $P \in \mathbb{R}[X]$, then the conjugate complex number $\bar{a}$ also is a root of P.*

Proof. Conjugating each side of the equality $P(a) = 0$, we obtain $P(\bar{a}) = 0$, since the coefficients of P are equal to their own conjugate. □

8.10. THEOREM. *Let $P \in \mathbb{R}[X]$ be a monic polynomial with real coefficients, which splits[†] into a product of linear factors over $\mathbb{C}$, so that*

$$P = (X - u_1) \cdots (X - u_n)$$

for some $u_1, \ldots, u_n \in \mathbb{C}$. Let $d \in \mathbb{R}$ be the discriminant of P. The equality $d = 0$ holds if and only if P has a root of multiplicity at least 2 in $\mathbb{C}$. If all the roots of P are real, then $d \geq 0$. The converse is true if $n = 2, 3$.

Proof. Since the calculations with the roots of the generic polynomial are also valid with the roots $u_1, \ldots, u_n$ of P (see the discussion following Definition 8.1), we have

$$d = \Delta(u_1, \ldots, u_n)^2 = \prod_{1 \leq i < j \leq n} (u_i - u_j)^2.$$

This readily shows that $d \geq 0$ if all the roots $u_1, \ldots, u_n$ are real. If P has a root of multiplicity at least 2, then $u_i - u_j = 0$ for some indices i, j with $i \neq j$, whence $d = 0$ since the product above has a zero factor. If the roots of P are all simple, then each factor is non-zero, whence $d \neq 0$.

For $n = 2$,

$$d = (u_1 - u_2)^2.$$

If u_1 is not real, then the preceding lemma shows that $u_2 = \overline{u_1}$. It follows that $\overline{(u_1 - u_2)} = -(u_1 - u_2)$, hence

$$(u_1 - u_2)^2 = -(u_1 - u_2)\overline{(u_1 - u_2)} = -|u_1 - u_2|^2,$$

and $d < 0$.

Similarly, for $n = 3$,

$$d = (u_1 - u_2)^2(u_1 - u_3)^2(u_2 - u_3)^2.$$

If one of the roots, u_1 say, is not real, then its conjugate $\overline{u_1}$ is among u_2, u_3. Without loss of generality, we can assume $\overline{u_1} = u_2$. Then $(X - u_1)(X - u_2) \in \mathbb{R}[X]$, whence

$$X - u_3 = \frac{P}{(X - u_1)(X - u_2)} \in \mathbb{R}[X].$$

[†]The fundamental theorem of algebra (Theorem 9.1, p. 115) will show that this is no restriction on P.

This shows $u_3 \in \mathbb{R}$. Now

$$\overline{(u_1 - u_2)} = -(u_1 - u_2), \qquad \overline{(u_1 - u_3)} = (u_2 - u_3),$$
$$\text{and} \quad \overline{(u_2 - u_3)} = (u_1 - u_3).$$

Therefore,

$$\overline{(u_1 - u_2)(u_1 - u_3)(u_2 - u_3)} = -(u_1 - u_2)(u_1 - u_3)(u_2 - u_3),$$

whence

$$d = -|(u_1 - u_2)(u_1 - u_3)(u_2 - u_3)|^2 < 0.$$

$\square$

8.11. Remarks. **(a)** For $n \geq 4$, the sign of d determines the number of real roots of P up to a multiple of 4, see Exercise 4.
(b) The first statement (about multiple roots) in the preceding theorem is valid over an arbitrary field. Thus, we now have two necessary and sufficient conditions for a (monic) polynomial to have at least one multiple root in some extension of the base field: its discriminant has to be zero or, equivalently, the polynomial has to have a non-constant common divisor with its derivative (see Proposition 5.19, p. 53). The equivalence of these two conditions can be verified directly by using Theorem 5.24. Indeed, the discriminant of a monic polynomial P is equal (up to sign) to the resultant of P and its derivative ∂P; see Exercise 6.

8.12. COROLLARY. *Let $p, q \in \mathbb{R}$. The equation*

$$X^3 + pX + q = 0$$

has three distinct real solutions if and only if $\left(\frac{p}{3}\right)^3 + \left(\frac{q}{2}\right)^2 < 0$.

Proof. This readily follows from the preceding theorem, since the discriminant d of $X^3 + pX + q$ is

$$d = -2^2 3^3 \left(\left(\frac{p}{3}\right)^3 + \left(\frac{q}{2}\right)^2\right).$$

(See equation (8.8), p. 107.) $\square$

This corollary shows that the "casus irreducibilis" of cubic equations (see §2.3(c)) is precisely the case where the equation has three distinct real roots.

Appendix: Euler's summation of the series of reciprocals of perfect squares

Around 1735, Euler succeeded in finding the sum of the series $\sum_{k=1}^{\infty} \frac{1}{k^2}$, thus achieving a result that had baffled Leibniz and Jacques Bernoulli (see Boyer [8, ch. 21, n° 4] or Goldstine [27, §3.2]). His method was to apply to a certain power series the relations between roots and coefficients of polynomials.

For the generic polynomial

$$X^n - s_1 X^{n-1} + s_2 X^{n-2} - \cdots + (-1)^n s_n = (X - x_1) \cdots (X - x_n),$$

we have seen that

$$s_1 = \sum x_1, \quad s_2 = \sum x_1 x_2, \quad \ldots, \quad s_{n-1} = \sum x_1 \cdots x_{n-1},$$
$$\text{and} \quad s_n = x_1 \cdots x_n.$$

Therefore, using the notation set up at the end of §8.1 (p. 100) for rational fractions,

$$\sum x_1^{-1} = s_{n-1} s_n^{-1}, \quad \sum (x_1 x_2)^{-1} = s_{n-2} s_n^{-1}, \quad \ldots,$$
$$\text{and} \quad \sum (x_1 \cdots x_{n-1})^{-1} = s_1 s_n^{-1}.$$

By the same calculations as for Newton's formulas in §4.2, we then obtain

$$\sum x_1^{-2} = \left(\sum x_1^{-1} \right)^2 - 2 \sum (x_1 x_2)^{-1} = (s_{n-1}^2 - 2s_{n-2} s_n) s_n^{-2},$$
$$\sum x_1^{-3} = \left(\sum x_1^{-1} \right) \left(\sum x_1^{-2} \right) - \left(\sum (x_1 x_2)^{-1} \right) \left(\sum x_1^{-1} \right) + 3 \sum (x_1 x_2 x_3)^{-1}$$
$$= (s_{n-1}^3 - 3s_{n-2} s_{n-1} s_n + 3s_{n-3} s_n^2) s_n^{-3},$$

etc.

Now, if $u_1, \ldots, u_n$ are the roots of an arbitrary (not necessarily monic) polynomial

$$a_n X^n + a_{n-1} X^{n-1} + \cdots + a_1 X + a_0, \quad (a_n \neq 0)$$

then $u_1, \ldots, u_n$ are the roots of the monic polynomial

$$X^n + a_{n-1} a_n^{-1} X^{n-1} + \cdots + a_1 a_n^{-1} X + a_0 a_n^{-1}.$$

Substituting $-a_{n-1}a_n^{-1}$, $a_{n-2}a_n^{-1}$, ..., $(-1)^n a_0 a_n^{-1}$ for s_1, s_2, ..., s_n in the calculations above, it follows (provided that $a_0 \neq 0$) that

$$u_1^{-1} + \cdots + u_n^{-1} = -a_1 a_0^{-1}, \tag{8.9}$$

$$u_1^{-2} + \cdots + u_n^{-2} = (a_1^2 - 2a_2 a_0)a_0^{-2}, \tag{8.10}$$

$$u_1^{-3} + \cdots + u_n^{-3} = (-a_1^3 + 3a_2 a_1 a_0 - 3a_3 a_0^2)a_0^{-3}. \tag{8.11}$$

As Euler pointed out, these calculations yield interesting results when applied to the sine function, considered as the "infinite polynomial"

$$\sin z = z - \frac{z^3}{3!} + \frac{z^5}{5!} - \frac{z^7}{7!} + \cdots$$

which has as roots 0, $\pm\pi$, $\pm 2\pi$, $\pm 3\pi$, ... Dividing the series by z to get rid of the root 0, and changing the variable to $x = z^2$, we obtain the series

$$1 - \frac{x}{3!} + \frac{x^2}{5!} - \frac{x^3}{7!} + \cdots$$

with roots π^2, $(2\pi)^2$, $(3\pi)^2$, ... Then equations (8.9), (8.10), (8.11), etc. with $n = \infty$, $a_0 = 1$, $a_1 = -\frac{1}{3!}$, $a_2 = \frac{1}{5!}$, $a_3 = -\frac{1}{7!}$, etc. yield

$$\sum_{k=1}^{\infty} (k\pi)^{-2} = \frac{1}{6}, \qquad \sum_{k=1}^{\infty} (k\pi)^{-4} \stackrel{.}{=} \frac{1}{90}, \qquad \sum_{k=1}^{\infty} (k\pi)^{-6} = \frac{1}{945}, \qquad \text{etc.}$$

Multiplying both sides by the appropriate power of π, we obtain

$$\sum_{k=1}^{\infty} \frac{1}{k^2} = \frac{\pi^2}{6}, \qquad \sum_{k=1}^{\infty} \frac{1}{k^4} = \frac{\pi^4}{90}, \qquad \sum_{k=1}^{\infty} \frac{1}{k^6} = \frac{\pi^6}{945}, \qquad \text{etc.}$$

That Euler's calculations are valid is of course not obvious, but they can be rigorously verified. The point is that the sine function can be expressed as an infinite product

$$\sin z = z \prod_{k=1}^{\infty} \left(1 - \frac{z^2}{k^2 \pi^2}\right) \qquad \text{for } z \in \mathbb{C}.$$

Note also that the above calculations do not yield any information on the values of Riemann's zeta function

$$\zeta(s) = \sum_{k=1}^{\infty} \frac{1}{k^s}$$

at *odd* integers s. Indeed, very little is known about these values. Until recently, it was not even known whether $\zeta(3)$ is rational or not. This question was answered in the negative by Roger Apéry in 1978 (see Van der Poorten [60]).

Exercises

1. *The following exercise provides an alternative procedure for calculating the discriminant of a polynomial.* Let

$$P(X) = X^n - s_1 X^{n-1} + \cdots + (-1)^n s_n = (X - x_1) \cdots (X - x_n),$$

let $\sigma_i = \sum_{j=1}^n x_j^i$ for $i = 1, 2, \ldots$ and let A and B be the $n \times n$ matrices

$$A = \begin{pmatrix} 1 & 1 & \cdots & 1 \\ x_1 & x_2 & \cdots & x_n \\ x_1^2 & x_2^2 & \cdots & x_n^2 \\ \vdots & \vdots & & \vdots \\ x_1^{n-1} & x_2^{n-1} & \cdots & x_n^{n-1} \end{pmatrix},$$

$$B = \begin{pmatrix} n & \sigma_1 & \sigma_2 & \cdots & \sigma_{n-1} \\ \sigma_1 & \sigma_2 & \sigma_3 & \cdots & \sigma_n \\ \sigma_2 & \sigma_3 & \sigma_4 & \cdots & \sigma_{n+1} \\ \vdots & \vdots & \vdots & & \vdots \\ \sigma_{n-1} & \sigma_n & \sigma_{n+1} & \cdots & \sigma_{2n-2} \end{pmatrix}.$$

(a) Show that $\det A = \prod_{i>j}(x_i - x_j)$. (The matrix A is called a *Vandermonde matrix*).

(b) Show that $B = AA^t$ (where A^t denotes the transpose of A).

(c) Derive from (a) and (b) that the discriminant of $P(X)$ is

$$D(s_1, \ldots, s_n) = \det B.$$

(d) Use this result to prove that the discriminant of the polynomial $X^n + pX + q$ is

$$(-1)^{\frac{n(n-1)}{2}} \left((-1)^{n-1}(n-1)^{n-1}p^n + n^n q^{n-1} \right).$$

2. Let x_1, x_2, x_3 be the roots of the cubic equation $X^3 + pX + q = 0$. Prove that the equation whose roots are $(x_1 - x_2)^2$, $(x_1 - x_3)^2$ and $(x_2 - x_3)^2$ is

$$Y^3 + 6pY^2 + 9p^2 Y + (4p^3 + 27q^2) = 0.$$

This equation is called the *equation of squared differences* of the given cubic equation.

3. Let

$$P(X) = X^4 - s_1 X^3 + s_2 X^2 - s_3 X + s_4$$
$$= (X - x_1)(X - x_2)(X - x_3)(X - x_4),$$

and let

$$u_1 = (x_1 + x_2)(x_3 + x_4), \qquad v_1 = x_1 x_2 + x_3 x_4,$$
$$u_2 = (x_1 + x_3)(x_2 + x_4), \qquad v_2 = x_1 x_3 + x_2 x_4,$$
$$u_3 = (x_1 + x_4)(x_2 + x_3), \qquad v_3 = x_1 x_4 + x_2 x_3.$$

(a) Show that the equation which has as roots u_1, u_2 and u_3 is

$$Q(Y) = Y^3 - 2s_2 Y^2 + (s_2^2 + s_1 s_3 - 4s_4)Y$$
$$- (s_1 s_2 s_3 - s_1^2 s_4 - s_3^2) = 0$$

and that the equation which has as roots v_1, v_2 and v_3 is

$$R(Z) = Z^3 - s_2 Z^2 + (s_1 s_3 - 4s_4)Z - (s_1^2 s_4 - 4s_2 s_4 + s_3^2) = 0.$$

(Compare equations (3.5) and (3.6) in chapter 3, pp. 23 and 24.)

(b) Show that the discriminants of P, Q and R are all equal. [Hint: Prove that $\Delta(X_1, X_2, X_3, X_4) = -\Delta(u_1, u_2, u_3) = -\Delta(v_1, v_2, v_3)$.]

4. Let $P \in \mathbb{R}[X]$ be a monic polynomial with real coefficients, which splits into a product of linear factors over $\mathbb{C}$, so that $P = (X - u_1) \cdots (X - u_n)$ for some $u_i \in \mathbb{C}$. Assume that the roots $u_1, \ldots, u_n$ of P are pairwise distinct and denote by r the number of real roots among $u_1, \ldots, u_n$ and by d the discriminant of P. Show that $n - r$ is an even integer, which is divisible by 4 if and only if $d > 0$.

5. Let $P = (X - x_1) \cdots (X - x_n)$ and $Q = (X - y_1) \cdots (X - y_m)$. Show that the resultant of P and Q is

$$\prod_{i=1}^{n} \prod_{j=1}^{m} (x_i - y_j) = \prod_{i=1}^{n} Q(x_i) = (-1)^n \prod_{j=1}^{m} P(y_j).$$

[Hint: Consider $x_1, \ldots, x_n, y_1, \ldots, y_m$ as indeterminates and use Theorem 5.24, p. 56.]

6. Let $P = (X - x_1) \cdots (X - x_n)$; let d denote the discriminant of P, and R the resultant of P and its derivative ∂P. Show that

$$R = \prod_{i \neq j} (x_i - x_j) = (-1)^{\frac{n(n-1)}{2}} d.$$

[Hint: Calculate $\partial P(x_i)$ and use Exercise 5.]

Chapter 9

The Fundamental Theorem of Algebra

9.1 Introduction

The title of this chapter refers to the following result:

9.1. THEOREM. *The number of roots of a non-zero polynomial over the field* $\mathbb{C}$ *of complex numbers, each root being counted with its multiplicity, is equal to the degree of the polynomial.*

Equivalently, by Theorem 5.15, the fundamental theorem of algebra asserts that every polynomial splits into a product of linear factors in $\mathbb{C}[X]$. There is also an equivalent formulation in terms of real polynomials only:

9.2. THEOREM. *Every (non-constant) real polynomial can be decomposed into a product of (real) polynomials of degree 1 or 2.*

The equivalence of these statements will be proved in Proposition 9.6 below.

Theorem 9.1 can be traced back to Girard, in a considerably looser form (see §4.2, p. 35). Indeed, Leibniz's proposed counter-example (in §7.2, p. 75) clearly shows how remote a proof still was at the beginning of the eighteenth century. Yet, during the first half of this century, de Moivre's work had prompted a deeper understanding of the operations on complex numbers and opened the way to the first attempts of proof.

Since the ultimate structure of $\mathbb{R}$ (or $\mathbb{C}$) is analytical, it is not surprising to us that the first idea of a proof, published in 1746 by Jean le Rond d'Alembert (1717–1783), used analytic techniques. However, an analytic proof for what was perceived as an algebraic theorem was hardly satisfactory, and Euler in 1749 tried a more algebraic method. Euler's idea was to prove the equivalent Theorem 9.2

by an induction argument on the highest power of 2 which divides the degree. Omitting several key details, Euler failed to carry out his program completely, so that his proof is only a sketch. Some simplifications in Euler's proof were subsequently suggested by Daviet François de Foncenex (1734–1799), and Lagrange eventually gave in 1772 a complete proof, elaborating on Euler's and de Foncenex's ideas and correcting all the flaws in their proofs.

All, but one. As Gauss noticed in 1799 in his inaugural dissertation [23, §12], a critical flaw remained. Lured by the custom inherited from seventeenth century mathematicians, Lagrange implicitly takes for granted the *existence* of n "imaginary" roots for any equation of degree n, and he thus only proves that the *form* of these imaginary roots is $a + b\sqrt{-1}$ with a, $b \in \mathbb{R}$. Gauss also shows that the same criticism can be addressed to the earlier proofs of Euler, de Foncenex and d'Alembert [23, §§6ff] and he then proceeds to give the first essentially complete proof of the fundamental theorem of algebra, along the lines of d'Alembert's proof.

In 1815, Gauss also found a way to mend the Euler–de Foncenex–Lagrange proof [25], and he subsequently gave two other proofs of the fundamental Theorem 9.1.

The proof we give is based upon the Euler–de Foncenex–Lagrange ideas. However, instead of following Gauss' correction, we use some ideas from the late nineteenth century to prove first Girard's "theorem" on the existence of imaginary roots (of which nothing is known, except that operations can be performed on these roots as if they were numbers). Having thus justified the *postulate* on which Euler's proof implicitly relied, we are then able to use Euler's arguments in a more direct way, to prove that the imaginary roots are of the form $a + b\sqrt{-1}$ (with a, $b \in \mathbb{R}$). We thus obtain a streamlined and almost completely algebraic proof of the fundamental theorem of algebra, also found in Samuel [52, p. 53].

9.2 Girard's theorem

In this section, we let F be an arbitrary field. The modern translation of Girard's intuition is as follows:

9.3. THEOREM. *For any non constant polynomial $P \in F[X]$, there is a field K containing F such that P splits over K into a product of linear factors, i.e.*

$$P = a(X - x_1) \cdots (X - x_n) \qquad in \ K[X].$$

The field K is constructed abstractly by successive quotients of polynomial rings by ideals.

We first recall that an *ideal* in a commutative ring A is a subgroup I of the additive group of A which is stable under multiplication by elements in A, i.e. such that $ax \in I$ for $a \in A$ and $x \in I$. In order to define the quotient ring of A by I, we set, for $a \in A$,

$$a + I = \{a + x \mid x \in I\}.$$

This is a subset of A, and from the hypothesis that I is a subgroup of the additive group of A, it easily follows that, for $a, b \in A$,

$$a + I = b + I \quad \text{if and only if} \quad a - b \in I. \tag{9.1}$$

We then set

$$A/I = \{a + I \mid a \in A\}.$$

The condition that I is an ideal ensures that the operations on A induce a ring structure on A/I, by

$$(a + I) + (b + I) = (a + b) + I \qquad \text{and} \qquad (a + I)(b + I) = ab + I.$$

The ring A/I is called the *quotient ring* of A by the ideal I.

The zero element in this quotient ring is $0 + I$ ($= I$), also denoted simply as 0; therefore, (9.1) shows that in A/I

$$a + I = 0 \quad \text{if and only if} \quad a \in I. \tag{9.2}$$

We shall need the following instance of this construction: let $A = F[X]$ and let I be the set (P) of multiples of a polynomial P,

$$(P) = \{PQ \mid Q \in F[X]\}.$$

It is readily verified that (P) is an ideal of $F[X]$, so that we can construct the quotient ring $F[X]/(P)$. This construction is essentially due to Kronecker (1823–1897), although a special case had been considered earlier by Cauchy. (In 1847, Cauchy represented $\mathbb{C}$ as the quotient ring $\mathbb{R}[X]/(X^2 + 1)$.)

The basic lemma required for the proof of Girard's theorem is the following:

9.4. LEMMA. *If $P \in F[X]$ is irreducible, then $F[X]/(P)$ is a field containing F and a root of P.*

Proof. In order to show that $F[X]/(P)$ is a field, it suffices to prove that every non-zero element $Q + (P)$ in $F[X]/(P)$ is invertible. Since $Q + (P) \neq 0$, it follows from (9.2) above that $Q \notin (P)$, i.e. that Q is not divisible by P. Since P is irreducible, P and Q are then relatively prime, whence, by Corollary 5.4, p. 47,

$$PP_1 + QQ_1 = 1$$

for some $P_1, Q_1 \in F[X]$. This relation shows that $QQ_1 - 1$ is divisible by P, so that, by (9.1) above,

$$QQ_1 + (P) = 1 + (P).$$

Therefore,

$$(Q + (P))(Q_1 + (P)) = 1 + (P),$$

which means that $Q_1 + (P)$ is the inverse of $Q + (P)$ in $F[X]/(P)$.

The map $a \mapsto a + (P)$ from F to $F[X]/(P)$ is injective since no non-zero element in F is divisible by P. Therefore, this map is an embedding of F in $F[X]/(P)$, and F is thus identified to a subfield of $F[X]/(P)$. We can thus consider P as a polynomial over $F[X]/(P)$, and it only remains to prove that $F[X]/(P)$ contains a root of P.

From the definition of the operations in $F[X]/(P)$ it follows that

$$P(X + (P)) = P(X) + (P).$$

Therefore, by (9.2) above,

$$P(X + (P)) = 0,$$

which means that $X + (P)$ is a root of P in $F[X]/(P)$. $\square$

Proof of Theorem 9.3. We decompose P into a product of irreducible factors in $F[X]$:

$$P = P_1 \cdots P_r.$$

Let s be the number of linear factors among $P_1, \ldots, P_r$. The integer s is thus the number of roots of P in F, each root being counted with its multiplicity. (Possibly $s = 0$, since P may have no root in F.)

We argue by induction on $(\deg P) - s$, noting that this number is equal to the degree of the product of the non-linear factors among $P_1, \ldots, P_r$.

If $(\deg P) - s = 0$, then each of the factors $P_1, \ldots, P_r$ is linear, and we can choose $K = F$.

If $(\deg P) - s > 0$, then at least one of the factors $P_1, \ldots, P_r$ has degree greater than or equal to 2. Assume for instance that $\deg P_1 \geq 2$, and let

$$F_1 = F[X]/(P_1).$$

Since P_1 has a root in F_1, the decomposition of P_1 over F_1 involves at least one linear factor, by Theorem 5.12 (p. 51), whence the number s_1 of linear factors in the decomposition of P into irreducible factors over F_1 is at least $s+1$. Therefore, $(\deg P) - s_1 < (\deg P) - s$, and the induction hypothesis implies that there is a field K containing F_1 such that P splits into linear factors over K. Since F_1 contains F, the field K also contains F and satisfies all the requirements. □

9.3 Proof of the fundamental theorem

Instead of proving Theorem 9.1 directly, we shall prove an equivalent formulation in terms of real polynomials. We first note for later reference the following easy special case of the fundamental Theorem 9.1:

9.5. LEMMA. *Every quadratic polynomial over $\mathbb{C}$ splits into a product of linear factors in $\mathbb{C}[X]$.*

Proof. It suffices to show that the roots of every quadratic equation with complex coefficients are complex numbers. This readily follows from the usual formula by radicals for the roots (§1.1), since, by Proposition 7.1 (p. 81), every complex number has a square root in $\mathbb{C}$. □

We now prove the equivalence of several formulations of the fundamental theorem.

9.6. PROPOSITION. *The following statements are equivalent:*

(a) *The number of roots of any non-zero polynomial over $\mathbb{C}$ is equal to its degree (each root being counted with its multiplicity).*

(b) *Every non-constant polynomial over $\mathbb{R}$ has at least one root in $\mathbb{C}$.*

(c) *Every non-constant real polynomial can be decomposed into a product of (real) polynomials of degree 1 or 2.*

Proof. (a) $\Rightarrow$ (b) This is clear.

(b) ⇒ (c) By Theorem 5.8 (p. 48), it suffices to show, assuming *(b)*, that every irreducible polynomial in $\mathbb{R}[X]$ has degree 1 or 2. Let P be an irreducible polynomial in $\mathbb{R}[X]$, and let $a \in \mathbb{C}$ be a root of P.

If $a \in \mathbb{R}$, then $X - a$ divides P in $\mathbb{R}[X]$, whence $\deg P = 1$ since, by definition, an irreducible polynomial cannot be divided by a non-constant polynomial of strictly smaller degree.

If $a \notin \mathbb{R}$, then $\overline{a} \neq a$, and $\overline{a}$ is also a root of P, by Lemma 8.9, p. 107. Therefore, by Proposition 5.10 (p. 50), P is divisible by $(X - a)(X - \overline{a})$ in $\mathbb{C}[X]$. But $(X - a)(X - \overline{a})$ lies in $\mathbb{R}[X]$ since

$$(X - a)(X - \overline{a}) = X^2 - (a + \overline{a})X + a\overline{a}.$$

Therefore, P is also divisible by $(X - a)(X - \overline{a})$ in $\mathbb{R}[X]$ (see Remark 5.5(b), p. 47), whence the same argument as above implies $\deg P = 2$.

(c) ⇒ (a) Let $P \in \mathbb{C}[X]$ be a non-constant polynomial. We extend to $\mathbb{C}[X]$ the complex conjugation map from $\mathbb{C}$ to $\mathbb{C}$ by setting $\overline{X} = X$; namely, we set

$$\overline{a_0 + a_1 X + \cdots + a_n X^n} = \overline{a_0} + \overline{a_1} X + \cdots + \overline{a_n} X^n.$$

The invariant elements are readily seen to be the polynomials with real coefficients. Therefore, $P\overline{P} \in \mathbb{R}[X]$ and it follows from the hypothesis *(c)* that

$$P\overline{P} = P_1 \cdot \ldots \cdot P_r$$

for some polynomials $P_1, \ldots, P_r \in \mathbb{R}[X]$ of degree 1 or 2.

By Lemma 9.5, the real polynomials of degree 2 split into products of linear factors in $\mathbb{C}[X]$, whence $P\overline{P}$ is a product of linear factors in $\mathbb{C}[X]$. Therefore, every irreducible factor of P in $\mathbb{C}[X]$ has degree 1, so that P splits into a product of linear factors in $\mathbb{C}[X]$, which proves *(a)*. □

As we noted in the introduction, every proof of the fundamental theorem of algebra uses at some point an analytical (or topological) argument, since $\mathbb{R}$ (or $\mathbb{C}$) cannot be completely defined without reference to some of its topological properties. The only analytical result we shall need in our proof is the following:

9.7. LEMMA. *Every real polynomial P of odd degree has at least one root in $\mathbb{R}$.*

Proof. Since $\deg P$ is odd, the polynomial function $P(\cdot) : \mathbb{R} \to \mathbb{R}$ changes sign when the variable runs from $-\infty$ to $+\infty$, so, by continuity, it must take the value 0 at least once. □

The continuity argument, according to which every continuous function which changes sign on an interval must take the value 0 at least once, may seem (and was for a long time considered as) evident by itself. It was first proved by Bolzano in 1817 (see Dieudonné [18, p. 340] or Kline [38, p. 952]), in an attempt to provide "arithmetical" proofs to the intuitive geometric arguments that Gauss used in his 1799 proof of the fundamental theorem.

Proof of the fundamental theorem. We shall prove the equivalent formulation (b) in Proposition 9.6, that every real non-constant polynomial has at least one root in $\mathbb{C}$.

Let $P \in \mathbb{R}[X]$ be a non-constant polynomial. Dividing P by its leading coefficient if necessary, we may assume that P is monic. We write the degree of P in the form

$$\deg P = n = 2^e m$$

where $e \geq 0$ and m is odd. If $e = 0$, then the degree of P is odd, and the preceding lemma shows that P has a root in $\mathbb{R}$. We then argue by induction on e, assuming that $e \geq 1$ and that the property holds when the exponent of the highest power of 2 which divides the degree of the polynomial is at most $e - 1$.

Let K be a field containing $\mathbb{C}$, over which P splits into a product of linear factors:

$$P = (X - x_1) \cdots (X - x_n).$$

(The existence of such a field K follows from Theorem 9.3.) For $c \in \mathbb{R}$ and for i, $j = 1, \ldots, n$ with $i < j$, let

$$y_{ij}(c) = (x_i + x_j) + cx_ix_j.$$

Let also

$$Q_c(Y) = \prod_{1 \leq i < j \leq n} \left(Y - y_{ij}(c)\right).$$

The coefficients of Q_c are the values of the elementary symmetric polynomials in the roots $y_{ij}(c)$. These coefficients are therefore the values of symmetric polynomials in $x_1, \ldots, x_n$ with real coefficients, hence they can be expressed in terms of the values of the elementary symmetric polynomials in $x_1, \ldots, x_n$, by the fundamental theorem of symmetric polynomials (Theorem 8.4, p. 99). Since the values of the elementary symmetric polynomials in $x_1, \ldots, x_n$ are the coefficients of P, which are real numbers, it follows that the coefficients of Q_c also are real numbers.

Moreover, the degree of Q_c is $\frac{n(n-1)}{2}$, whence

$$\deg Q_c = 2^{e-1}\big(m(2^e m - 1)\big),$$

and the integer between brackets is odd. We may therefore apply the induction hypothesis to conclude that Q_c has at least one root in $\mathbb{C}$, i.e. $y_{r(c),s(c)}(c) \in \mathbb{C}$ for some indices $r(c)$, $s(c)$. If we let the real number c run over the set of real numbers, the indices $r(c)$, $s(c)$ for which $y_{r(c),s(c)}(c) \in \mathbb{C}$ cannot be all distinct, since the set of indices is finite, while $\mathbb{R}$ is infinite. Therefore, we can find some distinct real numbers c_1, c_2 such that $r(c_1) = r(c_2)$ and $s(c_1) = s(c_2)$. Denoting by r and s these common indices, this means that

$$(x_r + x_s) + c_1 x_r x_s \in \mathbb{C} \qquad \text{and} \qquad (x_r + x_s) + c_2 x_r x_s \in \mathbb{C},$$

with c_1, $c_2 \in \mathbb{R}$ and $c_1 \neq c_2$. By subtraction, these relations imply

$$(c_1 - c_2) x_r x_s \in \mathbb{C},$$

whence $x_r x_s \in \mathbb{C}$. Comparing this result with the relations from which it has been derived, we obtain moreover $x_r + x_s \in \mathbb{C}$. This shows that the coefficients of the polynomial

$$X^2 - (x_r + x_s)X + x_r x_s$$

are complex numbers, and it then follows from Lemma 9.5 that its roots x_r and x_s are complex numbers. We have thus shown that at least one of the roots $x_1, \ldots, x_n$ of P in K is a complex number, as was required. $\qquad \square$

9.8. COROLLARY. *Over $\mathbb{C}$, the irreducible polynomials are the polynomials of degree 1. Over $\mathbb{R}$, the irreducible polynomials are the polynomials of degree 1, and the polynomials of degree 2 which have no real root.*

Proof. This readily follows from the fundamental Theorem 9.1 or the equivalent Theorem 9.2, since, by Theorem 5.12 (p. 51), the irreducible polynomials which have a root in the base field have degree 1. $\qquad \square$

Chapter 10

Lagrange

10.1 The theory of equations comes of age

In the second half of the eighteenth century, the algebraic theory of equations is ripe for new advances. All the more or less elementary facts on polynomials are well-known, and computational skills are very high, even by modern standards. Moreover, deeper insights on the ambiguity of roots of (complex) numbers become available through de Moivre's work. The relevance of these insights for the problem of solving equations by radicals is obvious (see the end of §7.2), and one may venture the hypothesis that de Moivre's work provided an important stimulus to new research in the algebraic theory of equations.

Whatever its origin, it is clear that the spirit of the most significant research in this period is completely different from that of Cardano and his contemporaries: no direct application to the solution of numerical equations is expected, and no reference to any practical problem is made, even allusively. The subject has become pure mathematics, and is pursued for its own interest. Within less than a century, in the hands of several mathematicians of genius, it will undergo a rapid development which will dramatically change the whole subject of algebra.

The earliest works in this line appear in the sixties of the eighteenth century, when Euler and Bezout devise various new methods to solve equations of degree at most 4, which can seemingly be extended to equations of higher degree. One of these methods, proposed by Bezout in 1765, is of particular interest because of its explicit use of roots of unity; it is in fact very close to a method of Euler, and has deep resemblances to Tschirnhaus' method.

The idea* is to eliminate the indeterminate Y between the two equations

$$X = a_0 + a_1 Y + a_2 Y^2 + \cdots + a_{n-1} Y^{n-1} \qquad (10.1)$$

$$Y^n = 1, \qquad (10.2)$$

producing an equation of degree n in X,

$$R_n(X) = 0,$$

as in Tschirnhaus' method (see §6.4). Dividing R_n by its leading coefficient if necessary, we can assume that R_n is monic. The properties of $R_n(X)$ imply that if x, y are related by equations (10.1) and (10.2), i.e. if

$$x = a_0 + a_1\omega + a_2\omega^2 + \cdots + a_{n-1}\omega^{n-1}$$

for some n-th root of unity ω, then x is a root of $R_n(X)$, whence $R_n(X)$ is divisible by $X - (a_0 + a_1\omega + \cdots + a_{n-1}\omega^{n-1})$. Regarding $a_0, a_1, \ldots, a_{n-1}$ as independent indeterminates, the values of x corresponding to the various n-th roots of unity ω are all different, so that, by Proposition 5.10 (p. 50),

$$R_n(X) = \prod_{\omega}(X - (a_0 + a_1\omega + \cdots + a_{n-1}\omega^{n-1})), \qquad (10.3)$$

where the product runs over the n different n-th roots of unity ω. The roots of $R_n(X) = 0$ are thus known.

Now, to solve an arbitrary monic equation $P(X) = 0$ of degree n, the method is to determine the parameters $a_0, a_1, \ldots, a_{n-1}$ in such a way that the polynomial $R_n(X)$ be identical to $P(X)$. The solutions of $P(X) = 0$ are then readily obtained in the form $a_0 + a_1\omega + \cdots + a_{n-1}\omega^{n-1}$.

Of course, whether it is possible to assign some value to $a_0, \ldots, a_{n-1}$ in such a way that R_n becomes identical to P is not clear at all, but this turns out to be the case for $n = 2, 3$ or 4, as we are about to see.

The method for constructing $R_n(X)$ by elimination of Y between equations (10.1) and (10.2) has already been discussed in §6.4. For the small values of n,

---

*according to Bezout's presentation; in Euler's work, equation (10.2) is replaced by $Y^n = b$ and in (10.1), one of the coefficients a_i is chosen to be 1. Tschirnhaus' method can be presented in a similar way, replacing equation (10.2) by $Y^n = b$ and (10.1) by $Y = a_0 + a_1 X + \cdots + a_{n-2}X^{n-2} + X^{n-1}$.

the following results are found:

$$R_2(X) = (X - a_0)^2 - a_1^2,$$
$$R_3(X) = (X - a_0)^3 - 3a_1a_2(X - a_0) - (a_1^3 + a_2^3),$$
$$R_4(X) = (X - a_0)^4 - 2(a_2^2 + 2a_1a_3)(X - a_0)^2 - 4a_2(a_1^2 + a_3^2)(X - a_0)$$
$$- (a_1^4 - a_2^4 + a_3^4 + 4a_1a_2^2a_3 - 2a_1^2a_3^2).$$

Alternatively, these results can be obtained from equation (10.3) by expanding the right-hand side.

To obtain the solutions of the cubic equation

$$X^3 + pX + q = 0 \tag{10.4}$$

(to which the general cubic equation can be reduced by a linear change of variable, see §2.2), it now suffices to assign values to a_0, a_1 and a_2 in such a way that $R_3(X)$ takes the form $X^3 + pX + q$. We thus choose $a_0 = 0$ and determine a_1 and a_2 by

$$-3a_1a_2 = p \tag{10.5}$$
$$-(a_1^3 + a_2^3) = q \tag{10.6}$$

(compare §2.2). The first equation gives the value of a_2 as a function of a_1; substituting this value in the second equation yields the following quadratic equation in a_1^3:

$$a_1^6 + qa_1^3 - \left(\frac{p}{3}\right)^3 = 0.$$

A root of this equation is easily found: one can choose for a_1 any cube root of $-\frac{q}{2} + \sqrt{\left(\frac{p}{3}\right)^3 + \left(\frac{q}{2}\right)^2}$ or $-\frac{q}{2} - \sqrt{\left(\frac{p}{3}\right)^3 + \left(\frac{q}{2}\right)^2}$. Letting then $a_2 = -\frac{p}{3a_1}$, it follows that equations (10.5) and (10.6) both hold, whence $R_3(X) = X^3 + pX + q$. Then, equation (10.3) shows that the solutions of equation (10.4) are of the form $\omega a_1 + \omega^2 a_2$, where ω runs over the set of cube roots of unity. If ζ denotes one of these cube roots other than 1, then the cube roots of unity are 1, ζ and ζ^2, and therefore the solutions of (10.4) are

$$a_1 + a_2, \qquad \zeta a_1 + \zeta^2 a_2 \qquad \text{and} \qquad \zeta^2 a_1 + \zeta a_2.$$

(Note that $\zeta^4 = \zeta$.)

Remark. The fact that one can choose $a_0 = 0$ obviously follows from the particular form of the proposed cubic equation (10.4), which lacks the term in X^2. The general case is in no way more difficult and could be treated in the same way, but the calculations are less transparent.

Similarly, for equations of degree 4 such as

$$X^4 + pX^2 + qX + r = 0, \qquad (10.7)$$

we seek values of a_0, a_1, a_2 and a_3 for which $R_4(X) = X^4 + pX^2 + qX + r$. As above, we choose $a_0 = 0$ and we are left with the equations

$$-2(a_2^2 + 2a_1a_3) = p, \qquad (10.8)$$

$$-4a_2(a_1^2 + a_3^2) = q, \qquad (10.9)$$

$$-(a_1^4 - a_2^4 + a_3^4 + 4a_1a_2^2a_3 - 2a_1^2a_3^2) = r. \qquad (10.10)$$

Substituting in the third equation $(a_1^2 + a_3^2)^2 - 2a_1^2a_3^2$ for $a_1^4 + a_3^4$, we get

$$-r = (a_1^2 + a_3^2)^2 - 4a_1^2a_3^2 + 4(a_1a_3)a_2^2 - a_2^4.$$

Equations (10.8) and (10.9) can then be used to eliminate a_1 and a_3 from this equation. The resulting equation is a cubic equation in a_2^2, from which a value of a_2 can be determined. Values for a_1 and a_3 are then easily found from equations (10.8) and (10.9), and the roots of the proposed quartic equation (10.7) are obtained in the form $a_1\omega + a_2\omega^2 + a_3\omega^3$, where ω runs over the set of 4-th roots of unity. Letting i denote (as usual) a square root of -1, the 4-th roots of unity are $1, i, -1, -i$ and the roots of the quartic equation (10.7) are

$$a_1 + a_2 + a_3, \quad ia_1 - a_2 - ia_3, \quad -a_1 + a_2 - a_3 \quad \text{and} \quad -ia_1 - a_2 + ia_3.$$

As noted above, the principle of this method, whence also its difficulty, is not very different from that of Tschirnhaus' method. To its credit, one can nevertheless observe that the method of Euler and Bezout leads to easier calculations, that it is somewhat more direct and, what is more significant for later researches, that Bezout's method stresses the importance of the roots of unity. Altogether, it does not represent a very substantial progress.

The first really important burst of activity in the theory of equations takes place only a few years later, around 1770, with the almost simultaneous publication of Lagrange's "Réflexions sur la résolution algébrique des équations" and Vandermonde's "Mémoire sur la résolution des équations," and of comparatively less important works such as Waring's "Meditationes Algebricae," which we already quoted in Chapter 8. Among all the works of this period, Lagrange's massive paper clearly is the most lucid and the most comprehensive. Therefore, it proved to be also the most influential. Moreover, Lagrange provides an almost unhoped-for link between the early stages of the theory of equations and the subsequent period, by first reviewing the various methods for equations of degree 3 and 4, and the attempts at equations of higher degree so far proposed, before making his own highly original observations.

We shall thus begin our study of the two critical works of this period with Lagrange's paper, and discuss Vandermonde's memoir in the next chapter.

10.2 Lagrange's observations on previously known methods

Lagrange's discussion of the previously known methods is not a mere summary, it is a vast unification and reassessment of these methods. His very explicit aim is to determine not only *how* these methods work, but *why*.

> I propose in this Memoir to examine the various methods found so far for the algebraic solution of equations, to reduce them to general principles, and to let see *a priori* why these methods succeed for the third and the fourth degree, and fail for higher degrees.
>
> This examination will have a double advantage: on one hand, it will shed a greater light on the known solutions of the third and the fourth degree; on the other hand, it will be useful to those who will want to deal with the solution of higher degrees, by providing them with various views to this end and above all by sparing them a large number of useless steps and attempts. [40, pp. 206–207]

The phrase *a priori* keeps recurring throughout Lagrange's work. It is the hallmark of his new fruitful methodology. Lagrange started from the rather obvious observation that the various methods for solving equations have a common feature: they all reduce the problem by some clever transformations to the solution of a certain auxiliary equation of smaller degree. *A posteriori*, when these clever transformations have been found, one can only ascertain that the method provides the required solutions from those of the auxiliary equation, but this does not yield any valuable insight into the solution of equations of higher degree. Indeed, the only evidence that supports the belief that Tschirnhaus', Euler's or Bezout's method could be applied to equations of higher degree is that the approach is the same in all cases, that the first calculations are parallel, and that it works for equations of degree 2, 3 or 4. This is rather scant evidence.

To find out *a priori* why a method works, Lagrange's highly original idea is to reverse the steps, and determine the roots of the auxiliary equations as functions of the roots of the proposed equation. The properties of the roots of the auxiliary equation then become apparent, and they clearly show why these roots provide the solution of the proposed equation.

As a first example, we take Cardano's method for cubic equations, which is the first method scrutinized by Lagrange. Lagrange begins with a careful description of the method. The cubic equation

$$X^3 + aX^2 + bX + c = 0$$

is first reduced to the form

$$X'^3 + pX' + q = 0$$

by the change of variable $X' = X + \frac{a}{3}$. Next, by setting $X' = Y + Z$, the equation becomes

$$(Y^3 + Z^3 + q) + (Y + Z)(3YZ + p) = 0.$$

The solutions of the cubic equation are then obtained from the solutions of the system

$$\begin{cases} Y^3 + Z^3 + q = 0 \\ 3YZ + p = 0. \end{cases}$$

From the second equation comes

$$Z = -\frac{p}{3Y} \tag{10.11}$$

and, substituting in the first equation, one gets

$$Y^3 - \left(\frac{p}{3Y}\right)^3 + q = 0,$$

whence

$$Y^6 + qY^3 - \left(\frac{p}{3}\right)^3 = 0. \tag{10.12}$$

This is the auxiliary equation, which Lagrange terms the "reduced" equation, on which Cardano's method depends. From this equation, the values of Y are easily obtained, since it is really a quadratic equation in Y^3. The corresponding values of Z are derived from (10.11), and the solutions of the initial equation are then

$$X = -\frac{a}{3} + Y + Z.$$

Since the reduced equation (10.12) has degree 6, it has six roots, so we end up with six roots for the initial equation of degree 3. In fact, it can be seen that these six values of X are pairwise equal, so that each root of the cubic equation is obtained twice. Indeed, let $y_1, y_2, \ldots, y_6$ be the six roots of (10.12). Since this equation is quadratic in Y^3, their cubes $y_1^3, \ldots, y_6^3$ take only two values v_1, v_2, whose product is $v_1 v_2 = -\left(\frac{p}{3}\right)^3$, since $-\left(\frac{p}{3}\right)^3$ is the constant term of (10.12). Changing the numbering if necessary, we may assume

$$y_1^3 = y_2^3 = y_3^3 = v_1 \quad \text{and} \quad y_4^3 = y_5^3 = y_6^3 = v_2.$$

So, y_1, y_2, y_3 (resp. y_4, y_5, y_6) are the various cube roots of v_1 (resp. v_2). Therefore, denoting by ω a cube root of unity other than 1, we may assume

$$y_2 = \omega y_1, \quad y_3 = \omega^2 y_1 \quad \text{and} \quad y_5 = \omega y_4, \quad y_6 = \omega^2 y_4.$$

Since $v_1 v_2 = -\left(\frac{p}{3}\right)^3$, there are some determinations of the cube roots of v_1 and v_2 which multiply up to $-\frac{p}{3}$. Assume for instance (renumbering $y_1, \ldots, y_6$ if necessary) that

$$y_1 y_4 = -\frac{p}{3}.$$

Then

$$y_2 y_6 = \omega^3 y_1 y_4 = -\frac{p}{3}$$

and similarly

$$y_3 y_5 = -\frac{p}{3}.$$

Therefore, if we denote by z_i the value of Z which corresponds to y_i by (10.11), we have

$$z_1 = y_4, \quad z_2 = y_6, \quad z_3 = y_5, \quad z_4 = y_1, \quad z_5 = y_3 \quad \text{and} \quad z_6 = y_2,$$

and it follows that $y_i + z_i$ takes only three different values, namely

$$y_1 + y_4, \quad y_2 + y_6 = \omega y_1 + \omega^2 y_4 \quad \text{and} \quad y_3 + y_5 = \omega^2 y_1 + \omega y_4,$$

which yield the three roots of the initial cubic equation in the form

$$x_1 = -\frac{a}{3} + y_1 + y_4,$$

$$x_2 = -\frac{a}{3} + \omega y_1 + \omega^2 y_4, \qquad (10.13)$$

$$x_3 = -\frac{a}{3} + \omega^2 y_1 + \omega y_4.$$

So, we see *a posteriori* how the reduced equation provides the solution of the initial cubic equation.

To understand *a priori* why it does, Lagrange determines $y_1, \ldots, y_6$ as functions of x_1, x_2 and x_3. This is fairly easy: it suffices to solve the system (10.13) for y_1 and y_4, and the other y_i's are multiples of y_1 or y_4 by ω or ω^2. It is even easier if one notices that, since ω is a cube root of unity other than 1, it is a root of

$$\frac{X^3 - 1}{X - 1} = X^2 + X + 1,$$

so that $\omega^2 + \omega + 1 = 0$. Therefore, multiplying the second equation of (10.13) by ω^2, the third by ω and adding to the first, one obtains

$$x_1 + \omega^2 x_2 + \omega x_3 = -\frac{a}{3}(1 + \omega + \omega^2) + 3y_1 + (1 + \omega + \omega^2)y_4,$$

whence

$$y_1 = \frac{1}{3}(x_1 + \omega^2 x_2 + \omega x_3).$$

Likewise, one obtains

$$y_4 = \frac{1}{3}(x_1 + \omega x_2 + \omega^2 x_3)$$

and the other roots of the reduced equation are easily obtained by multiplying y_1 or y_4 by ω or ω^2. One thus gets

$$y_2 = \frac{1}{3}(\omega x_1 + x_2 + \omega^2 x_3),$$

$$y_3 = \frac{1}{3}(\omega^2 x_1 + \omega x_2 + x_3),$$

$$y_5 = \frac{1}{3}(\omega x_1 + \omega^2 x_2 + x_3),$$

$$y_6 = \frac{1}{3}(\omega^2 x_1 + x_2 + \omega x_3).$$

Thus, the roots of the reduced equation are all the expressions obtained from $\frac{1}{3}(x_1 + \omega x_2 + \omega^2 x_3)$ by permutation of x_1, x_2 and x_3, and the purpose of solving the reduced equation is to determine some (whence all) of these expressions.

From this observation, Lagrange draws some clever conclusions. First, it explains why the reduced equation has degree 6. Indeed, since the coefficients of the reduced equation are functions of the coefficients of the proposed equation, which are the elementary symmetric polynomials in x_1, x_2, x_3, it follows that these coefficients are symmetric. Therefore, if some expression of x_1, x_2, x_3 is a root of the reduced equation, every other expression obtained from this one by permutation of x_1, x_2, x_3 also is a root of this equation. Since y_4 takes six different values by permutation of x_1, x_2, x_3, these six values are the roots of the reduced equation, which has therefore degree 6.

Moreover, it explains why the reduced equation is a quadratic equation in Y^3: this is because y_4^3 takes only two values by permutation of x_1, x_2, x_3. Indeed, since $\omega^3 = 1$, one has for instance

$$\omega x_1 + \omega^2 x_2 + x_3 = \omega(x_1 + \omega x_2 + \omega^2 x_3)$$

and it follows that

$$(\omega x_1 + \omega^2 x_2 + x_3)^3 = (x_1 + \omega x_2 + \omega^2 x_3)^3.$$

Likewise, we obtain

$$(x_1 + \omega x_2 + \omega^2 x_3)^3 = (\omega x_1 + \omega^2 x_2 + x_3)^3 = (\omega^2 x_1 + x_2 + \omega x_3)^3$$

and

$$(x_1 + \omega^2 x_2 + \omega x_3)^3 = (\omega^2 x_1 + \omega x_2 + x_3)^3 = (\omega x_1 + x_2 + \omega^2 x_3)^3.$$

Therefore, the two values of y_4^3 are

$$\left(\frac{1}{3}\right)^3 (x_1 + \omega x_2 + \omega^2 x_3)^3 \quad \text{and} \quad \left(\frac{1}{3}\right)^3 (x_1 + \omega^2 x_2 + \omega x_3)^3,$$

which are the roots of a quadratic equation.

The general result behind these arguments is the following:

10.1. PROPOSITION. *Let f be a rational fraction in n indeterminates $x_1, \ldots, x_n$. If f takes m different values*[†] *when the indeterminates $x_1, \ldots, x_n$ are per-*

[†]Properly speaking, one should say "if the permutations of $x_1, \ldots, x_n$ in f give rise to m different rational fractions." However, Lagrange's use of the term "values of a rational fraction" will be retained in the sequel since it is more suggestive and should not cause any confusion.

muted in all possible ways, then f is a root of a monic equation $\Theta = 0$ of degree m, whose coefficients are symmetric in $x_1, \ldots, x_n$, whence expressible as functions of the elementary symmetric polynomials (by the fundamental theorem of symmetric fractions, Theorem 8.3, p. 99). Moreover, if f is a root of another equation $\Phi = 0$ with coefficients symmetric in $x_1, \ldots, x_n$, then

$$\deg \Phi \geq m.$$

Proof. Let $f_1, f_2, \ldots, f_m$ be the various values of f obtained by permutation of $x_1, \ldots, x_n$ (with $f = f_1$, say), and let

$$\Theta(Y) = (Y - f_1) \cdots (Y - f_m).$$

Since every permutation of $x_1, \ldots, x_n$ permutes the f_i's among themselves, the coefficients of Θ, which are the elementary symmetric polynomials in $f_1, \ldots, f_m$, are not altered when the indeterminates $x_1, \ldots, x_n$ are permuted. Therefore, these coefficients are symmetric in $x_1, \ldots, x_n$, and the equation $\Theta = 0$ satisfies the required properties.

If $\Phi(Y)$ is another polynomial with symmetric coefficients such that $\Phi(f) = 0$, then, for any $i = 1, \ldots, m$, the permutation of $x_1, \ldots, x_n$ which gives to f the value f_i transforms $\Phi(f)$ into $\Phi(f_i)$, since it does not change the coefficients of Φ. Thus, $\Phi(f_i) = 0$ for $i = 1, \ldots, m$, whence Φ has m different roots and it follows that $\deg \Phi \geq m$ (and, in fact, Φ is divisible by Θ, by Proposition 5.10, p. 50). $\square$

For instance, the polynomial $x_1 + \omega x_2 + \omega^2 x_3$ takes six different values by permutation of x_1, x_2, x_3. It is therefore a root of an equation of degree 6 with symmetric coefficients, and of no equation of smaller degree. On the other hand, $(x_1 + \omega x_2 + \omega^2 x_3)^3$ takes only two different values, whence it is a root of a quadratic equation.

After Cardano's method, Lagrange investigates Tschirnhaus' method. If the change of variable

$$Y = b_0 + b_1 X + X^2$$

transforms a given cubic equation in X with roots x_1, x_2, x_3 into an equation like

$$Y^3 = c,$$

which has as roots $\sqrt[3]{c}$, $\omega\sqrt[3]{c}$ and $\omega^2\sqrt[3]{c}$ (where, as above, ω denotes a cube root of unity other than 1), then we can assume

$$x_1^2 + b_1 x_1 + b_0 = \sqrt[3]{c},$$
$$x_2^2 + b_1 x_2 + b_0 = \omega\sqrt[3]{c}, \qquad (10.14)$$
$$x_3^2 + b_1 x_3 + b_0 = \omega^2\sqrt[3]{c}.$$

Multiplying the second equation by ω, the third by ω^2 and adding to the first, we obtain

$$(x_1^2 + \omega x_2^2 + \omega^2 x_3^2) + b_1(x_1 + \omega x_2 + \omega^2 x_3) + b_0(1 + \omega + \omega^2) =$$
$$\sqrt[3]{c}(1 + \omega + \omega^2).$$

Since $1 + \omega + \omega^2 = 0$, it follows that

$$b_1 = -\frac{x_1^2 + \omega x_2^2 + \omega^2 x_3^2}{x_1 + \omega x_2 + \omega^2 x_3}. \qquad (10.15)$$

This rational fraction takes only two values under the permutations of x_1, x_2, x_3, namely

$$b_1 = -\frac{x_1^2 + \omega x_2^2 + \omega^2 x_3^2}{x_1 + \omega x_2 + \omega^2 x_3} \quad \text{and} \quad b_1' = -\frac{x_1^2 + \omega^2 x_2^2 + \omega x_3^2}{x_1 + \omega^2 x_2 + \omega x_3}.$$

Therefore, b_1 can be determined by solving the quadratic equation

$$Y^2 - (b_1 + b_1')Y + b_1 b_1' = 0.$$

The coefficients of this quadratic equation are symmetric in x_1, x_2, x_3, and can therefore be calculated from the coefficients of the proposed equation. On the other hand, adding the equations (10.14) and taking into account the fact that $1 + \omega + \omega^2 = 0$, we obtain

$$(x_1^2 + x_2^2 + x_3^2) + b_1(x_1 + x_2 + x_3) + 3b_0 = 0.$$

Since $x_1^2 + x_2^2 + x_3^2$ and $x_1 + x_2 + x_3$ are symmetric in x_1, x_2, x_3, they can be calculated from the coefficients of the proposed equation. This last equation then shows that b_0 can be rationally calculated from b_1 and the coefficients of the proposed equation. Similarly, multiplying the equations (10.14), we obtain

$$c = (x_1^2 + b_1 x_1 + b_0)(x_2^2 + b_1 x_2 + b_0)(x_3^2 + b_1 x_3 + b_0).$$

Since this expression is symmetric in x_1, x_2, x_3, it can be rationally calculated from b_1, b_0 and the coefficients of the proposed equation. Once b_0, b_1 and c have been calculated, the roots x_1, x_2 and x_3 can be rationally calculated as explained in §6.4 (p. 71). Therefore, Tschirnhaus' method for the cubic equation requires only the solution of a quadratic equation, beyond rational calculations; ultimately, this is because the rational fraction b_1 in (10.15) takes only two values.

Thereafter, Lagrange successively scrutinizes Euler and Bezout's methods, and the various methods for equations of degree 4. Each time, he shows how the roots of the auxiliary equations can be expressed in terms of those of the proposed equation, and he observes that the number of values of these expressions is less than the degree of the proposed equation.

For degree 4, Ferrari's method reduces the quartic equation

$$X^4 + pX^2 + qX + r = 0$$

to

$$\left(X^2 + \frac{p}{2} + u\right)^2 = \left(\sqrt{2u}X - \frac{q}{2\sqrt{2u}}\right)^2$$

by the choice of a suitable u (see §3.2). This last equation splits into two quadratic equations

$$X^2 + \frac{p}{2} + u = \sqrt{2u}X - \frac{q}{2\sqrt{2u}} \quad \text{and} \quad X^2 + \frac{p}{2} + u = -\left(\sqrt{2u}X - \frac{q}{2\sqrt{2u}}\right),$$

which yield the roots x_1, x_2, x_3, x_4 of the proposed quartic equation. Renumbering the roots if necessary, we may assume that x_1 and x_2 are the roots of the first quadratic equation and that x_3 and x_4 are the roots of the second one. Then, since the constant term is the product of the roots, it follows that

$$x_1x_2 = \frac{p}{2} + u + \frac{q}{2\sqrt{2u}} \quad \text{and} \quad x_3x_4 = \frac{p}{2} + u - \frac{q}{2\sqrt{2u}},$$

whence

$$x_1x_2 + x_3x_4 = p + 2u. \tag{10.16}$$

Since p is the coefficient of X^2 in the quartic equation, it is the second symmetric polynomial in x_1, x_2, x_3, x_4, i.e.

$$p = x_1x_2 + x_1x_3 + x_1x_4 + x_2x_3 + x_2x_4 + x_3x_4.$$

Substituting for p in (10.16), the following value of u is found:

$$u = -\tfrac{1}{2}(x_1 + x_2)(x_3 + x_4).$$

This expression takes only three values when x_1, x_2, x_3 and x_4 are permuted. This explains why u is a root of a cubic equation.

Lagrange concludes his review with the attempted applications of Tschirnhaus', Euler's and Bezout's methods to equations of higher degree. He then observes that, according to Bezout's method (see §10.1), the roots of the proposed equation of degree n are obtained in the form $a_0 + a_1\omega + a_2\omega^2 + \cdots + a_{n-1}\omega^{n-1}$, where ω runs over the set of n-th roots of unity. If ζ is a primitive n-th root of unity, then by Proposition 7.11 (p. 89) the n-th roots of unity are $1, \zeta, \zeta^2, \ldots, \zeta^{n-1}$. Substituting successively $1, \zeta, \zeta^2, \ldots, \zeta^{n-1}$ for ω, we obtain the following expressions of the roots:

$$x_1 = a_0 + a_1 + a_2 + \cdots + a_{n-1},$$
$$x_2 = a_0 + a_1\zeta + a_2\zeta^2 + \cdots + a_{n-1}\zeta^{n-1},$$
$$x_3 = a_0 + a_1\zeta^2 + a_2\zeta^4 + \cdots + a_{n-1}\zeta^{2(n-1)},$$
$$\cdots$$
$$x_n = a_0 + a_1\zeta^{n-1} + a_2\zeta^{2(n-1)} + \cdots + a_{n-1}\zeta^{(n-1)^2},$$

whence, in general,

$$x_i = \sum_{j=0}^{n-1} a_j\zeta^{(i-1)j} \qquad \text{for } i = 1, \ldots, n. \tag{10.17}$$

This system is easily solved for $a_0, a_1, \ldots, a_{n-1}$: to obtain the value of a_k, it suffices to multiply each of these equations by a suitable power of ζ so that the coefficient of a_k be 1, and to add the equations thus produced. We obtain

$$\sum_{i=1}^{n} \zeta^{-(i-1)k} x_i = \sum_{j=0}^{n-1} a_j \left(\sum_{i=1}^{n} \zeta^{(j-k)(i-1)} \right). \tag{10.18}$$

If $j \neq k$, then ζ^{j-k} is an n-th root of unity different from 1. Therefore, ζ^{j-k} is a root of

$$\frac{X^n - 1}{X - 1} = X^{n-1} + X^{n-2} + \cdots + X + 1,$$

whence

$$\sum_{i=1}^{n} \zeta^{(j-k)(i-1)} = 0.$$

Therefore, in the right-hand side of (10.18), all the terms vanish except the term corresponding to the index $j = k$, which is na_k. Thus, equation (10.18) yields

$$a_k = \frac{1}{n}\Big(\sum_{i=1}^{n}\zeta^{-(i-1)k}x_i\Big). \tag{10.19}$$

It is easily seen that, if $x_1, \ldots, x_n$ are considered as independent indeterminates, all the values of a_k obtained by the various permutations of $x_1, \ldots, x_n$ are distinct. Therefore, a_k is a root of an equation of degree $n!$. However, Lagrange shows[‡] that a_k^n takes only $(n-1)!$ values. Moreover, if n is prime, then a_k^n is a root of an equation of degree $n-1$ whose coefficients can be determined from the solution of a single equation of degree $(n-2)!$. Thus, for $n = 5$, the determination of a_k^5 still requires the solution of an equation of degree $3! = 6$.

If n is not prime, the result is more complicated. Using the same arguments as in the case where n is prime, Lagrange shows that if $n = pq$ with p prime, and if k is divisible by q, then a_k^p is a root of an equation of degree $p - 1$ whose coefficients depend on a single equation of degree $\frac{n!}{(p-1)p(q!)^p}$. For $n = 4$, it follows that a_2^2 can be found by solving an equation of degree $\frac{4!}{1.2.(2!)^2} = 3$, but for $n = 6$ the determination of a_3^2 requires the solution of an equation of degree $\frac{6!}{1.2.(3!)^2} = 10$. These results led Lagrange to doubt the possibility of solving algebraically the general equations of degree 5 or higher, although he cautiously avoided to reject this possibility too categorically. The conclusion of his investigations of previously known methods is given in Article 86 [40, pp. 355–357], which is quoted here extensively to give an idea of Lagrange's leisurely style of writing.

> As should be clear from the analysis that we have just given of the main known methods for the solution of equations, all these methods reduce to the same general principle, namely to find functions of the roots of the proposed equation which are such: 1° that the equation or the equations by which they are given, i.e. of which they are the roots (equations that are usually called the *reduced* equations), happen to be of a degree smaller than that of the proposed equation, or at least decomposable in other equations of a degree smaller than this one; 2° that the

---

[‡]Lagrange's arguments are elementary, but can be more easily explained when they are related to some subsequent results of Lagrange and with the help of an appropriate notation. Therefore, we postpone the proof of this result to the next section (see Proposition 10.8, p. 149).

values of the sought roots can be easily deduced from them.

The art of solving equations thus consists of discovering functions of the roots which have the above-mentioned properties; but is it always possible to find such functions, for equations of any degree, i.e. for any number of roots? That is a question which seems very difficult to decide in general.

As to equations which do not exceed the fourth degree, the simplest functions which yield their solution can be represented by the general formula

$$x_1 + \omega x_2 + \omega^2 x_3 + \cdots + \omega^{n-1} x_n,$$

x_1, x_2, x_3, ..., x_n being the roots of the proposed equation, which is assumed to be of degree n, and ω being an arbitrary root other than 1 of the equation

$$\omega^n - 1 = 0,$$

i.e. an arbitrary root of the equation

$$\omega^{n-1} + \omega^{n-2} + \omega^{n-3} + \cdots + 1 = 0,$$

as follows from what has been shown in the first two sections about the solution of equations of the third and the fourth degree. [...]

It thus seems that one could conclude from this by induction that every equation of any degree will also be solvable with the help of a reduced equation whose roots are represented by the same formula

$$x_1 + \omega x_2 + \omega^2 x_3 + \omega^3 x_4 + \cdots.$$

However, after what has been proved in the preceding section about the methods of MM. Euler and Bezout, which readily lead to such reduced equations, one has, it seems, the occasion to convince oneself beforehand that this conclusion will be defective from the fifth degree on; hence it follows that, if the algebraic solution of the equations of degree higher than four is not impossible, it must rely on some functions of the roots, other than the above.

The polynomials of the form

$$x_1 + \omega x_2 + \omega^2 x_3 + \cdots + \omega^{n-1} x_n,$$

where ω is an n-th root of unity, were subsequently christened *Lagrange resolvents*. As we have seen, they originate from the works of Euler and Bezout, and it will be clear later that they play a prominent role in Galois' theory of equations. For the convenience of later reference, we recapitulate the formula which yields the roots of equations of any degree n in terms of Lagrange resolvents.

FORMULA. *If $t(\omega)$ denotes the Lagrange resolvent*

$$t(\omega) = x_1 + \omega x_2 + \omega^2 x_3 + \cdots + \omega^{n-1} x_n,$$

where ω is an n-th root of unity, then for $i = 1, \ldots, n$,

$$x_i = \frac{1}{n}\left(\sum_\omega \omega^{-(i-1)} t(\omega)\right),$$

where the sum runs over all the n-th roots of unity.

This was shown above: see equations (10.17) and (10.19).

10.3 First results of group theory and Galois theory

In the final section of his paper, Lagrange draws from his investigations some general conclusions concerning the degree of the equations by which functions of the roots of a given equation can be determined. Proposition 10.1 above (p. 131) is a first instance of Lagrange's observations, but in his conclusions Lagrange goes much farther. In effect, he begins to calculate with permutations of the roots, obtaining first results in group theory and Galois theory.

Remarkably enough, these results were achieved without even devising a notation for permutations, which were indeed very new objects to calculate with. Unfortunately, this makes Lagrange's arguments sometimes hard to follow. To facilitate our exposition of Lagrange's results, we shall not refrain from using modern notation. Thus, for any integer $n \geq 1$, we denote by S_n the symmetric group on $\{1, \ldots, n\}$, i.e. the group of permutations of $1, \ldots, n$. For $\sigma \in S_n$ and for any rational fraction f in n indeterminates $x_1, \ldots, x_n$, we set

$$\sigma\big(f(x_1, \ldots, x_n)\big) = f(x_{\sigma(1)}, \ldots, x_{\sigma(n)}).$$

So, S_n can be considered as the group of permutations of $x_1, \ldots, x_n$, and S_n acts on the rational fractions in $x_1, \ldots, x_n$ by permuting the indeterminates. For any rational fraction f, we denote by $I(f)$ the subgroup of permutations $\sigma \in S_n$ which leave f invariant, (sometimes called the *isotropy group* of f), i.e.

$$I(f) = \left\{ \sigma \in S_n \mid \sigma\big(f(x_1, \ldots, x_n)\big) = f(x_1, \ldots, x_n) \right\}.$$

We generally denote by $|E|$ the number of elements in a finite set E, which is called the *order* of E, if E is a group. Thus, for instance

$$|S_n| = n!.$$

In his Article 97, Lagrange proves the following theorem:

10.2. THEOREM. *Let $f = f(x_1, \ldots, x_n)$ be a rational fraction in n indeterminates. The number m of different values that f takes under the permutations of $x_1, \ldots, x_n$ is equal to the quotient of $n!$ by the number of permutations which leave f invariant,*

$$m = \frac{n!}{|I(f)|}.$$

Proof. Let $f_1, \ldots, f_m$ be the various values of f (with $f = f_1$, say). For $i = 1, \ldots, m$, let $I(f \mapsto f_i)$ be the set of permutations $\sigma \in S_n$ such that $\sigma(f) = f_i$; thus,

$$I(f \mapsto f_1) = I(f).$$

Now, fix some $i = 1, \ldots, m$ and some $\sigma \in I(f \mapsto f_i)$. If $\tau \in I(f)$, then

$$\sigma \circ \tau(f) = \sigma(f) = f_i,$$

whence $\sigma \circ \tau \in I(f \mapsto f_i)$. Conversely, if $\rho \in I(f \mapsto f_i)$, then $\sigma^{-1} \circ \rho \in I(f)$. Since

$$\rho = \sigma \circ (\sigma^{-1} \circ \rho),$$

it follows that every element in $I(f \mapsto f_i)$ is of the form $\sigma \circ \tau$ where $\tau \in I(f)$. Therefore, composition on the left with σ defines a bijection from $I(f)$ onto $I(f \mapsto f_i)$, hence

$$|I(f)| = |I(f \mapsto f_i)|.$$

Since every permutation in S_n maps f onto one of its values $f_1, \ldots, f_m$, we have a decomposition of S_n as a disjoint union

$$S_n = \bigcup_{i=1}^{m} I(f \mapsto f_i),$$

whence

$$|S_n| = \sum_{i=1}^{m} |I(f \mapsto f_i)|.$$

Since $|S_n| = n!$ and since each term in the right-hand side is equal to $|I(f)|$, this last equation yields

$$n! = m \cdot |I(f)|.$$

$\square$

Here now are Lagrange's own words (in free translation, and with slight notational changes to fit the notation of this section). To understand the following quotation, one needs to know that Θ is the polynomial

$$\Theta(t) = \prod_{\sigma \in S_n} \Big(t - \sigma\big(f(x_1, \ldots, x_n) \big) \Big),$$

whose roots are the values of f under the permutations of $x_1, \ldots, x_n$ (compare Proposition 10.1, p. 131).

> Although the equation $\Theta = 0$ must in general be of degree $1 \cdot 2 \cdot 3 \cdot \ldots \cdot n = n!$, which is equal to the number of permutations of $x_1, \ldots, x_n$, yet if it happens that the function be such that it does not receive any change by some or several permutations, then the equation in question will necessarily reduce to a smaller degree.
>
> Assume, for instance, that the function $f(x_1, x_2, x_3, x_4, \ldots)$ be such that it keeps the same value when x_1 is changed into x_2, x_2 into x_3 and x_3 into x_1, so that
>
> $$f(x_1, x_2, x_3, x_4, \ldots) = f(x_2, x_3, x_1, x_4, \ldots),$$
>
> it is clear that the equation $\Theta = 0$ will already have two equal roots; but I am going to prove that with this hypothesis all the other roots will be pairwise equal too. Indeed, let us consider

an arbitrary root of the same equation, which be represented by the function $f(x_4, x_3, x_1, x_2, \ldots)$, as this one derives from the function $f(x_1, x_2, x_3, x_4, \ldots)$ by changing x_1 into x_4, x_2 into x_3, x_3 into x_1, x_4 into x_2, it follows that it will have to keep the same value when we change in it x_4 into x_3, x_3 into x_1 and x_1 into x_4; so that we shall also have

$$f(x_4, x_3, x_1, x_2, \ldots) = f(x_3, x_1, x_4, x_2, \ldots).$$

Therefore, in this case, the quantity Θ will be equal to a square θ^2 and consequently the equation $\Theta = 0$ will be reduced to this one $\theta = 0$, which will have dimension $\frac{n!}{2}$.

Likewise, one will prove that if the function f is by its own nature such that it keeps the same value when two or three or a greater number of different permutations are made among the roots x_1, x_2, x_3, x_4, $\ldots$, x_n, the roots of the equation $\Theta = 0$ will be equal three by three or four by four or, etc.; so that the quantity Θ will be equal to a cube θ^3 or to a square-square θ^4 or, etc., and therefore the equation $\Theta = 0$ will reduce to this one $\theta = 0$, whose degree will be equal to $\frac{n!}{3}$, or equal to $\frac{n!}{4}$, or, etc.

To see the link between Lagrange's argument in the quotation above and the preceding proof, let $f = f(x_1, x_2, x_3, x_4, \ldots)$ and $f_i = f(x_4, x_3, x_1, x_2, \ldots)$, and let σ be the permutation $x_1 \mapsto x_4$, $x_2 \mapsto x_3$, $x_3 \mapsto x_1$, $x_4 \mapsto x_2$, $\ldots$ so that

$$\sigma(f) = f_i, \qquad \text{i.e. } \sigma \in I(f \mapsto f_i).$$

Lagrange's observation is that if $\tau\colon x_1 \mapsto x_2 \mapsto x_3 \mapsto x_1$ is in $I(f)$, then $\sigma \circ \tau\colon x_1 \mapsto x_3$, $x_2 \mapsto x_1$, $x_3 \mapsto x_4$, $x_4 \mapsto x_2$, $\ldots$ is such that

$$\sigma \circ \tau(f) = f_i, \qquad \text{i.e. } \sigma \circ \tau \in I(f \mapsto f_i).$$

This is indeed the crucial step in the proof.

The theorem which is often referred to as "Lagrange's theorem" nowadays deals with the order (i.e. the number of elements) of subgroups of a group. It is stated as follows:

10.3. THEOREM. *Let H be a subgroup of a finite group G. Then $|H|$ divides $|G|$.*

Proof. For $g \in G$, define the (left) coset gH by

$$gH = \{gh \mid h \in H\}.$$

We readily have $|gH| = |H|$, since multiplication by g defines a bijection from H onto gH. Since $g = g1 \in gH$, it is clear that every element of G is in some coset. Moreover, if two cosets have a common element, then they are equal. Indeed, if there exists an element $x \in G$ such that $x \in g_1H \cap g_2H$, let $x = g_1h_1 = g_2h_2$ for some $h_1, h_2 \in H$, then every element $g_1h \in g_1H$ can be written as

$$g_1h = g_2(h_2h_1^{-1}h) \in g_2H,$$

so that $g_1H \subset g_2H$. Interchanging the indices 1 and 2, we obtain $g_2H \subset g_1H$, whence $g_1H = g_2H$.

This shows that the group G decomposes into a disjoint union of cosets. Since the number of elements of each of these cosets is equal to $|H|$, it follows that $|H|$ divides $|G|$ (and the quotient of $|G|$ by $|H|$ is the number of different cosets in a decomposition of G, which is called the *index* of H in G). $\square$

Although the pattern of this proof is quite similar to that of Theorem 10.2 (observe that $I(f \mapsto f_i)$ is a left coset of $I(f)$), Lagrange did not reach this generality, nor did he need to. His primary concern was to obtain some information on the number of values of functions, whence, by Proposition 10.1 above (p. 131), on the degree of the equation by which a given function of the roots can be determined.

In this respect, his achievement is even more stunning: he proves a "relative version" of Proposition 10.1 above, which can be seen as a part of the fundamental theorem of Galois theory for the splitting field of the general polynomial:

> Now, as soon as the value of a given function of the roots $x_1, \ldots, x_n$ has been found, either by the solution of the equation $\theta = 0$ or otherwise, I claim that the value of another arbitrary function of the same roots can be found, and that, generally speaking, simply by a linear equation, except for some particular cases which demand an equation of the second degree, or of the third, etc. This Problem seems to me to be one of the most important of the theory of equations, and the general Solution that we are going to give will shed a new light on this part of Algebra. [40, Art. 100]

Lagrange's result can be stated more precisely as follows:

10.4. THEOREM. *Let f and g be two rational fractions in n indeterminates x_1, $\ldots$, x_n. If f takes m different values by the permutations which leave g invariant, then f is a root of an equation of degree m whose coefficients are rational fractions in g and in the elementary symmetric polynomials $s_1, \ldots, s_n$.*

In particular, if f is invariant by the permutations which leave g invariant, then f is a rational fraction in g and $s_1, \ldots, s_n$.

Proof. We begin with the special case above, i.e. we first assume $m = 1$. Then, let $g_1, \ldots, g_r$ be the different values of g under the permutations of $x_1, \ldots, x_n$ (with $g_1 = g$, say). Let $f_1 = f$, $f_2, \ldots, f_r$ be the corresponding values of f, in the sense that if a permutation of $x_1, \ldots, x_n$ gives to g the value g_i (for some i), then it gives to f the value f_i. The possibility of defining such a correspondence from the values of g to the values of f follows from the hypothesis that f is invariant under the permutations which leave g invariant. Indeed, this hypothesis means that $I(g) \subset I(f)$; thus, if σ and ρ both give to g the value g_i, then the proof of Theorem 10.2 shows that $\rho = \sigma \circ \tau$ for some $\tau \in I(g)$, and the hypothesis ensures that $\tau \in I(f)$, whence $\rho(f) = \sigma(f)$. We may therefore define $f_i = \sigma(f)$ for *any* $\sigma \in S_n$ such that $\sigma(g) = g_i$.

Consider then the following expressions, which are denoted by $a_0, \ldots, a_{r-1}$:

$$a_0 = f_1 + f_2 + \cdots + f_r$$
$$a_1 = f_1 g_1 + f_2 g_2 + \cdots + f_r g_r$$
$$a_2 = f_1 g_1^2 + f_2 g_2^2 + \cdots + f_r g_r^2 \qquad (10.20)$$
$$\cdots$$
$$a_{r-1} = f_1 g_1^{r-1} + f_2 g_2^{r-1} + \cdots + f_r g_r^{r-1}.$$

From the definition of f_i and g_i, it follows that every permutation of $x_1, \ldots, x_n$ merely permutes the terms in $a_0, \ldots, a_{r-1}$, so that each of these expressions is symmetric in $x_1, \ldots, x_n$ and can therefore be calculated as a rational fraction in $s_1, \ldots, s_n$, by the fundamental theorem of symmetric fractions (Theorem 8.3, p. 99).

Now, the idea is to solve system (10.20) for $f_1, \ldots, f_r$. However, the usual elimination method would yield f_1 in terms of $a_0, \ldots, a_{r-1}$ (thus, eventually, in terms of $s_1, \ldots, s_n$) but also of $g_1, \ldots, g_r$, while we need an expression of f_1 in terms of $s_1, \ldots, s_n$ and g_1 only.

Lagrange then uses the following trick: let

$$\theta(Y) = (Y - g_1)(Y - g_2) \cdots (Y - g_r) = Y^r + b_{r-1} Y^{r-1} + \cdots + b_0.$$

Dividing this polynomial by $Y - g_1$, we obtain

$$\psi(Y) = (Y - g_2) \cdots (Y - g_r) = Y^{r-1} + c_{r-2} Y^{r-2} + \cdots + c_0$$

and the coefficients $c_0, c_1, \ldots, c_{r-2}$ are rational fractions in $b_0, b_1, \ldots, b_{r-1}$ and g_1, as is easily seen by carrying out explicitly the division of $\theta(Y)$ by $Y - g_1$. Now, the coefficients $b_0, \ldots, b_{r-1}$ of θ are symmetric in $g_1, \ldots, g_r$ whence also in $x_1, \ldots, x_n$, and can therefore be calculated in terms of $s_1, \ldots, s_n$. Thus, $c_0, c_1, \ldots, c_{r-2}$ are rational fractions in g_1 and $s_1, \ldots, s_n$.

Multiplying the first equation of (10.20) by c_0, the second by c_1, the third by c_2, etc. and the last by 1, and adding the equations thus obtained, we obtain an equation in which the coefficient of f_i (for $i = 1, \ldots, r$) is the polynomial $\psi(Y)$ calculated at $Y = g_i$, namely

$$a_0 c_0 + a_1 c_1 + \cdots + a_{r-1} = f_1 \psi(g_1) + f_2 \psi(g_2) + \cdots + f_r \psi(g_r).$$

Since $\psi(g_2) = \cdots = \psi(g_r) = 0$, we thus end up with an expression of f_1 as a rational fraction in g and $s_1, \ldots, s_n$,

$$f_1 = (a_0 c_0 + a_1 c_1 + \cdots + a_{r-1}) \psi(g_1)^{-1},$$

as was required.

This proves Theorem 10.4 in the special case where $m = 1$, but the general case follows easily. Indeed, assume now that f takes m values $f_1, \ldots, f_m$ under the permutations which leave g invariant. Then f is a root of the equation

$$(Y - f_1) \cdots (Y - f_m) = 0,$$

and this equation satisfies the required properties since its coefficients are symmetric in $f_1, \ldots, f_m$, whence invariant under the permutations which leave g invariant, whence, by the special case above, rational fractions in g and $s_1, \ldots, s_n$. $\qquad\qquad\square$

Lagrange's result is even more general than the above, since he also considers the case where $x_1, \ldots, x_n$ are related by some algebraic relations (this occurs when $x_1, \ldots, x_n$ are the roots of some particular equation instead of the general equation of degree n), but the theorem above gives the flavor of Lagrange's proof and covers the essential part of the applications that Lagrange had in mind, since his purpose was to investigate the solution of general equations.

After Lagrange's preceding result (Theorem 10.4), the solution of general equations is much enlightened indeed. The strategy appears as follows: to solve the general equation of degree n, one has to find a (finite) sequence of rational fractions $V_0, V_1, \ldots, V_r$ in n indeterminates $x_1, \ldots, x_n$ such that the first function V_0 is symmetric in $x_1, \ldots, x_n$, the last function V_r is one of the roots, say $V_r = x_1$, and for $i = 1, \ldots, r$, the function V_i satisfies either

(1) $V_i^n = V_{i-1}$, or

(2) the number of values of V_i under the permutations which leave V_{i-1} invariant is strictly less than n.

In case (1), the function V_i can be calculated from the preceding ones by extracting an n-th root, and in case (2) it can be found by solving an equation of degree less than n, by Theorem 10.4. Since the last function is a root of the proposed equation, it means that a root can be found by successive extractions of roots and solutions of equations of lower degree. The sequence $V_0, V_1, \ldots, V_r$ indicates in which order the calculations can be arranged. The other roots can be found likewise, substituting for $V_0, V_1, \ldots, V_r$ similar functions. (More precisely, for any $\sigma \in S_n$, the root $\sigma(V_r)$ can be found by the sequence $\sigma(V_0) = V_0, \sigma(V_1), \ldots, \sigma(V_r)$.) For $n = 2$, one chooses

$$V_0 = (x_1 - x_2)^2,$$
$$V_1 = x_1 - x_2,$$
$$V_2 = x_1.$$

Thus, V_1 can be found by extracting the square root of V_0, and V_2 can then be found rationally, since $V_2 = \frac{1}{2}\left(V_1 + (x_1 + x_2)\right)$.

For $n = 3$, one can choose for V_0 any symmetric function, next (denoting by ω a cube root of unity other than 1)

$$V_1 = (x_1 + \omega x_2 + \omega^2 x_3)^3,$$
$$V_2 = x_1 + \omega x_2 + \omega^2 x_3,$$
$$V_3 = x_1.$$

Since V_1 takes only two values by all the permutations of x_1, x_2, x_3, it can be found by solving a quadratic equation. Next, V_2 is found by extracting a cube root of V_1 and finally, since V_3 is invariant under the permutations which leave V_2 invariant (as only the identity leaves V_2 invariant), it can be determined rationally from V_2, i.e. by solving an equation of degree 1. Likewise, x_2 and x_3 can be determined rationally from V_2.

For $n = 4$, one can choose for V_0 any symmetric function, next

$$V_1 = (x_1 + x_2)(x_3 + x_4),$$
$$V_2 = x_1 + x_2,$$
$$V_3 = x_1.$$

Indeed, V_1 can be determined by solving a cubic equation since it takes only three values by the permutations of x_1, x_2, x_3, x_4. Next, V_2 can be determined by a quadratic equation since it takes only two values under the permutations which leave V_1 invariant, and finally V_3 can be found by a quadratic equation since it takes only two values under the permutations which leave V_2 invariant. The root x_2 is then readily found, since it is the other root of the quadratic equation which yields the value of V_3 ($= x_1$), and the other roots x_3 and x_4 are found by similar calculations: $x_3 + x_4$ is the other root of the equation which yields the value of V_2, and x_3, x_4 are the roots of a quadratic equation. This is the pattern which is suggested by Ferrari's solution of quartic equations. Of course, other choices are possible; for instance, using Lagrange's resolvents, one could choose

$$V_1 = (x_1 - x_2 + x_3 - x_4)^2,$$
$$V_2 = x_1 - x_2 + x_3 - x_4,$$
$$V_3 = x_1.$$

The first function V_1 is the root of a cubic equation, and V_2, V_3 are obtained by solving successively two quadratic equations.

> These are, if I am not mistaken, the genuine principles of the solution of equations and the analysis which is most suitable to lead to it; everything is reduced, as is seen, to a kind of calculus of combinations, by which the results to which one is led are found *a priori*. It would be opportune to apply it to the equations of the fifth degree and higher degrees, whose solution is so far unknown; but this application requires a too large amount of researches and combinations, whose success is, for that matter, still very dubious, for us to tackle this problem now; we hope however to come back to it at another time, and we will be content to have here set the foundations of a theory which seems to us new and general. [40, Art. 109]

To conclude this chapter, we now apply the results above to sketch a proof of the properties of "Lagrange resolvents," which we have pointed out in §10.2, at least for the case where n is prime. We shall need the following result on the existence of rational fractions which are invariant under a prescribed group:

10.5. PROPOSITION. *For any subgroup G in S_n, there exists a rational fraction f in n indeterminates such that $I(f) = G$.*

Proof. Choose a monomial m which is not invariant under any (non-trivial) permutation of the indeterminates, for instance $m = x_1 x_2^2 x_3^3 \cdots x_n^n$, and let

$$f = \sum_{\sigma \in G} \sigma(m).$$

Since for any $\tau \in G$ the set of products $\{\tau \circ \sigma \mid \sigma \in G\}$ is G, it follows that

$$\sum_{\sigma \in G} \tau \circ \sigma(m) = \sum_{\sigma \in G} \sigma(m),$$

whence

$$\tau(f) = f \qquad \text{for any } \tau \in G.$$

Therefore, $G \subset I(f)$.

On the other hand, if $\rho \notin G$, then the monomial $\rho(m)$ appears in $\rho(f)$ but not in f, so $\rho(f) \neq f$. Thus, $G = I(f)$. $\qquad\square$

Henceforth, to simplify notation a little, we index the indeterminates from 0. We shall thus consider S_n as the group of permutations of $\{0, 1, \ldots, n-1\}$, and we now define certain permutations which have interesting properties in relation with Lagrange's resolvents.

For any integer k relatively prime to n, we denote by

$$\sigma_k \colon \{0, 1, \ldots, n-1\} \to \{0, 1, \ldots, n-1\}$$

the map defined as follows: for any $i \in \{0, 1, \ldots, n-1\}$, the image $\sigma_k(i)$ is the unique integer j between 0 and $n-1$ such that $ik - j$ is divisible by n (i.e. $ik \equiv j \bmod n$, using a notation which will be introduced in §12.2). In other words, j is the remainder of the division of ik by n.

10.6 PROPOSITION. *For any integer k relatively prime to n, the map σ_k is a permutation of $\{0, 1, \ldots, n-1\}$.*

Proof. By Theorem 7.8 (p. 86), it is possible to find integers ℓ, m such that

$$k\ell + mn = 1. \tag{10.21}$$

For any $i \in \{0, 1, \ldots, n-1\}$, the definition of $\sigma_k(i)$ shows that $ik - \sigma_k(i)$ is divisible by n, hence $ik\ell - \sigma_k(i)\ell$ is divisible by n. Adding imn, which is clearly divisible by n, we see that $i(k\ell + mn) - \sigma_k(i)\ell$ is divisible by n. By (10.21), it follows that

$$i - \sigma_k(i)\ell \text{ is divisible by } n. \tag{10.22}$$

This last relation means that $\sigma_\ell(\sigma_k(i)) = i$, so that $\sigma_\ell \circ \sigma_k$ is the identity on $\{0, 1, \ldots, n-1\}$. Interchanging k and ℓ in the above discussion, it follows that $\sigma_k \circ \sigma_\ell$ also is the identity on $\{0, 1, \ldots, n-1\}$. Therefore, σ_ℓ and σ_k are reciprocal bijections of $\{0, 1, \ldots, n-1\}$ onto itself. $\square$

From now on, we assume that n *is a prime number*, so that σ_i is defined for any $i = 1, \ldots, n-1$. We denote by τ the cyclic permutation $0 \mapsto 1 \mapsto 2 \mapsto \cdots \mapsto n-1 \mapsto 0$ and by $\mathrm{GA}(n)$ the subgroup of S_n generated by $\sigma_1, \ldots, \sigma_{n-1}$ and τ. It can be shown that

$$|\mathrm{GA}(n)| = n(n-1)$$

(see Exercise 5). In fact, from a less elementary point of view, $\mathrm{GA}(n)$ can be identified to the group of affine transformations of the affine line over the field with n elements: τ generates the group of translations while $\sigma_1, \ldots, \sigma_{n-1}$ are homotheties.

Let V be a rational fraction in $x_0, x_1, \ldots, x_{n-1}$ such that $I(V) = \mathrm{GA}(n)$ (the existence of such a function is ensured by Proposition 10.5) and, for any n-th root of unity ω, let $t(\omega)$ denote the following Lagrange resolvent:

$$t(\omega) = x_0 + \omega x_1 + \omega^2 x_2 + \cdots + \omega^{n-1} x_{n-1}.$$

10.7 THEOREM. *Assume n is prime. If $\omega \neq 1$, then $t(\omega)^n$ is a root of an equation of degree $n-1$ whose coefficients are rational fractions in V and in the elementary symmetric polynomials in $x_0, \ldots, x_{n-1}$. Moreover, V is a root of an equation of degree $(n-2)!$ whose coefficients are rational fractions in the elementary symmetric polynomials.*

Proof. The fact that V is a root of an equation of degree $(n-2)!$ readily follows from Proposition 10.1 and Theorem 10.2, since $I(V)$ has $n(n-1)$ elements. To prove the rest, it suffices, by Theorem 10.4, to show that $t(\omega)^n$ takes $n-1$ values by the permutations which leave V invariant, i.e. by the permutations in $\mathrm{GA}(n)$.

First, we consider the action of σ_k:

$$\sigma_k(t(\omega)) = x_0 + \omega x_{\sigma_k(1)} + \omega^2 x_{\sigma_k(2)} + \cdots + \omega^{n-1} x_{\sigma_k(n-1)}.$$

Since $\omega^n = 1$, relation (10.22) yields

$$\omega^i = (\omega^\ell)^{\sigma_k(i)} \qquad \text{for } i = 0, 1, \ldots, n-1.$$

Therefore,

$$\sigma_k\big(t(\omega)\big) =$$
$$x_0 + (\omega^\ell)^{\sigma_k(1)} x_{\sigma_k(1)} + (\omega^\ell)^{\sigma_k(2)} x_{\sigma_k(2)} + \cdots + (\omega^\ell)^{\sigma_k(n-1)} x_{\sigma_k(n-1)},$$

which shows

$$\sigma_k\big(t(\omega)\big) = t(\omega^\ell). \tag{10.23}$$

Next, we consider the action of τ. Since

$$\tau\big(t(\omega)\big) = x_1 + \omega x_2 + \omega^2 x_3 + \cdots + \omega^{n-1} x_0,$$

we have

$$\tau\big(t(\omega)\big) = \omega^{-1} t(\omega),$$

for any n-th root of unity ω. Since $\omega^{-n} = 1$, this last equation yields

$$\tau\big(t(\omega)^n\big) = t(\omega)^n.$$

This result, together with (10.23), shows that under any product of the permutations $\sigma_1, \ldots, \sigma_{n-1}, \tau$ the function $t(\omega)^n$ takes one of the values $t(\omega)^n$, $t(\omega^2)^n$, $\ldots, t(\omega^{n-1})^n$, which are pairwise different if $\omega \neq 1$. Since $\mathrm{GA}(n)$ is generated by $\sigma_1, \ldots, \sigma_{n-1}, \tau$, this means that $t(\omega)^n$ takes $n - 1$ values under the permutations in $\mathrm{GA}(n)$, and the proof is complete. $\square$

Simpler arguments yield the number of values of $t(\omega)^n$:

10.8 PROPOSITION. *The function $t(\omega)^n$ takes $(n - 1)!$ values under the permutations of $x_0, \ldots, x_{n-1}$.*

Proof. Let k be the number of values of $t(\omega)^n$. At the end of the preceding proof, it was shown that $t(\omega)^n$ is invariant under τ, whence also under all the powers τ^2, $\tau^3, \ldots, \tau^{n-1}$. Thus, $\big|I\big(t(\omega)^n\big)\big| \geq n$, and Theorem 10.2 shows that

$$k \leq (n - 1)!.$$

On the other hand, it follows from Proposition 10.1 (p. 131) that $t(\omega)^n$ is a root of an equation $\Theta(Y) = 0$ of degree k. Thus, $t(\omega)$ is a root of $\Theta(Y^n) = 0$, which has degree kn; but since $t(\omega)$ takes $n!$ different values under the permutations of the variables, it cannot be a root of an equation of degree less than $n!$, by Proposition 10.1. Therefore, $kn \geq n!$, hence

$$k \geq (n-1)!.$$

$\square$

This proposition remains valid with the same proof when n is not prime, since the permutations σ_k were not used.

Exercises

1. As in Exercise 3 of Chapter 8, let

$$u_1 = (x_1 + x_2)(x_3 + x_4), \qquad v_1 = x_1 x_2 + x_3 x_4,$$
$$u_2 = (x_1 + x_3)(x_2 + x_4), \qquad v_2 = x_1 x_3 + x_2 x_4,$$
$$u_3 = (x_1 + x_4)(x_2 + x_3), \qquad v_3 = x_1 x_4 + x_2 x_3.$$

Show that v_1, v_2 and v_3 are rational fractions in u_1, u_2, u_3 with symmetric coefficients. Use this result to show how the cubic equation which has as roots v_1, v_2, v_3 is related to the equation with roots u_1, u_2, u_3. (Compare Exercise 3 of Chapter 8). Same questions with $w_1 = (x_1 - x_2 + x_3 - x_4)^2$, $w_2 = (x_1 + x_2 - x_3 - x_4)^2$, $w_3 = (x_1 - x_2 - x_3 + x_4)^2$ instead of v_1, v_2, v_3.

2. Use the arguments in the proof of Lagrange's theorem (Theorem 10.4, p. 142) to express $x_1 x_2$ as a rational fraction of $x_1 + x_2$ with coefficients symmetric in the three indeterminates x_1, x_2, x_3. Is this expression unique?

3. Find all the polynomials $f = ax_1 + bx_2 + cx_3$ (with a, b, $c \in \mathbb{C}$) which have the property that x_1, x_2, x_3 can be rationally expressed from f with symmetric coefficients and such that f^3 takes only two values by the permutations of x_1, x_2, x_3.

4. Let n be a prime number. For any n-th root of unity ω, let

$$t(\omega) = x_0 + \omega x_1 + \cdots + \omega^{n-1} x_{n-1}.$$

Show that $t(\omega^k) t(\omega)^{-k}$ is a rational fraction in $t(\omega)^n$ with symmetric coefficients, for any integer k.

5. Let n be a prime number and use the notation of Proposition 10.6 and after. Prove that $\tau \circ \sigma_i = \sigma_i \circ \tau^k$ for some k. Deduce that

$$GA(n) = \{\sigma_i \circ \tau^j \mid i = 1, \ldots, n-1 \text{ and } j = 0, \ldots, n-1\}$$

and that $|GA(n)| = n(n-1)$.

6. Show that for any group G and any subgroup H, the map $g \mapsto g^{-1}$ (which is an anti-automorphism of G) induces a bijection between the set of left cosets of H in G and the set of right cosets of H.

Chapter 11

Vandermonde

11.1 Introduction

Alexandre-Théophile Vandermonde (1735–1796) is not a mathematician in the same class as Lagrange or Euler. His contributions to mathematics were scarce and hardly influential. Ironically enough, he is most often remembered nowadays for a determinant which bears his name but is not to be found in his papers: Vandermonde determinants may have been so christened because someone misread indices for exponents (see Lebesgue [41, pp. 206–207]).

Nevertheless, his work, remarkably described by Lebesgue [41], shows that brilliant ideas and deep insights come not only from first-class mathematicians. Several of Lagrange's ideas were indeed discovered simultaneously or perhaps even a little earlier by Vandermonde. Most notably, Vandermonde performed calculations with permutations and singled out the functions known as Lagrange resolvents, but his exposition is less clear, less authoritative than Lagrange's.

Moreover, the delay in publication was such that Vandermonde's "Mémoire sur la résolution des équations" [59] appeared two years after the first part of Lagrange's "Réflexions sur la résolution algébrique des équations." Lagrange was already famous at that time, and Vandermonde's self-effacing comment (in a footnote added in proof)

> One will notice some conformities between this [Lagrange's] work and mine, of which I cannot feel but flattered. [59, p. 365]

did not help to secure notoriety for his paper. However, Vandermonde can be credited with a real breakthrough in the theory of equations: the solution of cyclotomic equations. This was definitely not obtained previously by Lagrange.

153

We shall divide up our discussion of Vandermonde's memoir into two parts: the discussion of general equations, which is somewhat analogous to Lagrange's, and the solution of cyclotomic equations.

11.2   The solution of general equations

Vandermonde's starting point is that the formula which yields the solutions of an equation in terms of the coefficients is necessarily ambiguous, since it must take as values the various roots. He then separates the solution into three "heads" [59, p. 370]:

1° To find a function of the roots, of which it can be said, in some sense, that it equals such of the roots that one wants.

2° To put this function in such a form that it be indifferent to interchange the roots in it.

3° To substitute in it the values of the sum of the roots, the sum of pairwise products, etc.

Consider for instance the solution of quadratic equations

$$X^2 - s_1 X + s_2 = 0,$$

with roots x_1, x_2. The function

$$F_2(x_1, x_2) = \frac{1}{2}\left((x_1 + x_2) + \sqrt{(x_1 - x_2)^2}\right)$$

satisfies the condition in 1°, since its value is x_1 or x_2, depending on the choice of the square root of $(x_1 - x_2)^2$, namely

$$\sqrt{(x_1 - x_2)^2} = \pm(x_1 - x_2).$$

Since moreover $F_2(x_1, x_2)$ is not altered when the roots x_1 and x_2 are interchanged, it is already in the form which is called for by 2°. Finally, 3° requires the evaluation of $F_2(x_1, x_2)$ in terms of s_1 and s_2. This is quite easy:

$$x_1 + x_2 = s_1 \qquad \text{and} \qquad (x_1 - x_2)^2 = s_1^2 - 4s_2,$$

whence

$$F_2(x_1, x_2) = \frac{1}{2}\left(s_1 + \sqrt{s_1^2 - 4s_2}\right).$$

Vandermonde first solves in full generality problem 3°. He thus proves the fundamental theorem of symmetric functions. which says that every symmetric function can be evaluated in terms of the elementary symmetric polynomials (Theorem 8.3, p. 99).

He then solves problem 1°, displaying the following formula:

$$F_n(x_1, \ldots, x_n) = \frac{1}{n}\left((x_1 + \cdots + x_n) + \sum_{i=1}^{n-1} \sqrt[n]{V_i^n}\right) \tag{11.1}$$

where

$$V_i = \rho_1^i x_1 + \cdots + \rho_n^i x_n$$

and $\rho_1, \ldots, \rho_n$ denote the n-th roots of unity (including 1).

To see that this function indeed answers to head 1°, we have to prove that for any $k = 1, \ldots, n$, some determination of the n-th roots $\sqrt[n]{V_i^n}$ can be chosen in such a way that $F_n(x_1, \ldots, x_n) = x_k$. This can be done as follows: choose $\sqrt[n]{V_i^n} = \rho_k^{-i} V_i$, i.e.

$$\sqrt[n]{V_i^n} = x_k + \sum_{j \neq k}(\rho_k^{-1}\rho_j)^i x_j.$$

Then

$$F_n(x_1, \ldots, x_n) = \frac{1}{n}\left(nx_k + \sum_{j \neq k}\left(\sum_{i=0}^{n-1}(\rho_k^{-1}\rho_j)^i\right)x_j\right). \tag{11.2}$$

Now, $\rho_k^{-1}\rho_j$ is an n-th root of unity, different from 1 if $k \neq j$, hence it is a root of

$$\frac{X^n - 1}{X - 1} = 1 + X + \cdots + X^{n-1}.$$

Therefore,

$$\sum_{i=0}^{n-1}(\rho_k^{-1}\rho_j)^i = 0$$

and equation (11.2) simplifies to

$$F_n(x_1, \ldots, x_n) = x_k.$$

Of course, if $n \geq 3$ the function $F_n(x_1, \ldots, x_n)$ also has other determinations besides $x_1, \ldots, x_n$, but this does not seem to matter to Vandermonde.

It is instructive to compare Vandermonde's formula (11.1) to Lagrange's formula, p. 138. It turns out that the functions V_i are none others than Lagrange resolvents. To establish this point, choose a primitive n-th root of unity ω; the various n-th roots of unity are then powers of ω, and we can set $\rho_k = \omega^{k-1}$ for $k = 1, \ldots, n$. Then

$$V_i = (\omega^0)^i x_1 + (\omega^1)^i x_2 + (\omega^2)^i x_3 + \cdots + (\omega^{n-1})^i x_n,$$

hence

$$V_i = x_1 + \omega^i x_2 + (\omega^i)^2 x_3 + \cdots + (\omega^i)^{n-1} x_n$$

and it follows that V_i is the Lagrange resolvent which was denoted by $t(\omega^i)$ in the formula of p. 138.

Problems 1° and 3° are thus completely solved by Vandermonde; the real stumbling-block is of course problem 2°. For $n = 3$, Vandermonde observes that, choosing $\rho_1 = 1$, $\rho_2 = \omega$ and $\rho_3 = \omega^2$, where ω is a cube root of unity other than 1, the functions involved in $F_3(x_1, x_2, x_3)$, which are

$$V_1^3 = (x_1 + \omega x_2 + \omega^2 x_3)^3 \quad \text{and} \quad V_2^3 = (x_1 + \omega^2 x_2 + \omega x_3)^3,$$

are not invariant under all the permutations of x_1, x_2, x_3, but every permutation either leaves V_1^3 and V_2^3 invariant or interchanges V_1^3 and V_2^3. Therefore, in order to make the function $F_3(x_1, x_2, x_3)$ invariant under all the permutations, it suffices to substitute for V_1^3 and V_2^3 an ambiguous function which takes the values V_1^3 and V_2^3. Such a function has been found previously in the solution of quadratic equations: it is

$$F_2(V_1^3, V_2^3) = \frac{1}{2}\left((V_1^3 + V_2^3) + \sqrt{(V_1^3 - V_2^3)^2}\right).$$

So, problem 2° is solved for $n = 3$.

Vandermonde argues similarly for $n = 4$, using $F_4(x_1, x_2, x_3, x_4)$. He also points out that in this case, since n is not a prime number, other functions can be chosen instead of $F_4(x_1, x_2, x_3, x_4)$, for instance

$$G_4(x_1, x_2, x_3, x_4) = \frac{1}{4}\left((x_1 + x_2 + x_3 + x_4) + \sqrt{W_1^2} + \sqrt{W_2^2} + \sqrt{W_3^2}\right)$$

where

$$W_1 = x_1 + x_2 - x_3 - x_4,$$
$$W_2 = x_1 - x_2 + x_3 - x_4,$$
$$W_3 = x_1 - x_2 - x_3 + x_4.$$

It is easy to put G_4 in such a form that it is not altered when x_1, x_2, x_3, x_4 are permuted, since every permutation interchanges W_1^2, W_2^2 and W_3^2. It therefore suffices to replace them by $F_3(W_1^2, W_2^2, W_3^2)$, which takes the values W_1^2, W_2^2 and W_3^2 and can be put in symmetric form, as previously observed.

For $n \geq 5$, the problem is that the functions V_i^n for $i = 1, \ldots, n-1$ are not interchanged among themselves when the indeterminates are permuted. Indeed, the function V_1^n takes $(n-1)!$ values under the permutations of the indeterminates (see Proposition 10.8, p. 149). Nevertheless, for $n = 5$ Vandermonde succeeds in reducing the determination of V_1^5 to the solution of an equation of degree 6 (compare Theorem 10.7, p. 148). For $n = 6$, he shows that his method requires the solution of an equation of degree 10 or 15.

Inconclusive as it is, this section is not devoid of interest, since it prompts Vandermonde to initiate fairly explicit calculations with permutations. He decomposes the symmetric polynomials (which he calls "types") into sums of "partial types" which are, in fact, sums of the values that a monomial takes under a subgroup of the symmetric group (especially, but not exclusively, cyclic subgroups). For instance, for three variables a, b, c, he denotes

$$[\begin{array}{ccc} \alpha & \beta & \gamma \\ \text{ii} & \text{iii} & \text{i} \end{array}] = a^\alpha b^\beta c^\gamma + a^\gamma b^\alpha c^\beta + a^\beta b^\gamma c^\alpha$$

(where α, β, γ are pairwise distinct integers). The Latin subscripts indicate that in the second term the exponents α, β, γ must be changed in such a way that γ takes the first place, α the second and β the third; the third term is obtained from the second as the second was obtained from the first, and so on for the next terms, as long as this process yields new monomials. The function thus produced is obviously invariant under the (cyclic) subgroup of S_3 generated by the permutation $a \mapsto b \mapsto c \mapsto a$. (Compare the proof of Proposition 10.5, p. 146.)

Sometimes, Vandermonde also uses a more general notation, which includes all the partial types which are invariant under the same group of permutations, but he stops short of devising a notation for permutations. For instance,

$$\begin{bmatrix} a & b & c & d & e \\ v & i & iv & ii & iii \end{bmatrix}$$

(where a, b, c, d, e are the indeterminates) is a generic notation for the various partial types which are invariant under the permutation which sets the letters a, b, c, d, e in the order b, d, e, c, a indicated by the Latin numerals, i.e. under the permutation $a \mapsto b \mapsto d \mapsto c \mapsto e \mapsto a$.

This notation allows Vandermonde to perform coherently some very complicated explicit calculations, but he cannot elude the conclusion that his method for equations of degree at least 5 leads to equations of ever higher degree, and that it may therefore not work eventually.

> That is all that the calculations taught me on this object, and I do not have enough faith in conjectures in such a thorny matter to dare try one here. I will only add that I have not found any partial type involving five letters which depends on an equation of the fourth or the third degree, and I am convinced that such a type does not exist. [59, p. 414]

However, that is not the end of the story. In the final two articles of his paper, Vandermonde briefly considers cyclotomic equations.

11.3   Cyclotomic equations

Recall from §7.3 (see Theorem 7.3, p. 83) that the problem of determining radical expressions for the roots of unity had been reduced to the solution by radicals of the cyclotomic equations

$$\Phi_p(X) = X^{p-1} + X^{p-2} + \cdots + X + 1 = 0$$

for p prime. Moreover, for p odd, de Moivre had shown that the change of variable $Y = X + X^{-1}$ converts $\Phi_p(X) = 0$ into an equation of degree $\frac{p-1}{2}$. Thus, for $p = 11$, the solution of the cyclotomic equation requires the solution of

$$Y^5 + Y^4 - 4Y^3 - 3Y^2 + 3Y + 1 = 0,$$

which de Moivre had been unable to solve by radicals. We also recall from Remark 7.6 (p. 85) that, since the roots of Φ_p are the complex numbers $e^{2k\pi i/11}$

for $k = 1, \ldots, 10$, it follows that the roots of the equation in Y are the values $2\cos\frac{2k\pi}{11}$ for $k = 1, \ldots, 5$.

Vandermonde in fact uses the (obviously equivalent) change of variable $Z = -(X + X^{-1})$, which yields the equation

$$Z^5 - Z^4 - 4Z^3 + 3Z^2 + 3Z - 1 = 0. \tag{11.3}$$

The roots of this equation, which are denoted by a, b, c, d, e, are chosen as

$$a = -2\cos\frac{2\pi}{11}, \qquad b = -2\cos\frac{4\pi}{11},$$
$$c = -2\cos\frac{6\pi}{11}, \qquad d = -2\cos\frac{8\pi}{11}, \qquad e = -2\cos\frac{10\pi}{11}.$$

It is useful to note, with Vandermonde, that the trigonometric formula

$$2\cos\alpha\cos\beta = \cos(\alpha + \beta) + \cos(\alpha - \beta) \tag{11.4}$$

yields relations between a, b, c, d and e. For instance, substituting $\frac{2\pi}{11}$ for α and β, we obtain

$$2\cos^2\frac{2\pi}{11} = \cos\frac{4\pi}{11} + \cos 0,$$

whence

$$a^2 = -b + 2.$$

Likewise, substituting $\frac{2\pi}{11}$ for α and $\frac{4\pi}{11}$ for β, we find

$$ab = -c - a,$$

and so on. Thus, substituting for α and β successively the various angles $\frac{2k\pi}{11}$ for $k = 1, \ldots, 5$, the trigonometric equation (11.4) yields linear expressions for the products of roots. Observing these expressions, Vandermonde draws an amazing conclusion, which enables him to find expressions by radicals for the eleventh roots of unity. Here are Vandermonde's own words, in the penultimate article of his paper [59, pp. 415–416]:

> In the particular cases where there are equations between the roots, the method just explained may be used to solve, without resorting to the general solution formulas. The equation $r^{11} -$

$1 = 0$ will provide us with an example: it leads (article VI) to this one

$$X^5 - X^4 - 4X^3 + 3X^2 + 3X - 1 = 0,$$

and denoting its roots by a, b, c, d, e, it will readily follow from article XI

$a^2 = -b + 2, \ b^2 = -d + 2, \ c^2 = -e + 2, \ d^2 = -c + 2,$
$e^2 = -a + 2,$
$ab = -a - c, \ bc = -a - e, \ cd = -a - d, \ de = -a - b,$
$ac = -b - d, \ bd = -b - e, \ ce = -b - c,$
$ad = -c - e, \ be = -c - d,$
$ae = -d - e,$

and all the partial types of the form $\begin{bmatrix} a & b & c & d & e \\ v & i & iv & ii & iii \end{bmatrix}$ will

have a purely rational value; thus, taking everywhere in article XXVIII $\begin{bmatrix} \alpha & \beta & \varepsilon & \delta & \gamma \\ v & i & iv & ii & iii \end{bmatrix}$ instead of $\begin{bmatrix} \alpha & \beta & \gamma & \delta & \varepsilon \\ v & iii & iv & i & ii \end{bmatrix}$

we will find [...]

$$X = \frac{1}{5}[1 + \Delta' + \Delta'' + \Delta''' + \Delta^{iv}]$$

with*

$$\Delta' = \sqrt[5]{\frac{11}{4}\left(89 + 25\sqrt{5} - 5\sqrt{-5 + 2\sqrt{5}} + 45\sqrt{-5 - 2\sqrt{5}}\right)},$$

$$\Delta'' = \sqrt[5]{\frac{11}{4}\left(89 + 25\sqrt{5} + 5\sqrt{-5 + 2\sqrt{5}} - 45\sqrt{-5 - 2\sqrt{5}}\right)},$$

$$\Delta''' = \sqrt[5]{\frac{11}{4}\left(89 - 25\sqrt{5} - 5\sqrt{-5 + 2\sqrt{5}} - 45\sqrt{-5 - 2\sqrt{5}}\right)},$$

*Oscar Luis Palacios-Vélez pointed out (private communication) that there are misprints in Vandermonde's formulas for Δ' and Δ''. They should be

$$\Delta' = \sqrt[5]{\frac{11}{4}\left(89 + 25\sqrt{5} - 5\sqrt{-5 - 2\sqrt{5}} + 45\sqrt{-5 + 2\sqrt{5}}\right)},$$

$$\Delta'' = \sqrt[5]{\frac{11}{4}\left(89 + 25\sqrt{5} + 5\sqrt{-5 - 2\sqrt{5}} - 45\sqrt{-5 + 2\sqrt{5}}\right)}.$$

$$\Delta^{iv} = \sqrt[5]{\frac{11}{4}\left(89 - 25\sqrt{5} + 5\sqrt{-5 + 2\sqrt{5}} + 45\sqrt{-5 - 2\sqrt{5}}\right)}.$$

Vandermonde's brilliant (but not quite explicit) observation is that the permutation $a \mapsto b \mapsto d \mapsto c \mapsto e \mapsto a$ preserves the relations between the roots. For instance, applying this permutation to the relation

$$a^2 = -b + 2,$$

i.e. changing a in b and b in d, we obtain

$$b^2 = -d + 2,$$

and this relation actually holds! (Compare Exercise 2).

This is very significant since the relations between the roots can be used to lower the degree of any polynomial in a, b, c, d, e, eventually providing a linear expression. Thus, suppose f is a polynomial in five variables (of any degree). Using the relations between the roots, we can eventually find

$$f(a, b, c, d, e) = Aa + Bb + Cc + Dd + Ee + F \qquad (11.5)$$

for some numbers A, B, $\ldots$, F which can be explicitly determined from the coefficients of f. Now, since the permutation $a \mapsto b \mapsto d \mapsto c \mapsto e \mapsto a$ preserves the relations which have been used in simplifying the expression of $f(a, b, c, d, e)$, we can perform the *same* simplification procedure changing a in b, b in d, d in c, c in e and e in a at each step, and we end up with

$$f(b, d, e, c, a) = Ab + Bd + Ce + Dc + Ea + F.$$

This point is somewhat delicate, since the expression (11.5) of $f(a, b, c, d, e)$ is *not* unique. Indeed, since the coefficient of Z^4 in (11.3) is the opposite of the sum of the roots, it follows that

$$a + b + c + d + e = 1. \qquad (11.6)$$

Therefore, we have for instance

$$Aa + Bb + Cc + Dd + Ee + F =$$
$$(A + F)a + (B + F)b + (C + F)c + (D + F)d + (E + F)e.$$

This does not matter, however, as long as we use the same procedure to simplify $f(a, b, c, d, e)$ and (after the permutation $a \mapsto b \mapsto d \mapsto c \mapsto e \mapsto a$) $f(b, d, e, c, a)$.

If we apply this observation to the Vandermonde (–Lagrange) resolvents

$$V_i(a, b, c, d, e)^5 = (\rho_1^i a + \rho_2^i b + \rho_3^i c + \rho_4^i d + \rho_5^i e)^5$$

where $\rho_1, \ldots, \rho_5$ are the 5-th roots of unity, we find

$$V_i(a, b, c, d, e)^5 = Aa + Bb + Cc + Dd + Ee + F \qquad (11.7)$$

where $A, B, \ldots, F$ are rational expressions in $\rho_1, \ldots, \rho_5$.

Applying the permutation four times, we obtain

$$
\begin{aligned}
V_i(b, d, e, c, a)^5 &= Ab + Bd + Ce + Dc + Ea + F, \\
V_i(d, c, a, e, b)^5 &= Ad + Bc + Ca + De + Eb + F, \\
V_i(c, e, b, a, d)^5 &= Ac + Be + Cb + Da + Ed + F, \\
V_i(e, a, d, b, c)^5 &= Ae + Ba + Cd + Db + Ec + F.
\end{aligned}
\qquad (11.8)
$$

Now, if we choose $\rho_1 = 1$, $\rho_2 = \omega$, $\rho_3 = \omega^3$, $\rho_4 = \omega^2$ and $\rho_5 = \omega^4$, where ω is some primitive 5-th root of unity, then

$$V_i(a, b, c, d, e) = a + \omega^i b + \omega^{2i} d + \omega^{3i} c + \omega^{4i} e,$$

and the permutation $a \mapsto b \mapsto d \mapsto c \mapsto e \mapsto a$ then leaves $V_i(a, b, c, d, e)^5$ invariant. Indeed, this permutation changes $V_i(a, b, c, d, e)$ into

$$V_i(b, d, e, c, a) = b + \omega^i d + \omega^{2i} c + \omega^{3i} e + \omega^{4i} a$$

and since $\omega^5 = 1$, we have

$$V_i(b, d, e, c, a) = \omega^{-i} V_i(a, b, c, d, e)$$

(compare the proof of Theorem 10.7, p. 148). Therefore,

$$V_i(b, d, e, c, a)^5 = V_i(a, b, c, d, e)^5.$$

Likewise, the left-hand sides of the equalities (11.7) and (11.8) are all equal to $V_i(a, b, c, d, e)^5$. Therefore, summing up all these equalities, we obtain

$$5 V_i(a, b, c, d, e)^5 = (A + B + C + D + E)(a + b + c + d + e) + 5F.$$

Using (11.6) we conclude

$$V_i(a, b, c, d, e)^5 = \frac{1}{5}(A + B + C + D + E) + F.$$

Since $A, B, \ldots, F$ can be rationally calculated from ω, which is already expressed by radicals (see §7.3), it follows that the functions V_i^5 can be expressed by radicals.

Therefore, a, b, c, d, and e can also be expressed by radicals. Using the formula $F_5(a, b, c, d, e)$ of §11.2 (and (11.6)), we obtain

$$a, b, c, d, e = \frac{1}{5}\left(1 + \sum_{i=1}^{4} \sqrt[5]{V_i^5}\right).$$

With hindsight, and with a view towards a possible generalization to cyclotomic equations of higher degree, the crucial steps in the calculations above appear to be

(a) the existence of relations among the roots, which can be used to reduce to degree 1 each polynomial expression in the roots;

(b) the existence of a cyclic permutation of the roots which preserves the relations above.

Given (a) and (b), we number the roots in such a way that the cyclic permutation be

$$x_1 \mapsto x_2 \mapsto x_3 \mapsto \cdots \mapsto x_n \mapsto x_1,$$

and the same arguments as above then show that the n-th power of each Lagrange resolvent of the form

$$t(\omega) = x_1 + \omega x_2 + \omega^2 x_3 + \cdots + \omega^{n-1} x_n,$$

for any n-th root of unity ω, is a rational expression in ω. Arguing inductively, we may assume that an expression by radicals has been found for ω. Then $t(\omega)^n$ can be expressed by radicals and the roots $x_1, \ldots, x_n$ can also be found by radicals, by Lagrange's formula (p. 138), namely

$$x_i = \frac{1}{n}\left(\sum_{\omega} \omega^{-(i-1)} t(\omega)\right).$$

Now, (a) is clear for the cyclotomic equations Φ_p for any prime p or, rather, for the equations obtained by de Moivre's change of variable $Y = X + X^{-1}$, whose roots are $2\cos\frac{2k\pi}{p}$ for $k = 1, \ldots, \frac{p-1}{2}$. Indeed, the same trigonometric equation (11.4) yields the required relations. But (b) is very far from clear! Yet Vandermonde simply states [59, p. 416]

Since, to solve the equation

$$X^m - X^{m-1} - (m-1)X^{m-2} + \text{etc.} = 0,$$

the question is at most to determine (article VI) the quantity
which is indifferently one of its roots, and by no means to arrange it in such a way that it be indifferent to interchange the
roots among themselves, this solution will always be very easy.

And he leaves it at that.

Yet, he had certainly noticed that the relations were not preserved by *any* permutation of the roots. The existence of a cyclic permutation which does preserve
the relations is a very remarkable, and quite mysterious, property of cyclotomic
equations, which should have awaken Vandermonde's curiosity. If he had investigated this property, he could have developed the theory of cyclotomy about thirty
years before Gauss.

Moreover, Vandermonde had pinpointed the very basic idea of Galois theory:
in order to determine the "structure" of an equation, deciding eventually whether
it is solvable by radicals, and more generally to evaluate its difficulty, one has
to look at the permutations of the roots; but one needs only to consider those
permutations which preserve the relations between the roots.[†]

This is a conspicuous example of a deep insight which was completely wasted.
To conclude this chapter, I could do no better than to quote Lebesgue [41, p. 222–
223]:

Surely, any man who discovers something truly important is left
behind by his own discovery; he himself hardly understands it,
and only by pondering over it for a long time. But Vandermonde
never came back to his algebraic investigations because he did
not realize their importance in the first place, and if he did not
understand them afterwards, it is precisely because he did not
reflect deeply on them; he was interested in everything, he was
busy with everything; he was not able to go slowly to the bot-
tom of anything. [...] To assess exactly what Vandermonde
saw, understood and what he did not catch, one would have to
reconstruct not only the mind of a man from the eighteenth century, but Vandermonde's mind, and at the moment when he had
a glimpse of genius and went ahead of his age. When trying
to do so, one will always give too much or too little credit to
Vandermonde.

[†] This restriction does not appear for general equations since their roots are independent indeterminates.
In this case, there is no relation to preserve, so every permutation is admissible.

Exercises

1. List all the terms in the partial types

$$[\begin{matrix} \alpha & \beta & \gamma & \delta & \varepsilon \\ \text{v} & \text{iii} & \text{iv} & \text{i} & \text{ii} \end{matrix}], \quad [\begin{matrix} \alpha & \varepsilon & \delta & \beta & \gamma \\ \text{v} & \text{iii} & \text{iv} & \text{i} & \text{ii} \end{matrix}], \quad [\begin{matrix} \alpha & \gamma & \beta & \varepsilon & \delta \\ \text{v} & \text{iii} & \text{iv} & \text{i} & \text{ii} \end{matrix}]$$

and $[\begin{matrix} \alpha & \delta & \varepsilon & \gamma & \beta \\ \text{v} & \text{iii} & \text{iv} & \text{i} & \text{ii} \end{matrix}]$. Show that the sum of all these partial types is also a partial type. What is the subgroup of S_5 which leaves this new partial type invariant? [59, Art. 24, p. 391]

2. Show that the permutation $a \mapsto b \mapsto c \mapsto d \mapsto e \mapsto a$ does not preserve the relations among $a = 2\cos\frac{2\pi}{11}$, $b = 2\cos\frac{4\pi}{11}$, $c = 2\cos\frac{6\pi}{11}$, $d = 2\cos\frac{8\pi}{11}$ and $e = 2\cos\frac{10\pi}{11}$.

3. Show that the permutation $a \mapsto b \mapsto c \mapsto a$ preserves the relations among $a = 2\cos\frac{2\pi}{7}$, $b = 2\cos\frac{4\pi}{7}$ and $c = 2\cos\frac{6\pi}{7}$.

4. Find a permutation of the numbers $2\cos\frac{2k\pi}{13}$ for $k = 1, \ldots, 6$, which preserves the relations among them.

Chapter 12

Gauss on Cyclotomic Equations

12.1 Introduction

The contributions of Carl Friedrich Gauss (1777–1855) to the theory of equations measure up to the outstanding advances he made in many other research areas. They occupy a special place in his work however, since they were among his earliest achievements. They can be divided up into two main topics: the fundamental theorem of algebra (1799) which we already discussed in Chapter 9, and the solution of cyclotomic equations.

Gauss' results on cyclotomic equations show how to complete Vandermonde's arguments to provide inductively expressions by radicals for the roots of unity (see Corollary 12.29, p. 195), but they far exceed this goal. In effect, they yield a thorough description of the possible reductions of cyclotomic equations of prime index to equations of smaller degree. Thus, in a brilliant way, they carry out for cyclotomic equations the program envisioned by Lagrange: to solve an equation by determining successively certain functions of the roots. As Gauss shows, the solution of $\Phi_p(X) = 0$ can be reduced to the solution of equations of degree equal to the prime factors of $p - 1$. In particular, the 17-th roots of unity can be determined by solving successively four quadratic equations, since $17 - 1 = 2^4$. As an application of this result, it follows that the regular polygon with 17 sides can be constructed by ruler and compass; this result was obtained by Gauss as early as 1796, and it is said to have been decisive in his vocation (see Bühler [9, p. 10]). This application will be discussed in an appendix to this chapter.

A definitive account of his results on cyclotomic equations was published by Gauss as the seventh and final section of his epoch-making treatise on number theory, "Disquisitiones Arithmeticae" (1801). The inclusion of such algebraic results in a book on number theory was commented by Gauss himself in the preface [24, p. 8]:

> The theory of the division of the circle, or of regular polygons, which is treated in section VII, does not belong *by itself* to Arithmetic, but its *principles* cannot be found but in Higher Arithmetic: this may appear to geometers as unexpected as the new truths that follow from it, and which they will see, I hope, with pleasure.

Accordingly, our review of Gauss' results will be preceded by some number-theoretic preliminaries. Thereafter, we divide the contents of the seventh section of the "Disquisitiones Arithmeticae" into three sections: first, we prove the irreducibility of cyclotomic equations of prime index, which is a key result in Gauss' investigations; next, we discuss the possible reductions of cyclotomic equations and we finish with the solvability by radicals of the cyclotomic equations and auxiliary equations. Some extra results which are needed to justify some of the steps in Gauss' proofs will be found in the final section.

It should be noted that we only review those results of section VII of the "Disquisitiones Arithmeticae" which directly concern the theory of equations. Several details which are meaningful in view of applications to number theory will be omitted.

12.2   Number-theoretic preliminaries

At the very beginning of the "Disquisitiones Arithmeticae" [24, Art. 2], Gauss introduces the following notation, which has gained wide acceptance and will be used repeatedly in the sequel: if a, b and n are integers and $n \neq 0$, one denotes

$$a \equiv b \mod n$$

whenever $a - b$ is divisible by n. The integers a and b are then said to be *congruent modulo n*. The explicit reference to the modulus n is sometimes omitted when no confusion is likely to arise. It is readily verified that this relation is an equivalence relation which is compatible with the sum and the product of integers, i.e., if $a_1 \equiv b_1 \mod n$ and $a_2 \equiv b_2 \mod n$, then $a_1 + a_2 \equiv b_1 + b_2 \mod n$ and $a_1 a_2 \equiv b_1 b_2 \mod n$.

In the sequel, we focus on the case where the modulus n is prime. This case has very distinctive features. We shall prove in particular the following result, which plays a key role in Gauss' investigations on cyclotomic equations:

12.1. THEOREM. *For any prime number p, there exists an integer g whose various powers g^0, g^1, g^2, ..., g^{p-2} are congruent to 1, 2, ..., p − 1 modulo p (not necessarily in that order).*

In the course of proving this theorem, we shall see that an integer g satisfies the condition of the theorem if and only if

$$g^{p-1} \equiv 1 \mod p \quad \text{and} \quad g^i \not\equiv 1 \mod p \quad \text{for } i = 1, \ldots, p-2.$$

Therefore, any such integer g is called a *primitive root of p*. (It would be more accurate, although not shorter, to call it a primitive $(p-1)$-st root of 1 modulo p.) For instance, 2 is a primitive root of 11, since modulo 11 we have

$$2^0 \equiv 1, \quad 2^1 \equiv 2, \quad 2^2 \equiv 4, \quad 2^3 \equiv 8, \quad 2^4 \equiv 5,$$
$$2^5 \equiv 10, \quad 2^6 \equiv 9, \quad 2^7 \equiv 7, \quad 2^8 \equiv 3, \quad 2^9 \equiv 6.$$

By contrast, 3 is not a primitive root of 11, as $3^5 \equiv 1 \mod 11$, whence $3^6 \equiv 3$, $3^7 \equiv 3^2, 3^8 \equiv 3^3, \ldots$ and the powers of 3 take modulo 11 only the values $3^0 \equiv 1$, $3^1 \equiv 3, 3^2 \equiv 9, 3^3 \equiv 5$ and $3^4 \equiv 4$.

The proof of Theorem 12.1 occupies the rest of this section. We closely follow one of the two proofs which Gauss included in the "Disquisitiones Arithmeticae" [24, Art. 55].

12.2. LEMMA ([24, ART. 14]). *Let a, b be integers and let p be a prime number. If*

$$ab \equiv 0 \mod p,$$

then a ≡ 0 mod p or b ≡ 0 mod p.

This amounts to the well-known fact that if a prime number divides a product of integers, then it divides one of the factors. To prove it, it suffices to mimic the proof of Lemma 5.9, p. 49, substituting integers for polynomials.

12.3. PROPOSITION ([24, ART. 43]). *Let p be a prime number and let a_0, a_1, ..., a_d be integers. If $a_d \not\equiv 0$ mod p, then the congruence equation*

$$a_d X^d + a_{d-1} X^{d-1} + \cdots + a_1 X + a_0 \equiv 0 \mod p \qquad (12.1)$$

has at most d incongruent solutions modulo p.

Proof. We argue by induction on d. Since the case $d = 0$ is trivial, we may assume inductively that every congruence equation of degree $d - 1$ has at most $d - 1$ solutions modulo p.

If equation (12.1) has $d + 1$ solutions $x_1, \ldots, x_{d+1}$ pairwise distinct modulo p, then the change of variable $Y = X - x_1$ transforms equation (12.1) into another equation of degree d,

$$a_d Y^d + a'_{d-1} Y^{d-1} + \cdots + a'_1 Y + a'_0 \equiv 0 \mod p$$

which has the same leading coefficient a_d and has the $d + 1$ solutions

$$0, \quad x_2 - x_1, \quad \ldots, \quad x_{d+1} - x_1.$$

Since 0 is a root, it follows that $a'_0 \equiv 0 \mod p$ and the equation in Y can be written

$$Y(a_d Y^{d-1} + a'_{d-1} Y^{d-2} + \cdots + a'_1) \equiv 0 \mod p.$$

Now, $x_2 - x_1, x_3 - x_1, \ldots, x_{d+1} - x_1$ are non-zero roots of this equation, hence, by Lemma 12.2, they are roots of the second factor $a_d Y^{d-1} + \cdots + a'_1$, which is a polynomial of degree $d - 1$. This contradicts the induction hypothesis. $\square$

12.4. Remark. As with any other equivalence relation, the congruence relation defines a partition of the set on which it is defined into equivalence classes. The congruence class modulo n of an integer m consists of all the integers which are congruent to m modulo n or, in other words, of all the integers of the form $kn+m$, for $k \in \mathbb{Z}$.

Since the congruence relation is compatible with the sum and the product of integers, these operations induce well-defined operations on the set of congruence classes of integers, and it is readily verified that this set inherits the commutative ring structure of $\mathbb{Z}$. The ring of congruence classes modulo n is denoted by $\mathbb{Z}/n\mathbb{Z}$. (Compare §9.2). This ring has only finitely many elements, namely the congruence classes modulo n of $0, 1, \ldots, n - 1$.

Lemma 12.2 asserts that $\mathbb{Z}/p\mathbb{Z}$ is a domain, for p prime, and the arguments in Proposition 12.3 show, more generally, that an equation of degree d with coefficients in a domain has at most d solutions in the domain.

In fact, since $\mathbb{Z}/n\mathbb{Z}$ is finite, it is easily seen that this ring is a field whenever it is a domain. Indeed, if $aX \equiv 0 \mod n$ implies $X \equiv 0 \mod n$, then the products ax are pairwise distinct modulo n when x runs over $0, 1, \ldots, n - 1$. Therefore, one of these products is congruent to 1 modulo n. This proves that a is invertible modulo n. Consequently, for p prime, the ring $\mathbb{Z}/p\mathbb{Z}$ is a field, which is denoted by $\mathbb{F}_p$.

For the proof of Theorem 12.1, we also need the following result, due to Pierre de Fermat (1601–1665), and proved by Gauss in [24, Art. 50]:

12.5. THEOREM (FERMAT). *Let p be any prime number and let a be an integer. If $a \not\equiv 0 \bmod p$, then $a^{p-1} \equiv 1 \bmod p$.*

Proof (Euler). We shall prove

$$a^p \equiv a \mod p \qquad \text{for every integer } a. \tag{12.2}$$

It will then follow that

$$a(a^{p-1} - 1) \equiv 0 \mod p \qquad \text{for every integer } a,$$

hence, by Lemma 12.2, $a^{p-1} - 1 \equiv 0 \bmod p$ whenever $a \not\equiv 0 \bmod p$.

The basic observation is that the binomial coefficients $\binom{p}{i} = \frac{p!}{i!(p-i)!}$ are all divisible by p for $i = 1, \ldots, p - 1$. Therefore,

$$(a + 1)^p \equiv a^p + 1 \mod p \qquad \text{for every integer } a,$$

and it easily follows by induction on a that (12.2) holds for every positive integer a. The property for negative a is then readily proved, since

$$(-a)^p \equiv -a^p \mod p$$

for every integer a and every prime p. (For $p = 2$, observe that $-1 \equiv 1 \bmod 2$.) $\qquad \square$

12.6. COROLLARY. *Let p be a prime number. The following conditions on an integer g are equivalent:*

(a) $g^{p-1} \equiv 1 \bmod p$ and $g^i \not\equiv 1 \bmod p$ for $i = 1, \ldots, p - 2$;
(b) the powers $g^0, g^1, \ldots, g^{p-2}$ take the values $1, 2, \ldots, p - 1$ modulo p.

Proof. (a) $\Rightarrow$ (b) If $g^{p-1} \equiv 1 \bmod p$, then $g \not\equiv 0 \bmod p$ and it follows from Lemma 12.2 that the values modulo p of the powers $g^0, g^1, \ldots, g^{p-2}$ range in $\{1, 2, \ldots, p - 1\}$. Therefore, to prove *(b)*, it suffices to show that the powers g^i are pairwise distinct modulo p for $i = 0, 1, \ldots, p - 2$.

Assume on the contrary

$$g^i \equiv g^j \mod p$$

for some integers i, j between 0 and $p - 2$, and $i < j$. Then,

$$g^i(1 - g^{j-i}) \equiv 0 \quad \mod p,$$

hence, by Lemma 12.2,

$$g^{j-i} \equiv 1 \quad \mod p.$$

Since $j - i$ is an integer between 1 and $p - 2$, this relation contradicts *(a)*.

(b) $\Rightarrow$ *(a)* From *(b)*, it clearly follows that $g \not\equiv 0 \mod p$, whence

$$g^{p-1} \equiv 1 \quad \mod p,$$

by Theorem 12.5. Moreover, condition *(b)* ensures that $g^i \not\equiv g^0$ for $i = 1, \ldots$, $p - 2$, hence

$$g^i \not\equiv 1 \quad \mod p \qquad \text{for } i = 1, \ldots, p - 2.$$

(Compare Proposition 7.11, p. 89.) $\square$

As previously noted, every integer g satisfying the equivalent conditions *(a)* and *(b)* in Corollary 12.6 is called a *primitive root of p*.

We now introduce the following technical definition: for any prime number p and any integer a relatively prime to p, the *exponent (modulo p) of* a is the smallest positive integer e such that $a^e \equiv 1 \mod p$. Thus, the integers of exponent $p - 1$, which will be shown to exist for every prime p, are the primitive roots of p. (Compare Definitions 7.7, p. 86.)

12.7. LEMMA. *Let e be the exponent (modulo a prime number p) of an integer a relatively prime to p, and let m be an integer. Then $a^m \equiv 1 \mod p$ if and only if e divides m. In particular, e divides $p - 1$.*

Proof. The same arguments as in the proof of Lemma 7.9, p. 88, apply. $\square$

Of course, it is not clear *a priori* that there exist integers of exponent e for every divisor e of $p - 1$. As a first step in the proof of Theorem 12.1, we now show the existence of such integers in the case where e is the power of a prime number.

12.8. LEMMA ([24, ART. 55]). *Let p and q be prime numbers. If some power q^m of q divides $p - 1$, then there exists an integer of exponent q^m (modulo p).*

Proof. By Proposition 12.3, the congruence equation $X^{(p-1)/q} \equiv 1 \bmod p$ has at most $(p-1)/q$ solutions (modulo p). Therefore, one can find an integer x which is not a root of this equation and is relatively prime to p. Let then $a = x^{(p-1)/q^m}$. By Fermat's theorem (Theorem 12.5), we have $x^{p-1} \equiv 1 \bmod p$, hence

$$a^{q^m} \equiv 1 \quad \bmod p.$$

Lemma 12.7 then shows that the exponent of a divides q^m. On the other hand,

$$a^{q^{m-1}} = x^{(p-1)/q} \not\equiv 1 \quad \bmod p,$$

so the exponent of a does not divide q^{m-1}. Since q is prime, the only (positive) integer which divides q^m but not q^{m-1} is q^m, and the exponent of a is therefore equal to q^m. $\qquad\square$

Proof of Theorem 12.1. Let

$$p - 1 = q_1^{m_1} \cdots q_r^{m_r}$$

be the decomposition of $p - 1$ into a product of prime factors, where $q_1, \ldots, q_r$ are pairwise distinct prime numbers. By Lemma 12.8, one can find integers a_1, $\ldots$, a_r of respective exponents $q_1^{m_1}, \ldots, q_r^{m_r}$ modulo p. To prove the theorem, we show that the product $a_1 \cdots a_r$ has exponent $p - 1$, and is therefore a primitive root of p.

Let e be the exponent of $a_1 \cdots a_r$. Lemma 12.7 shows that e divides $p - 1$. If $e \neq p - 1$, then e lacks at least one of the prime factors of $p - 1$ and divides therefore $(p - 1)/q_i$ for some $i = 1, \ldots, r$. Assume for instance e divides $(p - 1)/q_1$. Then, by Lemma 12.7,

$$(a_1 \cdots a_r)^{(p-1)/q_1} \equiv 1 \quad \bmod p. \tag{12.3}$$

Now, since the exponent $q_i^{m_i}$ of a_i divides $(p - 1)/q_1$ for $i = 2, \ldots, r$, we have

$$a_i^{(p-1)/q_1} \equiv 1 \quad \bmod p \qquad \text{for } i = 2, \ldots, r.$$

Therefore, equation (12.3) yields

$$a_1^{(p-1)/q_1} \equiv 1 \quad \bmod p.$$

This congruence shows, by Lemma 12.7, that the exponent $q_1^{m_1}$ of a_1 divides $(p - 1)/q_1$. Therefore, $p - 1$ is divisible by $q_1^{m_1+1}$. This is a contradiction, which shows that is was absurd to assume $e \neq p - 1$. (Alternatively, the claim that

$e = p - 1$ can also be proved by the same arguments as in Proposition 7.10, p. 88.)   □

12.9. Remarks. (a) If g is a primitive root of an odd prime p, then $g^{(p-1)/2} \equiv -1 \bmod p$. Indeed, since

$$\left(g^{(p-1)/2}\right)^2 \equiv g^{p-1} \equiv 1 \quad \bmod p,$$

it follows that $g^{(p-1)/2}$ is a root of the congruence equation

$$X^2 \equiv 1 \quad \bmod p.$$

Now, this equation has two roots modulo p (see Proposition 12.3), which are 1 and -1, so

$$g^{(p-1)/2} \equiv \pm 1 \quad \bmod p.$$

Since g is primitive, no power of exponent smaller than $p - 1$ is congruent to 1; therefore, $g^{(p-1)/2} \equiv 1 \bmod p$ is impossible and it follows that $g^{(p-1)/2} \equiv -1 \bmod p$.

(b) Fermat's theorem (Theorem 12.5) can also be proved by the following elaboration on Lagrange's theorem (Theorem 10.3, p. 141): for any element a of a (multiplicative) group G, define the *order* (or the *exponent*) of a as the smallest positive integer e such that $a^e = 1$ in G. If the order of a is finite, and denoted by e, then it is easily seen that the set

$$S = \{1, a, a^2, \ldots, a^{e-1}\} \subset G$$

is a subgroup of G, and the arguments in Corollary 12.6 or Proposition 7.11 (p. 89) show that the elements $1, a, \ldots, a^{e-1}$ are pairwise distinct, whence $|S| = e$. By Lagrange's theorem, it follows that e divides $|G|$ (if G is finite). In particular, $a^{|G|} = 1$ for all $a \in G$, if G is finite.

Fermat's theorem follows by applying this result to the multiplicative group $\mathbb{F}_p^\times = \mathbb{F}_p \setminus \{0\}$ of the field with p elements. (Compare Remark 12.4.)

(c) Tracing back through the proof of Theorem 12.1, it appears that the only information on $\mathbb{F}_p$ which was needed (besides the fact that $\mathbb{F}_p^\times$ is an abelian group) was that the equation $X^{(p-1)/q} = 1$ does not have more than $(p-1)/q$ solutions. Therefore, the arguments in this proof provide the following result: if a finite abelian group G is such that for every integer n the equation $X^n = 1$ has at most n solutions in G, then G is cyclic, i.e. G is generated by a single element,

$$G = \{1, a, a^2, \ldots, a^{|G|-1}\} \qquad \text{for some } a \in G.$$

In particular, every finite subgroup of the multiplicative group of a field is cyclic.

12.3 Irreducibility of the cyclotomic polynomials of prime index

The aim of this section is to provide a proof and develop some of the consequences of the following theorem:

12.10. THEOREM. *For every prime p, the cyclotomic polynomial*

$$\Phi_p(X) = X^{p-1} + X^{p-2} + \cdots + X + 1$$

is irreducible over the field of rational numbers.

This theorem was first proved by Gauss in Article 341 of the "Disquisitiones Arithmeticae." Since then, it has been generalized, and the proofs have been simplified by several mathematicians, including Eisenstein, Dedekind, Kronecker, Mertens, Landau and Schur. Instead of following Gauss' own proof, which requires a careful analysis of several cases, we shall follow Eisenstein's ideas, which are simpler and in some sense more general, in that they provide a useful sufficient condition for the irreducibility of polynomials over $\mathbb{Q}$. In §12.6, we prove some generalizations of this theorem, after Dedekind and Kronecker. The proofs given there yield an alternative proof of Theorem 12.10. The starting point of all the proofs is a result known as "Gauss' lemma" [24, Art. 42]:

12.11. LEMMA (GAUSS). *If a monic polynomial in $\mathbb{Q}[X]$ divides a monic polynomial with integral coefficients, then its coefficients are all integral.*

Proof. Let $f = X^n + a_{n-1}X^{n-1} + \cdots + a_1 X + a_0 \in \mathbb{Q}[X]$ be a monic polynomial which divides a monic polynomial $P \in \mathbb{Z}[X]$, and let $g \in \mathbb{Q}[X]$ be the quotient,

$$fg = P \in \mathbb{Z}[X].$$

Since P and f are monic, g is monic too. Let

$$g = X^m + b_{m-1}X^{m-1} + \cdots + b_1 X + b_0 \in \mathbb{Q}[X].$$

We have to prove that the coefficients $a_0, a_1, \ldots, a_{n-1}$ are all integers. Suppose the contrary and let d be the least common multiple of the denominators of a_0,

$\ldots, a_{n-1}$; then

$$f = \frac{1}{d}(dX^n + a'_{n-1}X^{n-1} + \cdots + a'_1 X + a'_0)$$

where $d, a'_{n-1}, \ldots, a'_0$ are relatively prime integers. Similarly, let

$$g = \frac{1}{e}(eX^m + b'_{m-1}X^{m-1} + \cdots + b'_1 X + b'_0)$$

where $e, b'_{m-1}, \ldots, b'_0$ are relatively prime integers. Let p be a prime number which divides d. Since $d, a'_{n-1}, \ldots, a'_0$ are relatively prime, there is a greatest index k such that p does not divide a'_k. Let also ℓ be the greatest index such that p does not divide b'_ℓ (if p does not divide e, let $\ell = m$ and $b'_m = e$). Since $fg = P \in \mathbb{Z}[X]$, it follows that

$$(dX^n + a'_{n-1}X^{n-1} + \cdots + a'_0)(eX^m + b'_{m-1}X^{m-1} + \cdots + b'_0) \in de\mathbb{Z}[X].$$

In particular, the coefficient of $X^{k+\ell}$ in the product is divisible by p since p divides d, i.e.

$$\sum_{i+j=k+\ell} a'_i b'_j \equiv 0 \mod p.$$

Since $a'_i \equiv 0 \mod p$ for $i > k$ and $b'_j \equiv 0 \mod p$ for $j > \ell$, we have

$$\sum_{i+j=k+\ell} a'_i b'_j \equiv a'_k b'_\ell \mod p.$$

Therefore, the previous equation yields

$$a'_k b'_\ell \equiv 0 \mod p.$$

This is a contradiction, since p does not divide a'_k nor b'_ℓ. $\square$

12.12. THEOREM (EISENSTEIN). *Let P be a monic polynomial with integral coefficients,*

$$P = X^t + c_{t-1}X^{t-1} + \cdots + c_1 X + c_0 \in \mathbb{Z}[X].$$

If there is a prime number p which divides c_i for $i = 0, \ldots, t-1$ but such that p^2 does not divide c_0, then P is irreducible over $\mathbb{Q}$.

Proof. Assume on the contrary that

$$P = fg$$

for some non-constant polynomials $f, g \in \mathbb{Q}[X]$, of degree n and m respectively. Since P is monic, we may assume that f and g are both monic, whence, by Gauss' lemma (Lemma 12.11), that f and g have integral coefficients. Let

$$f = X^n + a_{n-1}X^{n-1} + \cdots + a_1X + a_0 \in \mathbb{Z}[X]$$

and

$$g = X^m + b_{m-1}X^{m-1} + \cdots + b_1X + b_0 \in \mathbb{Z}[X].$$

Since $a_0b_0 = c_0$, it follows that p divides a_0 or b_0, but not both since p^2 does not divide c_0. Since f and g are interchangeable, we may assume without loss of generality that p divides a_0 but not b_0.

Let then i be the largest index such that p divides a_i; thus, p divides a_k for $k \leq i$ but not for $k = i + 1$ (possibly, $i = n - 1$; we then let $a_{i+1} = 1$). Now, the idea, as in the proof of Gauss' lemma, is to look at a well-chosen coefficient in the product fg; namely, we consider the coefficient of X^{i+1}, which is

$$c_{i+1} = a_{i+1}b_0 + a_ib_1 + a_{i-1}b_2 + \cdots + a_0b_{i+1}. \tag{12.4}$$

(We let $b_m = 1$ and $b_j = 0$ if $j > m$). Since $i + 1 \leq n < t$, the hypothesis ensures that c_{i+1} is divisible by p; but since $a_i, a_{i-1}, \ldots, a_0$ are all divisible by p, it follows from (12.4) above that p divides $a_{i+1}b_0$. This is a contradiction since it was assumed that p does not divide a_{i+1} nor b_0. $\square$

Proof of Theorem 12.10. If $\Phi_p(X)$ is not irreducible, then there is a factorization

$$\Phi_p(X) = f(X)g(X)$$

for some non-constant polynomials f, g in $\mathbb{Q}[X]$. The change of variable $X = Y + 1$ converts this equation into

$$\Phi_p(Y + 1) = f(Y + 1)g(Y + 1).$$

Since $f(Y + 1)$ and $g(Y + 1)$ are non-constant polynomials in $\mathbb{Q}[Y]$, it follows that $\Phi_p(Y + 1)$ is reducible in $\mathbb{Q}[Y]$. But since

$$\Phi_p(Y + 1) = \frac{(Y + 1)^p - 1}{(Y + 1) - 1},$$

it follows by expanding the numerator that

$$\Phi_p(Y + 1) = Y^{p-1} + pY^{p-2} + \binom{p}{2}Y^{p-3} + \cdots + \binom{p}{2}Y + p$$

(where $\binom{p}{i} = \frac{p!}{i!(p-i)!}$ is the binomial coefficient), and this polynomial is easily seen to be irreducible by Eisenstein's criterion. Therefore, $\Phi_p(X)$ is irreducible. □

The importance of this theorem lies in the fact that it enables us to reduce to a standard form every rational expression in the p-th roots of unity. We denote by μ_p the set of p-th roots of unity, as in §7.3, and by $\mathbb{Q}(\mu_p)$ the set of complex numbers which are rational expressions in these p-th roots of unity. Thus, letting

$$\mu_p = \{\rho_1, \ldots, \rho_p\},$$

we have

$$\mathbb{Q}(\mu_p) = \left\{ \frac{f(\rho_1, \ldots, \rho_p)}{g(\rho_1, \ldots, \rho_p)} \mid f, g \in \mathbb{Q}[X_1, \ldots, X_p] \text{ and } g(\rho_1, \ldots, \rho_p) \neq 0 \right\}.$$

This set is clearly a subfield of $\mathbb{C}$, since it is closed under sums, differences, products and division by non-zero elements.

Recall from Proposition 7.11, p. 89 (and Corollary 7.14, p. 90) that if ζ is any p-th root of unity other than 1, then every p-th root of unity is a power of ζ with an exponent between 0 and $p - 1$. Thus

$$\mu_p = \{1, \zeta, \zeta^2, \ldots, \zeta^{p-1}\}$$

and the complex numbers which are denoted by $\rho_1, \ldots, \rho_p$ above are powers of ζ. Therefore, every rational expression in $\rho_1, \ldots, \rho_p$ is a rational expression in ζ, and conversely, so that

$$\mathbb{Q}(\mu_p) = \left\{ \frac{f(\zeta)}{g(\zeta)} \mid f, g \in \mathbb{Q}[X] \text{ and } g(\zeta) \neq 0 \right\}.$$

12.13. THEOREM. *Every element in $\mathbb{Q}(\mu_p)$ can be expressed in one and only one way as a linear combination with rational coefficients of the p-th roots of unity other than 1,*

$$a_1\zeta + a_2\zeta^2 + \cdots + a_{p-1}\zeta^{p-1} \qquad (a_i \in \mathbb{Q}).$$

Since some of the arguments used in the proof will be useful in various contexts, we shall quote them in full generality.

12.14. LEMMA. *Let P and Q be polynomials with coefficients in some field F, and assume P is irreducible in F[X]. If P and Q have a common root in some field K containing F, then P divides Q.*

(Compare Gauss [24, Art. 346].)

Proof. If P does not divide Q, then P and Q are relatively prime, since P is irreducible. Corollary 5.4 (p. 47) then shows that there exist polynomials U, V in $F[X]$ such that

$$P(X)U(X) + Q(X)V(X) = 1.$$

Substituting in this equality the common root u of P and Q for the indeterminate X, we obtain

$$P(u)U(u) + Q(u)V(u) = 1 \qquad \text{in } K.$$

Since $P(u) = Q(u) = 0$, this equality yields $0 = 1$ in K. This contradiction shows that P divides Q. $\qquad\square$

Now, consider a field F and an element u in a field K containing F. We generalize the above definition of $\mathbb{Q}(\mu_p)$, denoting by $F(u)$ the set of elements in K which are rational expressions of u with coefficients in F,

$$F(u) = \left\{ \frac{f(u)}{g(u)} \in K \mid f, g \in F[X] \text{ and } g(u) \neq 0 \right\}.$$

The set $F(u)$ is obviously closed under sums, differences, products and divisions by non-zero elements, and is therefore a subfield of K.

If u is a root of some non-zero polynomial $P \in F[X]$, then it is a root of some monic irreducible factor of P. Indeed, if $P = cP_1 \cdots P_r$ is the decomposition of P into prime factors according to Theorem 5.8 (p. 48), then the equation $P(u) = 0$ implies that $P_i(u) = 0$ for at least one index i. Therefore, substituting for P a suitable monic irreducible factor if necessary, we may assume P itself is irreducible and monic.

12.15. PROPOSITION. *If $u \in K$ is a root of an irreducible polynomial $P \in F[X]$ of degree d, then every element in $F(u)$ can be uniquely written in the form*

$$a_0 + a_1 u + a_2 u^2 + \cdots + a_{d-1} u^{d-1} \qquad \text{with } a_i \in F.$$

Proof. Let $f(u)/g(u)$ be an arbitrary element in $F(u)$. Since $g(u) \neq 0$, the polynomial g is not divisible by P, and is therefore relatively prime to P, since P is irreducible. By Corollary 5.4 (p. 47), there exist polynomials h and U such that

$$g(X)h(X) + P(X)U(X) = 1 \qquad \text{in } F[X].$$

Substituting u for the indeterminate X in this equation, and taking into account the fact that $P(u) = 0$, we get

$$g(u)h(u) = 1 \quad \text{in } K.$$

This shows that $f(u)/g(u)$ can be written as a *polynomial* expression in u,

$$\frac{f(u)}{g(u)} = f(u)h(u) \quad \text{in } K.$$

Now, let R be the remainder of the division of fh by P,

$$fh = PQ + R \quad \text{in } F[X], \quad \text{with } \deg R \le d - 1.$$

Since $P(u) = 0$, it follows that

$$f(u)h(u) = R(u) \quad \text{in } K,$$

and since $R \in F[X]$ is a polynomial of degree at most $d - 1$, we have converted an arbitrary rational expression $f(u)/g(u) \in F(u)$ into a polynomial expression of the type

$$a_0 + a_1 u + \cdots + a_{d-1} u^{d-1}$$

with $a_i \in F$.

To prove the uniqueness of this expression, assume

$$a_0 + a_1 u + \cdots + a_{d-1} u^{d-1} = b_0 + b_1 u + \cdots + b_{d-1} u^{d-1}$$

for some $a_0, \ldots, a_{d-1}, b_0, \ldots, b_{d-1} \in F$. Collecting all the terms on one side, we see that u is then a root of the polynomial

$$V(X) = (a_0 - b_0) + (a_1 - b_1)X + \cdots + (a_{d-1} - b_{d-1})X^{d-1} \in F[X].$$

From Lemma 12.14, it follows that P divides V, but since $\deg V \le d - 1$, this is impossible unless $V = 0$. Therefore,

$$a_0 - b_0 = a_1 - b_1 = \cdots = a_{d-1} - b_{d-1} = 0.$$

$\square$

12.16. Remark. There is only one monic irreducible polynomial $P \in F[X]$ which has u as a root. Indeed, if $Q \in F[X]$ is another polynomial with the same properties, then Lemma 12.14 shows that P divides Q and, reversing the roles of P and Q, that Q divides P also. Since P and Q are both monic, it follows that $P = Q$. Moreover, among the non-zero polynomials in $F[X]$ which have u as a root, P

is the polynomial of least degree since it divides all the others. Therefore, this (unique) monic irreducible polynomial $P \in F[X]$ which has u as a root is called the *minimum polynomial of u over F*.

Proof of Theorem 12.13. We already observed above that $\mathbb{Q}(\mu_p) = \mathbb{Q}(\zeta)$. Since ζ is a root of Φ_p, which is irreducible by Theorem 12.10 and has degree $p - 1$, it follows from Proposition 12.15 that every element $a \in \mathbb{Q}(\mu_p)$ can be uniquely expressed in the form

$$a = a_0 + a_1\zeta + a_2\zeta^2 + \cdots + a_{p-2}\zeta^{p-2} \tag{12.5}$$

for some $a_i \in \mathbb{Q}$. To obtain the required form, it now suffices to use the fact that

$$\Phi_p(\zeta) = 1 + \zeta + \zeta^2 + \cdots + \zeta^{p-1} = 0. \tag{12.6}$$

This equation shows that

$$a_0 = -a_0(\zeta + \zeta^2 + \cdots + \zeta^{p-1}),$$

whence, substituting in (12.5) above,

$$a = (a_1 - a_0)\zeta + (a_2 - a_0)\zeta^2 + \cdots + (a_{p-2} - a_0)\zeta^{p-2} + (-a_0)\zeta^{p-1}.$$

The uniqueness of this expression follows from the uniqueness of expression (12.5). Indeed, if

$$a_1\zeta + \cdots + a_{p-1}\zeta^{p-1} = b_1\zeta + \cdots + b_{p-1}\zeta^{p-1},$$

then, using (12.6) to eliminate ζ^{p-1}, we get

$$-a_{p-1} + (a_1 - a_{p-1})\zeta + \cdots + (a_{p-2} - a_{p-1})\zeta^{p-2} =$$
$$-b_{p-1} + (b_1 - b_{p-1})\zeta + \cdots + (b_{p-2} - b_{p-1})\zeta^{p-2}.$$

By the uniqueness of expression (12.5), it follows that the coefficients of $1, \zeta, \zeta^2,$ $\ldots, \zeta^{p-2}$ on both sides are equal, and consequently

$$a_{p-1} = b_{p-1}, \quad a_1 = b_1, \quad \ldots, \quad a_{p-2} = b_{p-2}.$$

$\square$

12.17. Remark. For later use, we note the expression of rational numbers in the form indicated by Theorem 12.13: it easily follows from (12.6) that any element $a \in \mathbb{Q}$ is expressed as

$$a = (-a)\zeta + (-a)\zeta^2 + \cdots + (-a)\zeta^{p-1}.$$

12.4   The periods of cyclotomic equations

Let p be a prime number and let $\zeta \in \mathbb{C}$ be a primitive p-th root of unity (i.e. a p-th root of unity other than 1, see Corollary 7.14, p. 90). If m and n are integers such that $m \equiv n \bmod p$, then $\zeta^m = \zeta^n$.

Now, let g be a primitive root of p. Since the integers $g^0, g^1, g^2, \ldots, g^{p-2}$ are congruent modulo p to $1, 2, 3, \ldots, p-1$ (in some order), it follows that the complex numbers

$$\zeta^{g^0}, \zeta^{g^1}, \zeta^{g^2}, \ldots, \zeta^{g^{p-2}}$$

are the same as

$$\zeta, \zeta^2, \zeta^3, \ldots, \zeta^{p-1},$$

in some order. Therefore, letting $\zeta_i = \zeta^{g^i}$ for $i = 0, \ldots, p-2$ to facilitate notations, the set of p-th roots of unity is

$$\mu_p = \{1, \zeta_0, \zeta_1, \ldots, \zeta_{p-2}\}.$$

It turns out that this new ordering $\zeta_0, \zeta_1, \ldots, \zeta_{n-2}$ of the primitive p-th roots of unity is exactly the ordering for which Vandermonde's arguments in §11.3 can be carried out. More precisely, the point is that the cyclic permutation

$$\sigma\colon \zeta_0 \mapsto \zeta_1 \mapsto \cdots \mapsto \zeta_{p-2} \mapsto \zeta_0,$$

extended to μ_p by setting $\sigma(1) = 1$, preserves the relations among the roots ζ_0, $\ldots, \zeta_{p-2}$. This is the essence of the following proposition:

12.18. PROPOSITION. *For $\rho, \omega \in \mu_p$,*

$$\sigma(\rho\omega) = \sigma(\rho)\sigma(\omega).$$

Proof. For $i = 0, \ldots, p-3$, we have $\sigma(\zeta_i) = \zeta_{i+1}$, hence, by definition of ζ_i and ζ_{i+1},

$$\sigma(\zeta_i) = \zeta_i^g.$$

This equation also holds for $i = p-2$, since $\zeta_{p-2}^g = \zeta^{g^{p-1}}$ and $g^{p-1} \equiv g^0 \bmod p$ by Fermat's theorem (Theorem 12.5, p. 171). It also holds trivially with 1 instead of ζ_i; thus,

$$\sigma(\rho) = \rho^g \qquad \text{for all } \rho \in \mu_p.$$

Since $\rho\omega \in \mu_p$ for $\rho, \omega \in \mu_p$, we have

$$\sigma(\rho\omega) = (\rho\omega)^g = \sigma(\rho)\sigma(\omega) \qquad \text{for } \rho, \omega \in \mu_p.$$

$\square$

For example, if $p = 11$, then we can choose $g = 2$, as we observed in the example after the statement of Theorem 12.1, p. 169. The corresponding ordering of the primitive 11-th roots of unity is the following:

$$\zeta_0 = \zeta, \quad \zeta_1 = \zeta^2, \quad \zeta_2 = \zeta^4, \quad \zeta_3 = \zeta^8, \quad \zeta_4 = \zeta^5,$$
$$\zeta_5 = \zeta^{10}, \quad \zeta_6 = \zeta^9, \quad \zeta_7 = \zeta^7, \quad \zeta_8 = \zeta^3, \quad \zeta_9 = \zeta^6$$

where we can choose for instance $\zeta = \cos\frac{2\pi}{11} + i\sin\frac{2\pi}{11}$.

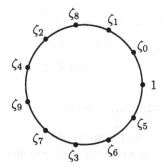

The values of $2\cos\frac{2k\pi}{11}$, which were denoted by a, b, c, d, e in §11.3, are thus given by

$$a = 2\cos\frac{2\pi}{11} = \zeta_0 + \zeta_5,$$
$$b = 2\cos\frac{4\pi}{11} = \zeta_1 + \zeta_6,$$
$$c = 2\cos\frac{6\pi}{11} = \zeta_3 + \zeta_8,$$
$$d = 2\cos\frac{8\pi}{11} = \zeta_2 + \zeta_7,$$
$$e = 2\cos\frac{10\pi}{11} = \zeta_4 + \zeta_9.$$

The permutation $\sigma\colon \zeta_0 \mapsto \zeta_1 \mapsto \cdots \mapsto \zeta_9 \mapsto \zeta_0$ induces the following permutation of $a, \ldots, e$:

$$a \mapsto b \mapsto d \mapsto c \mapsto e \mapsto a.$$

This is the permutation which played a crucial role in §11.3.

Thus, mimicking the arguments in our comments to Vandermonde's solution of $\Phi_{11}(X) = 0$, it is not hard to see that the cyclotomic equation $\Phi_p(X) = 0$ is solvable by radicals for every prime p. However, this result appears as secondary in Gauss' investigations and we postpone its proof to the next section. Gauss' primary concern is to decompose the solution of cyclotomic equations into the simplest steps as possible.

This decomposition is achieved as follows: for any two positive integers e, f such that

$$ef = p - 1,$$

Gauss defines e complex numbers which he calls the *periods of f terms*:

$$\eta_0 = \zeta_0 + \zeta_e + \zeta_{2e} + \cdots + \zeta_{e(f-1)},$$
$$\eta_1 = \zeta_1 + \zeta_{e+1} + \zeta_{2e+1} + \cdots + \zeta_{e(f-1)+1},$$
$$\eta_2 = \zeta_2 + \zeta_{e+2} + \zeta_{2e+2} + \cdots + \zeta_{e(f-1)+2},$$
$$\cdots$$
$$\eta_{e-1} = \zeta_{e-1} + \zeta_{2e-1} + \zeta_{3e-1} + \cdots + \zeta_{p-2}.$$

In particular, the periods of 1 term are the roots $\zeta_0, \zeta_1, \ldots, \zeta_{p-2}$, and the (unique) period of $p - 1$ terms is the sum of all the p-th roots of unity other than 1 or, in other words, the sum of all the roots of Φ_p. This period is therefore rational, it is the opposite of the first coefficient of Φ_p,

$$\zeta_0 + \zeta_1 + \cdots + \zeta_{p-2} = -1.$$

As a further example, for $p \geq 3$, the periods of two terms can be seen to be the values of $2 \cos \frac{2k\pi}{p}$ for $k = 1, \ldots, \frac{p-1}{2}$. This was already shown above for $p = 11$, but can be proved in general by considering the form of these periods,

$$\eta_j = \zeta_j + \zeta_{j + \frac{p-1}{2}}.$$

By definition of the indexing, we have

$$\zeta_{j + \frac{p-1}{2}} = (\zeta_j)^{g^{(p-1)/2}}$$

and since $g^{(p-1)/2} \equiv -1 \bmod p$ by Remark 12.9(a), it follows that

$$\eta_j = \zeta_j + \zeta_j^{-1}.$$

Therefore, the $\frac{p-1}{2}$ periods of two terms are the roots of the equation of degree $\frac{p-1}{2}$ obtained from $\Phi_p(X) = 0$ by setting $Y = X + X^{-1}$. This proves the claim, in view of Remark 7.6, p. 85.

As Gauss shows, the periods of f terms thus defined have the following remarkable properties:

12.19. PROPERTY. *Any period of f terms can be determined rationally from any other period of f terms.*

12.20. PROPERTY. *If f and g are two divisors of $p - 1$ and if f divides g, then any period of f terms is a root of an equation of degree g/f whose coefficients are rational expressions of a period of g terms.*

These properties will be proved below, see Corollary 12.24 (p. 190) and Corollary 12.26 (p. 191).

Thus, the periods can be used to provide remarkable examples of the step-by-step solution of equations as envisioned by Lagrange. Fix a sequence of integers:

$$f_0 = p - 1, \quad f_1, \quad \ldots, \quad f_{r-1}, \quad f_r = 1$$

such that f_i divides f_{i-1} for $i = 1, \ldots, r$, and define V_i to be a period of f_i terms (arbitrarily chosen) for $i = 0, \ldots, r$. Then V_0 is rational and for $i = 1, \ldots, r$, the complex number V_i can be determined by solving an equation of degree f_{i-1}/f_i whose coefficients are rational expressions in V_{i-1}. Since V_r is a period of 1 term, this process eventually yields a primitive p-th root of unity. The other p-th roots of unity are then readily obtained as powers of this one.

The choice of V_i among the periods of f_i terms does not affect essentially the solution, since Property 12.19 shows that the periods of f_i terms are rational expressions of each other.

Of course, it is not clear *a priori* that the equation which is used to determine V_i from V_{i-1} is solvable by radicals, since f_{i-1}/f_i might exceed 5, but Gauss further proves that these equations are indeed solvable by radicals for any value of f_{i-1}/f_i, including $p - 1$. (This case occurs if $r = 1$.) If one wants to deal with equations of the smallest degrees as possible, one can choose the sequence $f_0, f_1, \ldots, f_r$ in such a way that the successive quotients f_{i-1}/f_i are the prime factors which divide $p - 1$, but this is in no way compulsory.

Take for instance $p = 37$ and look at the lattice of divisors of $p - 1 = 2^2 \cdot 3^2$:

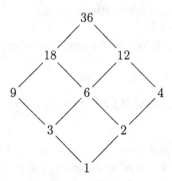

(In this diagram, a straight line indicates a relation of divisibility.) To every path going down from 36 to 1 (without going up at any step) corresponds a pattern of solution of $\Phi_{37}(X) = 0$ by successive equations, whose degrees are the successive quotients. For instance, if we choose the path 36, 12, 6, 1, then we first determine a period of 12 terms by an equation of degree $36/12 = 3$, next a period of 6 terms by an equation of degree $12/6 = 2$ and finally a period of 1 term, i.e. a primitive 37-th root of unity, by an equation of degree 6.

Instead of solving directly this last equation, one could determine a period of 3 terms by an equation of degree $6/3 = 2$ and a period of 1 term by an equation of degree 3. This amounts to refine the proposed path into 36, 12, 6, 3, 1.

For $p = 17$, the lattice of divisors of $p - 1 = 2^4$ is much simpler, it is

Thus, a primitive 17-th root of unity can be determined by solving successively four quadratic equations. This is the key fact which leads to the construction of the regular polygon with 17 sides by ruler and compass (see the appendix).

We now turn to the proof of Properties 12.19 and 12.20, which we adapt from

Gauss' own arguments with the added thrust of some elementary linear algebra.*
First, we define a map from the field $\mathbb{Q}(\mu_p)$ onto itself, extending by linearity the
map σ defined on μ_p. (See the definition of σ before Proposition 12.18, p. 182.)
We thus set

$$\sigma(a_0\zeta_0 + \cdots + a_{p-2}\zeta_{p-2}) = a_0\sigma(\zeta_0) + \cdots + a_{p-2}\sigma(\zeta_{p-2}),$$

i.e.

$$\sigma(a_0\zeta_0 + a_1\zeta_1 + \cdots + a_{p-3}\zeta_{p-3} + a_{p-2}\zeta_{p-2}) =$$
$$a_0\zeta_1 + a_1\zeta_2 + \cdots + a_{p-3}\zeta_{p-2} + a_{p-2}\zeta_0.$$

Theorem 12.13 shows that this is sufficient to define σ on the whole of $\mathbb{Q}(\mu_p)$.

12.21. PROPOSITION. *The map σ is a field automorphism of $\mathbb{Q}(\mu_p)$ which leaves
every element of $\mathbb{Q}$ invariant.*

Proof. That σ is bijective and that

$$\sigma(ua + vb) = u\sigma(a) + v\sigma(b)$$

for $a, b \in \mathbb{Q}(\mu_p)$ and $u, v \in \mathbb{Q}$ (i.e. that σ is $\mathbb{Q}$-linear) readily follow from the
definition of σ. Moreover, since by Remark 12.17 the rational numbers $a \in \mathbb{Q}$ are
written as

$$a = (-a)\zeta_0 + (-a)\zeta_1 + \cdots + (-a)\zeta_{p-2},$$

the definition also shows that every rational number is invariant under σ.

Thus, it only remains to prove

$$\sigma(ab) = \sigma(a)\sigma(b) \qquad \text{for all } a, b \in \mathbb{Q}(\mu_p).$$

This was already proved in Proposition 12.18 in the particular case where $a, b \in \mu_p$. From this case, the general case can be derived as follows: let

$$a = \sum_{i=0}^{p-2} a_i\zeta_i \qquad \text{and} \qquad b = \sum_{j=0}^{p-2} b_j\zeta_j$$

---

*Gauss' original arguments also use linear algebra, but expressed in an elementary way via systems of
linear equations, see [24, Art. 346].

with a_i, $b_j \in \mathbb{Q}$ for all i, j. Then

$$ab = \sum_{i,j=0}^{p-2} a_i b_j \zeta_i \zeta_j$$

whence, since σ is $\mathbb{Q}$-linear,

$$\sigma(ab) = \sum_{i,j=0}^{p-2} a_i b_j \sigma(\zeta_i \zeta_j).$$

On the other hand, we have

$$\sigma(a)\sigma(b) = \sum_{i,j=0}^{p-2} a_i b_j \sigma(\zeta_i)\sigma(\zeta_j).$$

Therefore, Proposition 12.18 shows that $\sigma(ab) = \sigma(a)\sigma(b)$.   □

Remark. The irreducibility of Φ_p was used above in an essential, but rather implicit, way. Indeed, that the map σ is well-defined on $\mathbb{Q}(\mu_p)$ results from the fact that the expression $a_0\zeta_0 + \cdots + a_{p-2}\zeta_{p-2}$ for the elements in $\mathbb{Q}(\mu_p)$ is unique; the proof of this fact, in Theorem 12.13, ultimately relies on the irreducibility of Φ_p.

Let now e and f be (positive) integers such that

$$ef = p - 1.$$

Denote by K_f the set of elements in $\mathbb{Q}(\mu_p)$ which are invariant under σ^e. Since σ, whence also σ^e, is a field automorphism of $\mathbb{Q}(\mu_p)$ which is the identity on $\mathbb{Q}$, the set K_f is clearly closed under sums, differences, products and divisions by non-zero elements, and contains $\mathbb{Q}$. In other words, K_f is a subfield of $\mathbb{Q}(\mu_p)$ containing $\mathbb{Q}$. Using the standard form of the elements in $\mathbb{Q}(\mu_p)$, a standard form for the elements of K_f is easily found, as the next proposition shows.

12.22. PROPOSITION. *Every element in K_f can be written in a unique way as a linear combination with rational coefficients of the e periods of f terms.*

Proof. Let a be an arbitrary element in $\mathbb{Q}(\mu_p)$, which we write as follows:

$$
\begin{aligned}
a = \ & a_0\zeta_0 + a_1\zeta_1 + \cdots + a_{e-1}\zeta_{e-1} \\
& + a_e\zeta_e + a_{e+1}\zeta_{e+1} + \cdots + a_{2e-1}\zeta_{2e-1} \\
& + \cdots \\
& + a_{e(f-1)}\zeta_{e(f-1)} + a_{e(f-1)+1}\zeta_{e(f-1)+1} + \cdots + a_{p-2}\zeta_{p-2}.
\end{aligned}
$$

Then, by definition of σ,

$$
\begin{aligned}
\sigma^e(a) = \ & a_0\zeta_e + a_1\zeta_{e+1} + \cdots + a_{e-1}\zeta_{2e-1} \\
& + a_e\zeta_{2e} + a_{e+1}\zeta_{2e+1} + \cdots + a_{2e-1}\zeta_{3e-1} \\
& + \cdots \\
& + a_{e(f-1)}\zeta_0 + a_{e(f-1)+1}\zeta_1 + \cdots + a_{p-2}\zeta_{e-1}.
\end{aligned}
$$

If $\sigma^e(a) = a$, then, by Theorem 12.13, the coefficient of ζ_i in the two expressions above are the same, for $i = 0, \ldots, p - 2$, hence

$$
\begin{array}{ccccccc}
a_0 & = & a_e & = & a_{2e} & = & \cdots & = & a_{e(f-1)}, \\
a_1 & = & a_{e+1} & = & a_{2e+1} & = & \cdots & = & a_{e(f-1)+1}, \\
& & & & \cdots & & & & \\
a_{e-1} & = & a_{2e-1} & = & a_{3e-1} & = & \cdots & = & a_{p-2}.
\end{array}
$$

Therefore, every element $a \in K_f$ can be written as

$$
\begin{aligned}
a = \ & a_0(\zeta_0 + \zeta_e + \cdots + \zeta_{e(f-1)}) \\
& + a_1(\zeta_1 + \zeta_{e+1} + \cdots + \zeta_{e(f-1)+1}) \\
& + \cdots \\
& + a_{e-1}(\zeta_{e-1} + \zeta_{2e-1} + \cdots + \zeta_{p-2}).
\end{aligned}
$$

This proves that a is a linear combination of the periods, since the expressions between brackets are the periods of f terms.

The uniqueness of this expression of a readily follows from Theorem 12.13, which asserts that every element in $\mathbb{Q}(\mu_p)$ can be written in only one way as a linear combination of $\zeta_0, \ldots, \zeta_{p-2}$. $\square$

12.23. PROPOSITION. *Let η be a period of f terms. Every element in K_f can be written as*

$$
a_0 + a_1\eta + a_2\eta^2 + \cdots + a_{e-1}\eta^{e-1}
$$

for some $a_0, \ldots, a_{e-1} \in \mathbb{Q}$.

Proof. Since K_f is a field containing $\mathbb{Q}$, it can be considered as a vector space over $\mathbb{Q}$ in a natural way: the vector space operations are induced by the operations in the field. To prove the proposition, it obviously suffices to show that $1, \eta, \ldots,$ η^{e-1} is a basis of K_f over $\mathbb{Q}$. In fact, it even suffices to prove that $1, \eta, \ldots, \eta^{e-1}$ are linearly independent over $\mathbb{Q}$, since Proposition 12.22 show that the e periods of f terms form a basis of K_f over $\mathbb{Q}$, hence that $\dim_{\mathbb{Q}} K_f = e$.

In order to prove this linear independence, suppose

$$a_0 + a_1 \eta + \cdots + a_{e-1} \eta^{e-1} = 0 \tag{12.7}$$

for some rational numbers $a_0, \ldots, a_{e-1}$. Then η is a root of the polynomial

$$P(X) = a_0 + a_1 X + \cdots + a_{e-1} X^{e-1}.$$

Applying σ, next σ^2, σ^3 and so on until σ^{e-1} to both sides of (12.7), and taking into account the fact that the coefficients a_i are invariant under σ, we observe that $\sigma(\eta), \sigma^2(\eta), \ldots, \sigma^{e-1}(\eta)$ are roots of $P(X)$ too. Now, $\eta, \sigma(\eta), \ldots, \sigma^{e-1}(\eta)$ are the e periods of f terms, which are pairwise distinct by Proposition 12.22. Since the polynomial $P(X)$ has degree at most $e - 1$, it cannot have as roots the e periods of f terms, unless it is the zero polynomial. Therefore,

$$a_0 = \cdots = a_{e-1} = 0,$$

and this proves the linear independence of $1, \eta, \ldots, \eta^{e-1}$.   □

12.24. COROLLARY. *If η and η' are periods of f terms, then*

$$\eta' = a_0 + a_1 \eta + \cdots + a_{e-1} \eta^{e-1}$$

for some rational numbers $a_0, \ldots, a_{e-1}$.

Proof. This readily follows from the proposition, since $\eta' \in K_f$.   □

This corollary proves Property 12.19 of the periods. In order to prove Property 12.20, we now introduce another pair of integers g, h such that

$$gh = p - 1,$$

and assume that f divides g. Then, denoting $k = g/f = e/h$, we have

$$\sigma^e = (\sigma^h)^k.$$

Therefore, every element invariant under σ^h is also invariant under σ^e, which means that

$$K_g \subset K_f.$$

12.25. PROPOSITION. *Let f and g be divisors of $p - 1$. If f divides g, then every element in K_f is a root of a polynomial of degree g/f with coefficients in K_g.*

Proof. For $a \in K_f$, we consider the polynomial

$$P(X) = (X - a)(X - \sigma^h(a))(X - \sigma^{2h}(a)) \cdots (X - \sigma^{h(k-1)}(a))$$

with the same notation as above. This polynomial has degree $k = g/f$, and its coefficients are the elementary symmetric polynomials in a, $\sigma^h(a)$, $\sigma^{2h}(a)$, ... , $\sigma^{h(k-1)}(a)$. Since

$$\sigma^h(\sigma^{h(k-1)}(a)) = \sigma^e(a) = a,$$

the map σ^h permutes a, $\sigma^h(a)$, ... , $\sigma^{h(k-1)}(a)$ among themselves and leaves therefore the coefficients of P invariant. This shows that the coefficients of P are in K_g. The polynomial P thus satisfies the required properties. $\square$

The proof of Property 12.20 can now be completed.

12.26. COROLLARY. *Let f and g be divisors of $p - 1$ and let η and ξ be periods of f and g terms respectively. If f divides g, then η is a root of a polynomial of degree g/f whose coefficients are rational expressions of ξ.*

Proof. Since $\xi \in K_g$ and $\eta \in K_f$, this corollary readily follows from Propositions 12.25 and 12.23. $\square$

It is instructive to note, with a view towards the modern framework of Galois theory, that the subfields K_f form a lattice of subfields of $\mathbb{Q}(\mu_p)$, which is anti-isomorphic to the lattice of divisors of $p - 1$, since $K_g \subset K_f$ if and only if f divides g. Thus, if for instance $p = 37$, the periods define the following lattice of subfields of $\mathbb{Q}(\mu_{37})$:

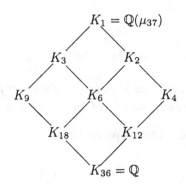

(A straight line indicates a relation of inclusion.)

12.5 Solvability by radicals

After his careful analysis of the periods of cyclotomic equations and their properties, Gauss shows in Art. 359–360 of "Disquisitiones Arithmeticae" that the equations by which the periods are determined can be solved by radicals. His exposition in this part is more sketchy and slurs at some points over a non-trivial difficulty which will be pinpointed below.

We use the notation of the preceding section. In particular, we let e, f and g, h be two pairs of integers such that

$$ef = gh = p - 1.$$

We assume that f divides g and set

$$k = \frac{g}{f} = \frac{e}{h}.$$

We denote by $\eta_0, \ldots, \eta_{e-1}$ (resp. $\xi_0, \ldots, \xi_{h-1}$) the periods of f (resp. g) terms,

$$\eta_i = \zeta_i + \zeta_{e+i} + \zeta_{2e+i} + \cdots + \zeta_{e(f-1)+i},$$
$$\xi_j = \zeta_j + \zeta_{h+j} + \zeta_{2h+j} + \cdots + \zeta_{h(g-1)+j}.$$

In Corollary 12.26, we have seen that, when the periods $\xi_0, \ldots, \xi_{h-1}$ are considered as known, then any period η_i can be determined by an equation of degree g/f. Our aim in this section is to show that this equation is solvable by radicals.

Consider for instance the equation which yields η_0. (The arguments for the other periods is exactly the same, but the notation is more complicated.) We denote this equation of degree k by $P(X) = 0$. Since the coefficients of P are in K_g, they are invariant under σ^h; hence, by repeatedly applying σ^h to both sides of the equation $P(\eta_0) = 0$ we find

$$P\big(\sigma^h(\eta_0)\big) = 0, \quad P\big(\sigma^{2h}(\eta_0)\big) = 0, \quad \ldots, \quad P\big(\sigma^{h(k-1)}(\eta_0)\big) = 0.$$

Therefore, the roots of P are η_0 and its images under $\sigma^h, \sigma^{2h}, \ldots, \sigma^{h(k-1)}$, which are $\eta_h, \eta_{2h}, \ldots, \eta_{h(k-1)}$.

In order to prove that $P(X) = 0$ is solvable by radicals, it suffices, after Lagrange's formula (p. 138), to show that the k-th power of the Lagrange resolvent

$$t(\omega) = \eta_0 + \omega\eta_h + \omega^2\eta_{2h} + \cdots + \omega^{k-1}\eta_{h(k-1)}$$

(where ω is a k-th root of unity) can be calculated from the periods of g terms.

12.27. PROPOSITION. *For every k-th root of unity ω, the complex number $t(\omega)^k$ has a rational expression in terms of ω and of the periods of g terms.*

Proof. First, we observe that, by Proposition 12.22, the product of any two periods of f terms can be expressed as a linear combination of the periods of f terms. We thus have relations among the periods, which can be used to reduce to 1 the degree of any polynomial expression in the periods. In particular,

$$
\begin{aligned}
t(\omega)^k &= \big(\eta_0 + \omega\eta_h + \cdots + \omega^{k-1}\eta_{h(k-1)}\big)^k \\
&= a_0\eta_0 + \cdots + a_{h-1}\eta_{h-1} \\
&\quad + a_h\eta_h + \cdots + a_{2h-1}\eta_{2h-1} \qquad\qquad (12.8)\\
&\quad + \cdots \\
&\quad + a_{h(k-1)}\eta_{h(k-1)} + \cdots + a_{e-1}\eta_{e-1}
\end{aligned}
$$

where the coefficients $a_0, \ldots, a_{e-1}$ are rational (in fact polynomial) expressions in ω over $\mathbb{Q}$.

Since the relations among the periods $\eta_0, \ldots, \eta_{e-1}$ are preserved under σ^h, by Proposition 12.21, we can replace η_0 by $\sigma^h(\eta_0) = \eta_h$, η_1 by $\sigma^h(\eta_1) = \eta_{h+1}$, etc. in the above calculation of $t(\omega)^k$. We thus find

$$
\begin{aligned}
(\eta_h + \omega\eta_{2h} + \cdots + \omega^{k-1}\eta_0)^k = \; & a_0\eta_h + \cdots + a_{h-1}\eta_{2h-1} \\
& + a_h\eta_{2h} + \cdots + a_{2h-1}\eta_{3h-1} \\
& + \cdots \\
& + a_{h(k-1)}\eta_0 + \cdots + a_{e-1}\eta_{h-1}.
\end{aligned}
\tag{12.9}
$$

This yields an expression of $\bigl(\sigma^h(t(\omega))\bigr)^k$. However, since

$$
\sigma^h(t(\omega)) = \omega^{-1}t(\omega),
$$

we have

$$
\bigl(\sigma^h(t(\omega))\bigr)^k = t(\omega)^k,
$$

so that (12.8) and (12.9) are two expressions of $t(\omega)^k$. Replacing in the initial calculation of $t(\omega)^k$ the period η_i by $\sigma^{2h}(\eta_i)$, next by $\sigma^{3h}(\eta_i), \ldots, \sigma^{h(k-1)}(\eta_i)$ (for $i = 0, \ldots, e-1$), we still find $k-2$ other expressions of $t(\omega)^k$. Inspection shows that the coefficients of a given period η_i in these various expressions are a_i, $a_{i+h}, a_{i+2h}, \ldots, a_{i+h(k-1)}$. Therefore, if we sum up all these expressions, we get

$$
\begin{aligned}
kt(\omega)^k = \; & (a_0 + \cdots + a_{h(k-1)})(\eta_0 + \cdots + \eta_{h(k-1)}) \\
& + (a_1 + \cdots + a_{h(k-1)+1})(\eta_1 + \cdots + \eta_{h(k-1)+1}) \\
& + \cdots \\
& + (a_{h-1} + \cdots + a_{e-1})(\eta_{h-1} + \cdots + \eta_{e-1}).
\end{aligned}
$$

Since $\eta_i + \eta_{h+i} + \cdots + \eta_{h(k-1)+i} = \xi_i$ for $i = 0, \ldots, h-1$, it follows that $t(\omega)^k$ is rationally expressed in terms of ω and $\xi_0, \ldots, \xi_{h-1}$, as

$$
t(\omega)^k = \frac{1}{k}\bigl((a_0 + \cdots + a_{h(k-1)})\xi_0 + \cdots + (a_{h-1} + \cdots + a_{e-1})\xi_{h-1}\bigr).
$$

$\square$

12.28. Remark. The above proof is quite similar to that of Gauss, but the final arguments are different. Gauss argues as follows: after observing that the right-hand sides of (12.8) and (12.9) are equal, since both are expressions of $t(\omega)^k$, he draws the conclusion that the coefficients of any given period are the same in both expressions, hence

$$
\begin{aligned}
a_0 &= a_h = a_{2h} = \cdots = a_{h(k-1)}, \\
a_1 &= a_{h+1} = a_{2h+1} = \cdots = a_{h(k-1)+1}, \\
&\quad \cdots \\
a_{h-1} &= a_{2h-1} = a_{3h-1} = \cdots = a_{e-1}.
\end{aligned}
$$

These equalities can be used to simplify (12.8) to

$$
\begin{aligned}
t(\omega)^k = \;& a_0(\eta_0 + \eta_h + \cdots + \eta_{h(k-1)}) \\
& + a_1(\eta_1 + \eta_{h+1} + \cdots + \eta_{h(k-1)+1}) \\
& + \cdots \\
& + a_{h-1}(\eta_{h-1} + \eta_{2h-1} + \cdots + \eta_{e-1}).
\end{aligned}
$$

This completes the proof of the proposition, since the expressions between brackets in the right-hand side are the periods of g terms.

However, the comparison of coefficients, which was also used in the proof of Proposition 12.22 above, is justified only insofar as the expression of an element as a linear combination of $\eta_0, \ldots, \eta_{e-1}$ (or, more generally, of $\zeta_0, \ldots, \zeta_{p-2}$) is known to be unique. This was shown in Theorem 12.13 (p. 178) for linear combinations with rational coefficients, which was sufficient to prove Proposition 12.22, but here the scalars are rational expressions of a k-th root of unity ω, so new arguments are needed.

From the proof of Theorem 12.13, it is clear that the crucial fact on which this uniqueness property ultimately relies is the irreducibility of Φ_p. Therefore, in order to justify Gauss' argument, we need to prove the irreducibility of Φ_p not only over the field $\mathbb{Q}$ of rational numbers, but over $\mathbb{Q}(\omega)$, where ω is a k-th root of unity for some integer k dividing $p - 1$. This will be done in the next section, see Corollary 12.33, p. 200.

To complete this section, we observe with Gauss [24, Art. 360] that the full generality of periods is not needed if we only aim to show that the roots of unity can be expressed by radicals.

12.29. COROLLARY. *For every integer n, the n-th roots of unity have expressions by radicals.*

Proof. We argue by induction on n. The corollary is trivial if $n = 1$ or 2, so we may assume that for every integer $k < n$ the k-th roots of unity are expressible by radicals. If n is not prime, then Theorem 7.3 (p. 83) and the induction hypothesis readily show that the n-th roots of unity can be expressed by radicals. We may thus assume that n is prime. We then order the n-th roots of unity other than 1 as at the beginning of §12.4 with the aid of a primitive root of n and we consider the Lagrange resolvent

$$t(\omega) = \zeta_0 + \omega\zeta_1 + \cdots + \omega^{n-2}\zeta_{n-2}$$

(where ω is an $(n-1)$-st root of unity). By the induction hypothesis, ω can be expressed by radicals. The preceding proposition (with $k = g = n - 1$) then shows that $t(\omega)^{n-1}$ has a rational expression in terms of ω, whence an expression by radicals. Lagrange's formula (p. 138) now yields expressions by radicals for the n-th roots of unity,

$$\zeta_i = \frac{1}{n-1}\left(\sum_\omega \omega^{-i}\,\sqrt[n-1]{t(\omega)^{n-1}}\right).$$

$\square$

12.6   Irreducibility of the cyclotomic polynomials

The aim of this section is to justify Gauss' argument (see Remark 12.28), by proving the irreducibility of the cyclotomic polynomial Φ_p over $\mathbb{Q}(\mu_k)$, when p is a prime number and k is an integer which is relatively prime to p.

A proof of this result was first published by Kronecker in 1854. The proof we give is inspired by some ideas of Dedekind (see Van der Waerden [61, §60], Weber [67, §174]). It holds in fact for any integer n instead of p. Its essential step is to prove the irreducibility of Φ_n over $\mathbb{Q}$, which was first established for non-prime n by Gauss in 1808 (see Bühler [9, p. 74]).

12.30. LEMMA. *Let f be a monic irreducible factor of Φ_n in $\mathbb{Q}[X]$ and let p be a prime number which does not divide n. If $\omega \in \mathbb{C}$ is a root of f, then ω^p also is a root of f, so*

$$f(\omega) = 0 \Rightarrow f(\omega^p) = 0.$$

Proof. Assume on the contrary that $f(\omega) = 0$ but $f(\omega^p) \neq 0$. Since Φ_n divides $X^n - 1$, we have

$$X^n - 1 = fg \tag{12.10}$$

for some monic polynomial $g \in \mathbb{Q}[X]$. Since $f(\omega) = 0$, it follows that $\omega^n = 1$, whence also, raising both sides to the p-th power,

$$(\omega^p)^n = 1.$$

In other words, ω^p is a root of $X^n - 1$. Since on the other hand it was assumed that $f(\omega^p) \neq 0$, equation (12.10) implies

$$g(\omega^p) = 0.$$

This last equality shows that ω is a root of $g(X^p)$. Therefore, by Lemma 12.14 (p. 178), $f(X)$ divides $g(X^p)$. Let $h(X) \in \mathbb{Q}[X]$ be a monic polynomial such that

$$g(X^p) = f(X)h(X). \tag{12.11}$$

Gauss' lemma (Lemma 12.11, p. 175) and equations (12.10) and (12.11) show that f, g and h have integral coefficients. Therefore, we may consider the polynomials $\overline{f}$, $\overline{g}$ and $\overline{h}$ whose coefficients are the congruence classes modulo p of the coefficients of f, g and h respectively, i.e. the images of these coefficients in $\mathbb{F}_p$ ($= \mathbb{Z}/p\mathbb{Z}$, see Remark 12.4, p. 170). By reduction modulo p, equations (12.10) and (12.11) yield

$$X^n - 1 = \overline{f}(X)\overline{g}(X) \qquad \text{in } \mathbb{F}_p[X] \tag{12.12}$$

and

$$\overline{g}(X^p) = \overline{f}(X)\overline{h}(X) \quad \text{in } \mathbb{F}_p[X]. \tag{12.13}$$

Now, Fermat's theorem (Theorem 12.5, p. 171) says that $a^p = a$ for all $a \in \mathbb{F}_p$. Therefore, if

$$\overline{g}(X) = a_0 + a_1 X + \cdots + a_{r-1}X^{r-1} + X^r,$$

we also have

$$\overline{g}(X) = a_0^p + a_1^p X + \cdots + a_{r-1}^p X^{r-1} + X^r,$$

whence

$$\overline{g}(X^p) = a_0^p + a_1^p X^p + \cdots + a_{r-1}^p X^{p(r-1)} + X^{pr}.$$

Since $(u+v)^p = u^p + v^p$ in $\mathbb{F}_p$ (because the binomial coefficients $\binom{p}{i}$ are divisible by p for $i = 1, \ldots, p-1$), it follows that

$$\bar{g}(X^p) = (a_0 + a_1 X + \cdots + a_{r-1} X^{r-1} + X^r)^p = \bar{g}(X)^p \qquad \text{in } \mathbb{F}_p[X].$$

Thus, equation (12.13) can be rewritten as

$$\bar{g}^p = \bar{f}\,\bar{h}$$

and this shows that $\bar{f}$ and $\bar{g}$ are not relatively prime. Let $\varphi(X) \in \mathbb{F}_p[X]$ be a non-constant common factor of $\bar{f}$ and $\bar{g}$. Equation (12.12) shows that φ^2 divides $X^n - 1$. Let

$$X^n - 1 = \varphi^2 \psi \qquad \text{in } \mathbb{F}_p[X].$$

Comparing the derivatives of both sides, we obtain

$$n X^{n-1} = \varphi \cdot (2\partial\varphi \cdot \psi + \varphi \cdot \partial\psi),$$

whence φ divides $X^n - 1$ and nX^{n-1}. This is impossible since $X^n - 1$ and nX^{n-1} are relatively prime in $\mathbb{F}_p[X]$. (It is here that the hypothesis that p does not divide n is needed.) This contradiction shows that the hypothesis $f(\omega^p) \neq 0$ was absurd. $\square$

12.31. THEOREM. *For every integer $n \geq 1$, the cyclotomic polynomial Φ_n is irreducible over $\mathbb{Q}$.*

Proof. Let f be a monic irreducible factor of Φ_n in $\mathbb{Q}[X]$. We shall prove that every root of Φ_n in $\mathbb{C}$ is a root of f. Since the roots of Φ_n are simple, it will then follow from Proposition 5.10 (p. 50) that Φ_n divides f, hence that $\Phi_n = f$, since f and Φ_n divide each other and are both monic.

Let ζ be a root of f. Then ζ is a root of Φ_n, which means that ζ is a primitive n-th root of unity. From Proposition 7.12 (p. 89) we recall that any other primitive n-th root of unity has the form ζ^k, where k is an integer relatively prime to n between 0 and n. Factoring k into (not necessarily distinct) prime factors

$$k = p_1 \cdots p_s,$$

we find, by successive applications of the preceding lemma,

$$f(\zeta) = 0 \Rightarrow f(\zeta^{p_1}) = 0 \Rightarrow f(\zeta^{p_1 p_2}) = 0 \Rightarrow \cdots$$
$$\Rightarrow f(\zeta^{p_1 \cdots p_{s-1}}) = 0 \Rightarrow f(\zeta^k) = 0.$$

Thus, f has as root every primitive n-th root of unity, i.e. every root of Φ_n. $\square$

12.32. THEOREM. *If m and n are relatively prime integers, then Φ_n is irreducible over $\mathbb{Q}(\mu_m)$.*

Proof. Let f be a monic irreducible factor of Φ_n in $\mathbb{Q}(\mu_m)[X]$ and let $\zeta \in \mathbb{C}$ be a root of f. Arguing as above, we see that it suffices to prove

$$f(\zeta^k) = 0$$

for every integer k relatively prime to n between 0 and n.

Let η be a primitive m-th root of unity. As observed before Theorem 12.13, p. 178, we have $\mathbb{Q}(\mu_m) = \mathbb{Q}(\eta)$, hence, by Proposition 12.15, every coefficient of f is a polynomial expression in η with rational coefficients. Therefore,

$$f(X) = \varphi(\eta, X)$$

for some polynomial $\varphi(Y, X) \in \mathbb{Q}[Y, X]$.

Let now $\rho = \zeta\eta$. Since m and n are relatively prime, it follows from Proposition 7.10 (p. 88) that ρ is a primitive mn-th root of unity. Moreover, since m and n are relatively prime, Theorem 7.8 (p. 86) shows that there exist integers r and s such that

$$mr + ns = 1.$$

Since $\zeta^n = 1$ and $\eta^m = 1$, this equation implies that

$$\zeta = \zeta^{mr} = \rho^{mr} \qquad \text{and} \qquad \eta = \eta^{ns} = \rho^{ns}.$$

Since $f(\zeta) = 0$, we have $\varphi(\eta, \zeta) = 0$, or

$$\varphi(\rho^{ns}, \rho^{mr}) = 0.$$

Lemma 12.14, p. 178, and the preceding theorem then show that $\Phi_{mn}(X)$ divides $\varphi(X^{ns}, X^{mr})$ and it follows that

$$\varphi(\omega^{ns}, \omega^{mr}) = 0 \tag{12.14}$$

for every primitive mn-th root of unity ω.

For any integer k relatively prime to n between 0 and n, let

$$\ell = kmr + ns.$$

Since $mr + ns = 1$, we have $mr \equiv 1 \bmod n$ and $ns \equiv 1 \bmod m$, whence

$$\ell \equiv k \quad \bmod n \qquad \text{and} \qquad \ell \equiv 1 \quad \bmod m. \tag{12.15}$$

It follows that $\zeta^\ell = \zeta^k$ and $\eta^\ell = \eta$, and since we already observed that $\zeta = \rho^{mr}$ and $\eta = \rho^{ns}$, we have

$$\rho^{\ell mr} = \zeta^k \qquad \text{and} \qquad \rho^{\ell ns} = \eta.$$

On the other hand, the congruences in (12.15) also show that ℓ is relatively prime to mn. Therefore, ρ^ℓ is a primitive mn-th root of unity, and equation (12.14) yields

$$\varphi(\rho^{\ell ns}, \rho^{\ell mr}) = 0,$$

i.e. $\varphi(\eta, \zeta^k) = 0$, or $f(\zeta^k) = 0$. $\qquad\qquad\qquad\qquad\qquad\qquad\qquad$ $\square$

12.33. COROLLARY. *Let p be a prime number and let k be an integer which divides $p - 1$. Let also $\zeta \in \mathbb{C}$ be a primitive p-th root of unity. Then every element in $\mathbb{Q}(\mu_k)(\mu_p)$ can be uniquely written in the form*

$$a_1\zeta + a_2\zeta^2 + \cdots + a_{p-1}\zeta^{p-1},$$

for some $a_1, \ldots, a_{p-1}$ in $\mathbb{Q}(\mu_k)$.

Proof. The hypothesis on k ensures that k is relatively prime to p, hence Φ_p is irreducible over $\mathbb{Q}(\mu_k)$, by the preceding theorem. The corollary then follows by the same arguments as in the proof of Theorem 12.13, p. 178. $\qquad\qquad$ $\square$

Appendix: Ruler and compass construction of regular polygons

We aim to find a process to construct regular polygons in the plane, using ruler and compass only. It is clear that this construction should be possible whenever the center of the polygon (i.e. the center of the circumscribed circle) and one of its vertices are arbitrarily chosen. Therefore, we may regard the center O and one of the vertices A as given, and we have to determine the other vertices. From the two given points O and A, new points can be constructed by a (finite) sequence of operations of the following types:

(1) draw a line through two points already determined,

(2) draw a circle with center a point already determined and radius the distance between two points already determined.

New points are determined as intersection points of the lines or circles drawn according to (1) and (2). The points which can be thus determined are called *constructible points*.

The problem is to decide for which values of n the vertices of the regular polygon with n sides, with center O, and A as one of the vertices, are constructible. To solve this problem, we first give an algebraic characterization of the constructible points, via their coordinates in a suitable basis, which we construct as follows: we consider the perpendicular to OA through O and denote by B one of the intersection points of this perpendicular and the circle with center O and radius OA:

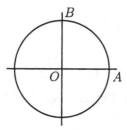

(Observe that the point B is constructible).

PROPOSITION. *A point in the plane can be constructed by ruler and compass from O and A if and only if its coordinates in the basis (OA, OB) can be obtained from 0 and 1 by a (finite) sequence of operations of the following types:*

   (i) rational operations,
   (ii) extraction of square roots.

Proof. First, we show that the points whose coordinates satisfy the condition above are constructible.

Since the perpendicular through a given point to a given line can be constructed by ruler and compass, a point with coordinates (a, b) is constructible if (and only if) the points $(a, 0)$ and $(0, b)$ are constructible. Moreover, since $(0, b)$ is the intersection of the axis OB with the circle with center O passing through $(b, 0)$, it suffices to consider points with coordinates $(u, 0)$. So, we have to prove that a point with coordinates $(u, 0)$ is constructible with ruler and compass if u is obtained from 0 and 1 by a sequence of operations *(i)* and *(ii)* above.

We argue inductively on the number of operations. Thus, we shall prove that if $(u, 0)$ and $(v, 0)$ are constructible, then $(u + v, 0)$, $(u - v, 0)$, $(uv, 0)$, $(uv^{-1}, 0)$ (assuming $v \neq 0$) and $(\sqrt{u}, 0)$ (assuming $u \geq 0$) are constructible.

This is clear for $(u + v, 0)$ and $(u - v, 0)$. In order to construct $(uv, 0)$ and $(uv^{-1}, 0)$, we consider the figure below:

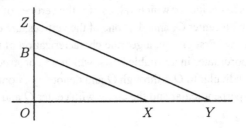

Since BX and YZ are parallel,

$$\frac{OX}{OB} = \frac{OY}{OZ}.$$

Since $B = (0, 1)$, it follows that, denoting $X = (x, 0)$, $Y = (y, 0)$ and $Z = (0, z)$,

$$x = yz^{-1}, \qquad \text{or} \qquad y = xz.$$

Therefore, if we regard X and Z as given, we can construct Y by drawing the parallel to BX through Z. This construction yields $(xz, 0)$ from $(x, 0)$ and $(0, z)$ (or, equivalently, $(z, 0)$). On the other hand, if we regard Y and Z as given, then we can obtain X by drawing the parallel to YZ through B; this yields $(yz^{-1}, 0)$ from $(y, 0)$ and $(0, z)$ (or $(z, 0)$).

To complete the proof of the "if" part, it only remains to show that $(\sqrt{u}, 0)$ can be constructed from $(u, 0)$ (assuming $u \geq 0$). This can be done as follows:

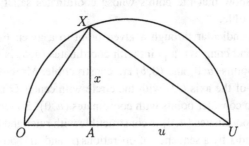

Let U be the point with coordinates $(1 + u, 0)$ and let X be one of the intersection points of the perpendicular to OU through A with the circle with diameter OU. We thus have $X = (1, x)$ for some x. Since the triangles OAX and XAU are similar, we have

$$\frac{AX}{OA} = \frac{AU}{AX},$$

whence

$$\frac{x}{1} = \frac{u}{x} \quad \text{or} \quad x = \sqrt{u}.$$

Since the point $(x, 0)$ can be easily determined from $(1, x)$, this construction yields $(\sqrt{u}, 0)$ from $(u, 0)$.

We have thus proved that the points whose coordinates are obtained from 0 and 1 by rational operations and extraction of square roots are constructible.

To prove the converse, we first observe that if a line passes through two points (a_1, b_1) and (a_2, b_2), then its equation has the form

$$\alpha X + \beta Y = \gamma$$

where α, β and γ are rational expressions of a_1, a_2, b_1 and b_2. (Specifically, $\alpha = b_2 - b_1$, $\beta = a_1 - a_2$ and $\gamma = a_1 b_2 - b_1 a_2$.) Likewise, the equation of a circle with center (a_1, b_1) and radius the distance between (a_2, b_2) and (a_3, b_3) is

$$(X - a_1)^2 + (Y - b_1)^2 = (a_2 - a_3)^2 + (b_2 - b_3)^2$$

hence it has the form

$$X^2 + Y^2 = \alpha X + \beta Y + \gamma$$

where α, β, γ are rational expressions of a_1, a_2, a_3, b_1, b_2, b_3. Now, direct calculations show that the coordinates of the intersection point of two lines

$$\alpha_1 X + \beta_1 Y = \gamma_1 \quad \text{and} \quad \alpha_2 X + \beta_2 Y = \gamma_2$$

are rational expressions of α_1, β_1, γ_1, α_2, β_2, γ_2. Thus, if a point is constructed as the intersection of two lines passing through given points, its coordinates are rational expressions of the coordinates of the given points.

Similarly, it can be seen that the coordinates of the intersection points of a line and a circle

$$\alpha_1 X + \beta_1 Y = \gamma_1 \quad \text{and} \quad X^2 + Y^2 = \alpha_2 X + \beta_2 Y + \gamma_2$$

are obtained by rational operations and extraction of a square root from $\alpha_1, \beta_1, \gamma_1,$ $\alpha_2, \beta_2, \gamma_2$. Therefore, if a point is constructed as the intersection of a line through given points and a circle with given center and with radius the distance between given points, then its coordinates are obtained from the coordinates of the given points by rational operations and extraction of a square root.

Finally, the intersection of two circles

$$X^2 + Y^2 = \alpha_1 X + \beta_1 Y + \gamma_1 \qquad \text{and} \qquad X^2 + Y^2 = \alpha_2 X + \beta_2 Y + \gamma_2$$

can be obtained as the intersection of the circle

$$X^2 + Y^2 = \alpha_1 X + \beta_1 Y + \gamma_1$$

and the line

$$\alpha_1 X + \beta_1 Y + \gamma_1 = \alpha_2 X + \beta_2 Y + \gamma_2,$$

hence the same conclusion as for the preceding case holds. These arguments show that the coordinates of the constructible points are obtained by operations *(i)* and *(ii)* from the coordinates of O and A, i.e. from 0 and 1. $\qquad\square$

This constructibility criterion seems to have been first published by Pierre Laurent Wantzel (1814–1848) in 1837, but it was undoubtedly known to Gauss (and presumably also to others) around 1796.

THEOREM. *If p is a prime number of the form $p = 2^m + 1$ (with $m \in \mathbb{N}$) then the regular polygon with p sides can be constructed with ruler and compass.*

Proof. Since $p - 1$ is a power of 2, the lattice of divisors of $p - 1$ is a chain

$$2^m = p - 1$$
$$|$$
$$2^{m-1}$$
$$|$$
$$2^{m-2}$$
$$|$$
$$\vdots$$
$$|$$
$$2$$
$$|$$
$$1.$$

Therefore, the results of §12.4 show that the periods of two terms can be determined by solving a sequence of quadratic equations. Since, as observed p. 184, the periods of two terms are the values $2 \cos \frac{2k\pi}{p}$, for $k = 1, \ldots, \frac{p-1}{2}$, and since the solution of a quadratic equation only requires rational operations and extraction of square roots, it follows that $\cos \frac{2\pi}{p}$ can be obtained from the integers (or even from 0 and 1) by rational operations and extraction of square roots. The preceding proposition then shows that the point with coordinates $(\cos \frac{2\pi}{p}, 0)$ is constructible.

The point $P = (\cos \frac{2\pi}{p}, \sin \frac{2\pi}{p})$ can then be obtained as the intersection of the circle with center O and radius OA with the perpendicular to OA through $(\cos \frac{2\pi}{p}, 0)$. The point P is a vertex of the regular polygon with p sides. In fact, it is one of the two vertices which is closest to A, and the other vertices can be found by reproducing the distance AP on the circle. $\qquad\square$

If a prime number p has the form $2^m + 1$, then it is easily seen that m is a power of 2. Indeed, if m is divisible by some odd integer k, then $2^m + 1$ is divisible by $2^{m/k} + 1$, as can be seen by letting $X = 2^{m/k}$ in the relation

$$X^k + 1 = (X + 1)(X^{k-1} - X^{k-2} + \cdots - X + 1).$$

Thus, the prime numbers which satisfy the hypothesis of the proposition are in fact of the form $p = 2^{2^n} + 1$ for some integer n. These prime numbers are called *Fermat primes*, after Pierre de Fermat, who conjectured that the number $F_n = 2^{2^n} + 1$ is prime for every integer n. For $n = 0, 1, 2, 3, 4$, this formula yields 3, 5, 17, 257 and 65537, which are indeed prime, but in 1732 Euler showed that $F_5 = 641 \cdot 6700417$. Since then, the numbers F_n have been shown to be composite for various values of n, and no new Fermat prime has been found. Although it has not been proved that no other Fermat prime exists, it is at least known that there is no such prime between 65538 and 10^{39456} (i.e. F_n is not prime for $5 \le n \le 16$).

COROLLARY. *The regular polygon with n sides can be constructed with ruler and compass if n is a product of distinct Fermat primes and of a power of 2.*

Proof. Since the regular polygon with n sides is constructible when n is a power of 2 (by repeated bisections of angles) or when n is a Fermat prime (by the preceding theorem), it suffices to show that when n_1 and n_2 are relatively prime integers such that the regular polygons with n_1 and n_2 sides are constructible, then the regular polygon with $n_1 n_2$ sides is constructible.

If n_1 and n_2 are relatively prime, Theorem 7.8 (p. 86) shows that there exist integers m_1 and m_2 such that $m_1 n_1 + m_2 n_2 = 1$. Multiplying both sides by

$\frac{2\pi}{n_1 n_2}$, we get

$$m_1 \frac{2\pi}{n_2} + m_2 \frac{2\pi}{n_1} = \frac{2\pi}{n_1 n_2}.$$

Therefore, the arc $\frac{2\pi}{n_1 n_2}$ can be constructed by reproducing a certain number of times the arcs $\frac{2\pi}{n_1}$ and $\frac{2\pi}{n_2}$, and it readily follows that the regular polygon with $n_1 n_2$ sides can be constructed from the regular polygons with n_1 and n_2 sides. $\square$

Remark. It can be proved that the converses of the theorem and of the corollary above also hold. Thus, the regular polygon with n sides can be constructed with ruler and compass if and only if n is a product of distinct Fermat primes and of a power of 2. This result is explicitly stated (without proof) by Gauss [24, Art. 366], but a smooth proof of these converses requires a detailed analysis of field degrees, which would carry us too far afield. Therefore, we refer the interested reader to Carrega [12, Chap. 4] or Stewart [56, Chap. 17] for a proof. We also refer to Hardy and Wright [29, §5.8] for an explicit geometric construction of the 17-gon with ruler and compass.

Exercises

1. Recall from Exercise 7 of Chapter 7 that for every integer $n \geq 2$, the number of integers which are relatively prime to n between 0 and n is denoted by $\varphi(n)$. Prove the following generalization (due to Euler) of Fermat's theorem (Theorem 12.5, p. 171): $a^{\varphi(n)} \equiv 1 \bmod n$ for every integer $n \geq 2$ and every integer a relatively prime to n.

2. Show that Girard's theorem (Theorem 6.1, p. 65) readily follows from Gauss' lemma (Lemma 12.11, p. 175).

3. Prove that the periods with an even number of terms are real numbers.

4. Prove that the set of periods of f terms does not depend on the choice of a primitive root of p nor of a primitive p-th root of unity. More precisely, let ζ and $\zeta' \in \mathbb{C}$ be primitive p-th roots of unity, let g, $g' \in \mathbb{Z}$ be primitive roots of p and let $p - 1 = ef$ for some positive integers e, f. Denote $\zeta_i = \zeta^{g^i}$ and $\zeta_i' = \zeta'^{g'^i}$ for $i = 0, \ldots, p - 2$. Show that for any $i = 0, \ldots, f - 1$, there is an integer j between 0 and $f - 1$ such that

$$\zeta_i + \zeta_{e+i} + \cdots + \zeta_{e(f-1)+i} = \zeta_j' + \zeta_{e+j}' + \cdots + \zeta_{e(f-1)+j}'.$$

5. With the notation of Proposition 12.22, p. 188, prove that if $K_g \subset K_f$, then f divides g. Moreover, show that in this case, $\dim_{K_g} K_f = g/f$.

6. *The following exercise provides complementary observations to the proof of Corollary 12.29, p. 195.* Let the notation be as in Corollary 12.29. Show that $t(\omega) \neq 0$ and that $t(\omega^k)t(\omega)^{-k}$ has a rational expression in terms of ω, for $k \in \mathbb{Z}$. (Compare Exercise 4 of Chapter 10.) Conclude that it suffices to extract a single $(n-1)$-st root to determine ζ_i.

7. By looking at the algebraic expression of 5-th roots of unity, find a construction of regular pentagons by ruler and compass. Find also a construction of regular 20-gons.

Chapter 13

Ruffini and Abel on General Equations

13.1 Introduction

Lagrange's investigations were primarily aimed at the solution of "general" equations, i.e. equations whose coefficients are letters, such as

$$X^n - s_1 X^{n-1} + s_2 X^{n-2} - \cdots + (-1)^n s_n = 0$$

(see Definition 8.1, p. 98). At about the same time when Gauss completed the solution of the class of particular equations which arise from the division of the circle (known as cyclotomic equations), Lagrange's line of investigation bore new fruits in the hands of Paolo Ruffini (1765–1822). In 1799, Ruffini published a massive two-volume treatise: "Teoria Generale delle Equazioni" [51, t. 1, pp. 1–324], in which he proves that the general equations of degree at least 5 are not solvable by radicals.

Ruffini's proof was received with skepticism by the mathematical community. Indeed, the proof was rather hard to follow through the 516 pages of his books. A few years after the publication, negative comments were made but, to Ruffini's dismay, no clear, focused objection was raised. Vague criticism was denying Ruffini the credit of having validly proved his claim. Negative reactions prompted Ruffini to simplify his proof, and he eventually came up with very clean arguments, but distrust of Ruffini's work did not subside. Typical in this respect is the following anecdote: in order to get a clear, motivated pronouncement from the French Academy of Sciences, Ruffini submitted a paper to the Academy in 1810. A year later, the referees (Lagrange, Lacroix and Legendre) had not yet given their conclusions. Ruffini then wrote to Delambre, who was secretary of the Academy, to withdraw his paper. In his reply, Delambre explains the referees' attitude:

Whatever decision Your Referees would have reached, they had
to work considerably either to motivate their approval or to re-
fute Your proof. You know how precious is time to realize also
how reluctant most geometers are to occupy themselves for a
long time with the works of each other, and if they would have
happened not to be of Your opinion, they would have had to be
moved by a quite powerful motive to enter the lists against a
geometer so learned and so skillful. [51, t. 3, p. 59].

At least, unconvincing as it was, Ruffini's proof seems to have completed the
reversal of the current opinion towards general equations: while the works of
Bezout and Euler around the middle of the eighteenth century were grounded on
the opinion that general equations were solvable, and that finding the solution of
the fifth degree equations was only a matter of clever transformations, the opposite
view became common in the beginning of the nineteenth century (see Ayoub [4,
p. 274]). Some comments of Gauss may also have been influential in this respect.
In his proof of the fundamental theorem of algebra, [23, §9], Gauss writes:

After the works of many geometers left very little hope of ever
arriving at the resolution of the general equation algebraically, it
appears more and more likely that this resolution is impossible
and contradictory.

He voiced again the same skepticism in Article 359 of "Disquisitiones Arithmeticae."

Ruffini's credit also includes advances in the theory of permutations, which
was crucial for his proof. Ruffini's results in this direction were soon generalized
by Cauchy. Incidentally, it is noteworthy that Cauchy was very appreciative of
Ruffini's work and that he supported Ruffini's claim that his proof was valid (see
[51, t. 3, pp. 88–89]). In fact, it now appears that Ruffini's proofs do have a
significant gap, which we shall point out below.

In 1824, a new proof was found by Niels-Henrik Abel (1802–1829) [1, n° 3],
independently of Ruffini's work. An expanded version of Abel's proof was published in 1826 in the first issue of Crelle's journal (the "Journal für die reine und
angewandte Mathematik") [1, n° 7]. This proof also contains some minor flaws
(see [1, vol. 2, pp. 292–293]), but it essentially settled the issue of solvability of
general equations.

Abel's approach is remarkably methodical. He explains it in some detail in the
introduction to a subsequent paper: "Sur la résolution algébrique des équations"

(1828) [1, n° 18].

> To solve these equations [of degree at most 4], a uniform method
> has been found, and it was believed that it could be applied to
> equations of arbitrary degree; but in spite of the efforts of a La-
> grange and other distinguished geometers, one was not able to
> reach this goal. This led to the presumption that the algebraic
> solution of general equations was impossible; but that could not
> be decided, since the method which was used could not lead to
> definite conclusions except in the case where the equations were
> solvable. Indeed, the purpose was to solve equations, without
> knowing whether this was possible. In this case, one could get
> the solution, although that was not sure at all; but if unfortu-
> nately the solution happened to be impossible, one could have
> sought it for ever without finding it. In order to obtain unfail-
> ingly something in this matter, it is therefore necessary to take
> another way. One has to cast the problem in such a form that it be
> always possible to solve, which can be done with any problem.
> Instead of seeking a relation of which it is not known whether
> it exists or not, one has to seek whether such a relation is in-
> deed possible. For instance, in the integral calculus, instead of
> trying by a kind of divination or by trial and error to integrate
> differential formulas, one has to look rather whether it is pos-
> sible to integrate them in this or that way. When a problem is
> thus presented, the statement itself contains the seed of the solu-
> tion and shows the way that is to be taken; and I think that there
> will be few cases where one could not reach more or less impor-
> tant propositions, even when one could not completely solve the
> question because the calculations would be too complicated.

The method which is thus advocated by Abel can be interpreted in the realm of
algebraic equations as a kind of generic method. One has to find the most general
form of the expected solution and work on it to investigate what kind of information can be obtained on this expression if it is a root of the general equation. Abel
thus proves, by an intricate inductive argument, that if an expression by radicals
is a root of the general equation of some degree, then every function of which it
is composed is a rational expression of the roots (see Theorem 13.13, p. 224, for
a precise statement). This fills a gap in Ruffini's proofs. Some delicate arguments
involving the number of values of functions under permutations of the variables

and, in particular, a theorem of Cauchy generalizing earlier results of Ruffini, complete the proof. This last part of the proof can be significantly streamlined by using arguments from the last of Ruffini's proofs, as Wantzel later noticed. In the following sections, we shall present this easy version, but we point out that this approach unfortunately downplays the advances in the theory of permutations (i.e. in the study of the symmetric group S_n) which were prompted by Ruffini's earlier work.

13.2   Radical extensions

Abel's calculations with expressions by radicals, which we discuss in this section and the following as a first step in the proof that general equations of degree higher than 4 are not solvable, can be adequately cast into the vocabulary of field extensions. This point of view will be used throughout since it is probably more enlightening for the modern reader.

An expression by radicals is constructed from some quantities which are regarded as known (usually the coefficients of an equation, in this context) by the four usual operations of arithmetic and the extraction of roots. This means that any such expression lies in a field obtained from the field of rational expressions in the known quantities by successive adjunctions of roots of some orders. In fact, it is clearly sufficient to consider roots of prime order, since if $n = p_1 \cdots p_r$ is the factorization of a positive integer n into prime factors, then

$$a^{1/n} = \left(\dots \left((a^{1/p_1})^{1/p_2} \right) \dots \right)^{1/p_r}.$$

This shows that an n-th root of any element a can be obtained by extracting a p_1-th root a^{1/p_1} of a, next a p_2-th root of a^{1/p_1} and so on. Moreover, it obviously suffices to extract p-th roots of elements which are not p-th powers, otherwise the base field is not enlarged. We thus come to the notion of a radical field extension. Before spelling out this notion in mathematical terms, we note that, in order to avoid some technical difficulties, we shall restrict attention throughout the chapter to fields of characteristic zero; in other words, we shall assume that $1+1+\cdots+1 \neq 0$ or that every field under consideration contains (an isomorphic copy of) the field $\mathbb{Q}$ of rational numbers. This is of course the classical case, which was the only case considered by Ruffini and Abel.

13.1. DEFINITIONS. A field R containing a field F is called a *radical extension of height* 1 of F if there exist a prime number p, an element $a \in F$ which is not a

p-th power in F and an element $u \in R$ such that

$$R = F(u) \qquad \text{and} \qquad u^p = a.$$

Such an element u is sometimes denoted by $a^{1/p}$ or $\sqrt[p]{a}$, and, accordingly, one sometimes writes

$$R = F(a^{1/p}) \qquad \text{or} \qquad R = F(\sqrt[p]{a}).$$

This is in fact an abuse of notation, since the element u is not uniquely determined by a and p. There are indeed p different p-th roots of a. Worse still, the field R itself is in general not uniquely determined by F, a and p. For instance, there are three subfields of $\mathbb{C}$ which qualify as $\mathbb{Q}(2^{1/3})$. (See however Exercises 4 and 5.) Therefore, the notation above will be used with caution.

Radical extensions of height h, for any positive integer h, are defined inductively as radical extensions of height 1 of radical extensions of height $h - 1$. More precisely, a field R containing a field F is called a *radical extension of height h* of F if there is a field R_1 between R and F such that R is a radical extension of height 1 of R_1 and R_1 is a radical extension of height $h - 1$ of F. Thus, in this case we can find a tower of extensions between R and F,

$$R \supset R_1 \supset R_2 \supset \cdots \supset R_{h-1} \supset F$$

such that, letting $R = R_0$ and $F = R_h$, we have for $i = 0, \ldots, h - 1$

$$R_i = R_{i+1}\left(a_i^{1/p_i}\right)$$

for some prime number p_i and some element $a_i \in R_{i+1}$ which is not a p_i-th power in R_{i+1}.

We simply term *radical extension* any radical extension of some (finite) height and, for completeness, we say that any field is a *radical extension of height* 0 of itself.

The definitions above are quite convenient to translate into mathematically amenable terms questions concerning expressions by radicals. For instance, to say that a complex number z has an expression by radicals means that there is a radical extension of the field $\mathbb{Q}$ of rational numbers containing z. More generally, we shall say that an element v of a field L *has an expression by radicals over some field F contained in L* if there is a radical extension of F containing v.

Likewise, we say that a polynomial equation $P(X) = 0$ over some field F is *solvable by radicals over F* if there is a radical extension of F containing a root of P. In the case of general equations

$$P(X) = (X - X_1) \cdots (X - X_n) = X^n - s_1 X^{n-1} + \cdots + (-1)^n s_n = 0,$$

we are concerned with radical expressions involving only the coefficients $s_1, \ldots,$ s_n, so the base field F will be the field of rational fractions in $s_1, \ldots, s_n$ (which can be considered as independent indeterminates, according to Remark 8.8(a), p. 105). To be more precise, we have to specify a field of reference in which the rational fractions are allowed to take their coefficients. A logical choice is of course the field $\mathbb{Q}$ of rational numbers, but in fact, since we are aiming at a negative result, the reference field can be chosen arbitrarily large. Indeed, we shall prove that if an equation is solvable by radicals over some field F, then it is solvable by radicals over every field L containing F; therefore, if the general equation of degree n is not solvable over $\mathbb{C}(s_1, \ldots, s_n)$, it is not solvable over $\mathbb{Q}(s_1, \ldots, s_n)$ either.

Of course, Ruffini and Abel did not address in these terms the problem of assigning a reference field, but their free use of roots of unity suggests that all the roots of unity are at their disposal in the base field. The choice $F = \mathbb{C}(s_1, \ldots, s_n)$ seems therefore close in spirit to Ruffini's and Abel's work.

The hypothesis that the base field contains all the roots of unity also has a technical advantage, in that it allows more flexibility in the treatment of radical extensions, as the next result shows:

13.2. PROPOSITION. *Let R be a field containing a field F. If R has the form $R = F(u)$ for some element u such that $u^n \in F$ for some integer n, and if F contains a primitive n-th root of unity (hence all the n-th roots of unity, since the other roots are powers of this one), then R is a radical extension of F.*

In other words, in the definition of radical extensions, we need not require that the exponent n be a prime number, nor that u^n be not the n-th power of an element in F, provided that F contains a primitive n-th root of unity.

Proof. We argue by induction on n. If $n = 1$, then $u \in F$, hence $R = F$ and R is then a radical extension of height 0 of F. We may thus assume that $n \geq 2$ and that the proposition holds when the exponent of u is at most $n - 1$.

If n is not prime, let $n = rs$ for some (positive) integers $r, s < n$. By the induction hypothesis, $F(u)$ is a radical extension of $F(u^r)$ and $F(u^r)$ is a radical extension of F, since u^r satisfies $(u^r)^s \in F$. Therefore, $F(u)$ is a radical extension of F, since it is clear from the definition that the property of being radical is transitive, namely, in a tower of extensions $F \subset K \subset L$, if L is a radical extension of K and K is a radical extension of F, then L is a radical extension of F.

If n is prime, we consider two cases, according to whether u^n is or is not the n-th power of an element in F. If it is not, then R is a radical extension of F, by definition. If it is, let

$$u^n = b^n$$

for some $b \in F$. If $b = 0$, then $u = 0$ and $R = F$, a radical extension of height 0 of F. If $b \neq 0$, then the preceding equation yields

$$\left(\frac{u}{b}\right)^n = 1,$$

hence u/b is an n-th root of unity. Since the n-th roots of unity are all in F, it follows that $u/b \in F$, hence $u \in F$ and again $R = F$, a radical extension of height 0 of F. $\qquad\square$

As an application, we have the following result, which will be useful later through its corollary:

13.3. PROPOSITION. *Let R and L be subfields of a field K, both containing a subfield F. Assume F contains the field $\mathbb{C}$ of complex numbers, so that all the roots of unity are in F. If R is a radical extension of F, then there is a radical extension S of L containing R and contained in K.*

Proof. We argue by induction on the height of R. If this height is zero, then $R = F$ and we can choose $S = L$. We may thus let the height of R be $h \geq 1$ and assume that the proposition holds for radical extensions of height at most $h - 1$. By definition of radical extensions of height h, we can find inside R a radical extension R_1 of F of height $h - 1$ and an element u such that

$$R = R_1(u) \qquad \text{and} \qquad u^p \in R_1$$

for some prime number p. By the induction hypothesis, there is a radical extension S_1 of L in K which contains R_1. Then $u^p \in S_1$ and Proposition 13.2 shows that $S_1(u)$ is a radical extension of L. This extension is contained in K, since $u \in K$ and $S_1 \subset K$, and it contains R, since $R = R_1(u)$ and $R_1 \subset S_1$. It thus satisfies the required conditions. $\qquad\square$

13.4. COROLLARY. *Let $v_1, \ldots, v_n$ be elements of a field K containing a field F. Assume that F contains $\mathbb{C}$ and that each of $v_1, \ldots, v_n$ lies in a radical extension* *of F contained in K. Then there is a single radical extension of F in K which contains all of $v_1, \ldots, v_n$.*

Proof. We argue by induction on n. There is nothing to prove if $n = 1$, so we may assume that $n \geq 2$ and that the corollary holds for $n - 1$ elements. Hence, there is a radical extension L of F in K which contains $v_1, \ldots, v_{n-1}$. Let R be a radical extension of F in K containing v_n. The preceding proposition shows that there is a radical extension S of L in K containing R. Since S contains both L and R, it contains $v_1, \ldots, v_n$. Since moreover S is a radical extension of L, which is a radical extension of F, it is a radical extension of F.   □

So far, we have dealt only with the case where roots of unity are in the base field. In order to reduce more general situations to this case, we have to use Gauss' result that every root of unity has an expression by radicals. Since we now have a formal definition for "expression by radicals," it seems worthwhile to spell out how Gauss' arguments actually fit in this framework.

13.5. PROPOSITION. *For any integer n and any field F, the n-th roots of unity lie in a radical extension of F.*

Proof. It suffices to show that a primitive n-th root of unity ζ lies in a radical extension of F, since the other n-th roots of unity are powers of ζ and lie therefore in the same radical extension as ζ.

We argue by induction on n. For $n = 1$, we have $\zeta = 1$, hence ζ lies in F, which is a radical extension of height 0 of itself. We may thus assume that $n \geq 2$ and that the proposition holds for roots of unity of exponent less than n.

If n is not prime, let $n = rs$ for some (positive) integers $r, s < n$. Then ζ^r is an s-th root of unity. By the induction hypothesis, we can find a radical extension R_1 of F containing ζ^r. By the induction hypothesis again, we can find a radical extension R_2 of R_1 (hence also of F) which contains a primitive r-th root of unity. Then, since $\zeta^r \in R_2$, it follows from Proposition 13.2 that $R_2(\zeta)$ is a radical extension of R_2, hence of F. The proposition is thus proved in this case.

If n is prime, then we have to use Gauss' results. First, we can find a radical extension R_1 of F which contains the $(n - 1)$-st roots of unity, by the induction hypothesis. We then consider the Lagrange resolvents $t(\omega)$ as in the proof of Corollary 12.29, p. 195. By Proposition 12.27, p. 193, we have

$$t(\omega)^{n-1} \in R_1$$

for every $(n - 1)$-st root of unity ω. Therefore Proposition 13.2 shows that

$R_1\big(t(\omega)\big)$ is a radical extension of R_1. Adjoining successively all the Lagrange resolvents $t(\omega)$, we find a radical extension R_2 of R_1, whence of F, which contains $t(\omega)$ for all $\omega \in \mu_{n-1}$. From Lagrange's formula (p. 138) it follows that ζ can be rationally calculated from the Lagrange resolvents, hence $\zeta \in R_2$ and the proof is complete. □

We now aim to prove the afore-mentioned fact that solvability of an equation by radicals over some field F implies solvability by radicals over any larger field L. This fact may seem obvious, since every expression by radicals involving elements of F is an expression by radicals involving elements of L. However, it needs a careful justification. The point is that, in building radical extensions or expressions by radicals, we allow only extractions of p-th roots of elements which are not p-th powers in F, but these elements could become p-th powers in the larger field L.

13.6. LEMMA. *Let L be a field containing a field F. For any radical extension R of F, there is a radical extension S of L such that R can be identified to a subfield of S.*

Proof. We argue by induction on the height h of R. If $h = 0$, then $R = F$ and we can choose $S = L$.

If $h = 1$, let $R = F(u)$ where u is such that $u^p = a$ for some element $a \in F$ which is not a p-th power in F. Let also K be a field containing L and over which the polynomial $X^p - a$ splits into a product of linear factors. (The existence of such a field K follows from Girard's theorem (Theorem 9.3, p. 116).) Since u is one of the roots of $X^p - a$, it can be identified with an element in K, and every rational fraction in u with coefficients in F, i.e. every element in R, is then identified with an element in K. We may thus henceforth assume that R is contained in K.

If a is not a p-th power in L, then $L(u)$ is a radical extension of height 1 of L, and this extension contains R since it contains F and u. It thus fulfills the required conditions.

If a is a p-th power in L, then let $b \in L$ be a p-th root of a,

$$b^p = a.$$

Since the p-th powers of u and b are equal, it follows that

$$\left(\frac{u}{b}\right)^p = 1.$$

Therefore, u/b is a p-th root of unity, and Proposition 13.5 shows that there is a radical extension S of L which contains u/b. Since $b \in L$, it follows that $u \in S$, hence $R \subset S$ and the proof is complete in the case where the height h of R is 1.

If $h \geq 2$, the lemma readily follows from the preceding case and the induction hypothesis. Indeed, we can find in R a subfield R_1 which is a radical extension of height $h - 1$ of F and such that R is a radical extension of height 1 of R_1. By the induction hypothesis, we may assume that R_1 is contained in a radical extension S_1 of L and, by the case $h = 1$ already considered, R can be identified to a subfield of a radical extension S of S_1. The field S is then a radical extension of L and it satisfies the condition of the lemma. $\qquad\Box$

13.7. THEOREM. *Let P be a polynomial with coefficients in a field F. If $P(X) = 0$ is solvable by radicals over F, then it is solvable by radicals over every field L containing F.*

Proof. Let R be a radical extension of F containing a root r of P. The preceding lemma shows that we may assume R is contained in some radical extension S of L. The radical extension S then contains the root r, hence $P(X) = 0$ is solvable by radicals over L. $\qquad\Box$

The following special case of the theorem is particularly relevant for this chapter:

13.8. COROLLARY. *If the general equation of degree n*

$$P(X) = (X - x_1) \cdots (X - x_n) = X^n - s_1 X^{n-1} + \cdots + (-1)^n s_n = 0$$

is not solvable by radicals over $\mathbb{C}(s_1, \ldots, s_n)$, then it is not solvable by radicals over $\mathbb{Q}(s_1, \ldots, s_n)$ either.

We may thus henceforth assume that the base field contains all the roots of unity.

13.3   Abel's theorem on natural irrationalities

Any proof that the general equation of some degree is not solvable by radicals obviously proceeds *ad absurdum*. Thus, we assume by way of contradiction that there is a radical extension R of $\mathbb{C}(s_1, \ldots, s_n)$ which contains a root x_i of the general equation

$$(X - x_1) \cdots (X - x_n) = X^n - s_1 X^{n-1} + s_2 X^{n-2} - \cdots + (-1)^n s_n = 0.$$

The first step in Abel's proof (which was missing in Ruffini's proofs) is to show that R can be supposed to lie inside $\mathbb{C}(x_1, \ldots, x_n)$. This means that the irrationalities which occur in an expression by radicals for a root of the general equation of degree n can be chosen to be *natural*, as opposed to *accessory* irrationalities, which designate the elements of extensions of $\mathbb{C}(s_1, \ldots, s_n)$ outside $\mathbb{C}(x_1, \ldots, x_n)$ (see Ayoub [4, p. 268]). (The terms "natural" and "accessory" irrationalities were coined by Kronecker.)

The aim of this section is to prove this result, following Abel's approach in [1, n° 7, §2].

13.9. LEMMA. *Let p be a prime number and let a be an element of some field F, which is not a p-th power in F.*

(a) *For $k = 1, \ldots, p - 1$, the k-th power a^k is not a p-th power in F either.*

(b) *The polynomial $X^p - a$ is irreducible over F.*

Proof. (a) If k is an integer between 1 and $p - 1$, then it is relatively prime to p, whence by Theorem 7.8 (p. 86) we can find integers ℓ and q such that $pq + k\ell = 1$. Then

$$a = (a^q)^p (a^k)^\ell.$$

Therefore, if $a^k = b^p$ for some $b \in F$, then we have

$$a = (a^q b^\ell)^p,$$

in contradiction with the hypothesis that a is not a p-th power in F. This contradiction proves (a).

(b) Let P and Q be polynomials in $F[X]$ such that

$$X^p - a = PQ.$$

We may assume that P and Q are monic, and we have to prove that P or Q is the constant polynomial 1. Let K be an extension of F over which $X^p - a$ splits into a product of linear factors. (The existence of such a field follows from Girard's theorem (Theorem 9.3, p. 116).) Since the roots of $X^p - a$ are the p-th roots of a, which are obtained from any of them by multiplication by the various p-th roots of unity (see §7.3), we have in $K[X]$

$$\prod_{\omega \in \mu_p} (X - \omega u) = PQ$$

where $u \in K$ is one of the p-th roots of a in K. This equation shows that P and Q split in $K[X]$ into products of factors $X - \omega u$. More precisely, μ_p decomposes into a union of disjoint subsets I and J such that

$$P = \prod_{\omega \in I}(X - \omega u) \quad \text{and} \quad Q = \prod_{\omega \in J}(X - \omega u).$$

Consider then the constant term of P, which we denote by b. The above factorization of P shows that

$$b = \left(\prod_{\omega \in I} \omega\right)(-u)^k$$

where k denotes the number of elements of I. Since $\omega^p = 1$ for any $\omega \in I$, we get by raising both sides of the preceding equality to the p-th power

$$((-1)^k b)^p = a^k.$$

Part *(a)* of the lemma then shows that $k = 0$ or $k = p$. In the first case $P = 1$ and in the second $P = X^p - a$, whence $Q = 1$. $\square$

Let now R be a radical extension of height 1 of some field F. By definition, this means that there exists an element $u \in R$ such that $R = F(u)$ and $u^p = a$ for some element $a \in F$ which is not a p-th power in F. Using the preceding lemma, we can give a standard form to the elements of R.

13.10. COROLLARY. *Every element $v \in R$ can be written in a unique way as*

$$v = v_0 + v_1 u + v_2 u^2 + \cdots + v_{p-1} u^{p-1},$$

for some elements $v_0, v_1, \ldots, v_{p-1} \in F$.

Proof. This readily follows from Proposition 12.15 (p. 179), by the preceding lemma. $\square$

In fact, when $v \in R$ is given beforehand outside F, then the element u can be chosen in such a way that $v_1 = 1$ in the expression above, as we now show:

13.11 LEMMA. *Let R be a radical extension of height 1 of some field F and let $v \in R$. If $v \notin F$, then the element $u \in R$ such that $R = F(u)$ and $u^p \in F$ can be chosen in such a way that*

$$v = v_0 + u + v_2 u^2 + \cdots + v_{p-1} u^{p-1}$$

for some $v_0, v_1, \ldots, v_{p-1} \in F$.

Proof. Let u' be an element of R such that $R = F(u')$ and $u'^p = a'$ for some element $a' \in F$ which is not a p-th power in F. By Corollary 13.10, we may write

$$v = v_0' + v_1' u' + v_2' u'^2 + \cdots + v_{p-1}' u'^{p-1}$$

for some $v_0', \ldots, v_{p-1}' \in F$. These elements are not all zero since $v \notin F$. Let k be an index between 1 and $p-1$ such that $v_k' \neq 0$, and let

$$u = v_k' u'^k. \tag{13.1}$$

Raising both sides of this equation to the p-th power, we get

$$u^p = v_k'^p a'^k.$$

This shows that u satisfies the equation $u^p = a$ with $a = v_k'^p a'^k \in F$.

If a is the p-th power of an element in F, then the last equation shows that a'^k also is a p-th power in F. But then it follows from Lemma 13.9(a) that a' itself is a p-th power in F, which contradicts the hypothesis on u'. Therefore, a is not a p-th power in F.

Since $u \in R$, we obviously have $F(u) \subset R$. In order to prove that $R = F(u)$, it thus suffices to show that every element in R has a rational expression in u with coefficients in F. We first show that the powers of u' have such expressions. For any $i = 0, \ldots, p-1$, we get by raising both sides of equation (13.1) to the i-th power

$$u^i = v_k'^i u'^{ki}. \tag{13.2}$$

Now, recall the permutation σ_k of $\{0, 1, \ldots, p-1\}$ which maps every integer i between 0 and $p-1$ to the unique integer $\sigma_k(i)$ between 0 and $p-1$ such that

$$\sigma_k(i) \equiv ik \mod p$$

(see Proposition 10.6, p. 147). By definition of $\sigma_k(i)$, there is an integer m such that

$$ik - \sigma_k(i) = pm,$$

hence

$$u'^{ik} = (u'^p)^m u'^{\sigma_k(i)}.$$

Therefore, recalling that $u'^p = a'$ and letting $b_i = (v_k'^i a'^m)^{-1}$ for $i = 0, \ldots,$ $p - 1$, we get from equation (13.2)

$$b_i u^i = u'^{\sigma_k(i)}$$

for $i = 0, \ldots, p - 1$. Now, every $x \in R$ has an expression

$$x = \sum_{i=0}^{p-1} x_i u'^i,$$

with $x_i \in F$ for $i = 0, \ldots, p - 1$, which can be alternatively written as

$$x = \sum_{i=0}^{p-1} x_{\sigma_k(i)} u'^{\sigma_k(i)},$$

as σ_k is a permutation of $\{0, \ldots, p - 1\}$. Substituting $b_i u^i$ for $u'^{\sigma_k(i)}$, we obtain

$$x = \sum_{i=0}^{p-1} x_{\sigma_k(i)} b_i u^i.$$

This shows that every element in R has a rational expression in u with coefficients in F, whence $R = F(u)$. For the given $v \in R$, the coefficient of u in this expression is 1, since taking $i = 1$ in the calculations above, we find $\sigma_k(1) = k$ and $m = 0$, whence $b_1 = v_k'^{-1}$. This completes the proof. $\square$

13.12. LEMMA. *We keep the same notation as in Lemma 13.11 and assume moreover that F contains a primitive p-th root of unity ζ (whence all the p-th roots of unity, since the others are powers of ζ). If v is a root of an equation with coefficients in F, then R contains p roots of this equation, and $u, v_0, v_2, \ldots, v_{p-1}$ are rational expressions of these roots with coefficients in $\mathbb{Q}(\zeta)$.*

Proof. Let $P \in F[X]$ be such that $P(v) = 0$. Using the expression of v as in Lemma 13.11, we derive from P another polynomial Q with coefficients in F,

$$Q(Y) = P(v_0 + Y + v_2 Y^2 + \cdots + v_{p-1} Y^{p-1}) \in F[Y].$$

This definition is designed so that the equation $P(v) = 0$ yields $Q(u) = 0$. On the other hand, u is also a root of the polynomial $Y^p - a$, which is irreducible by Lemma 13.9(b). Therefore, Lemma 12.14 (p. 178) shows that $Y^p - a$ divides

$Q(Y)$, and it follows that every root of $Y^p - a$ is a root of $Q(Y)$. Since the roots of $Y^p - a$ are the p-th roots of a, which are of the form $\zeta^i u$, for $i = 0, \ldots, p-1$, we have

$$Q(\zeta^i u) = 0 \qquad \text{for } i = 0, \ldots, p-1. \tag{13.3}$$

Let then

$$z_i = v_0 + \zeta^i u + v_2 \zeta^{2i} u^2 + \cdots + v_{p-1} \zeta^{(p-1)i} u^{p-1}$$

for $i = 0, \ldots, p-1$. Equation (13.3) yields

$$P(z_i) = 0 \qquad \text{for } i = 0, \ldots, p-1,$$

which proves that R contains p roots of P. To complete the proof, we now show that $u, v_0, v_2, \ldots, v_{p-1}$ are rational expressions of $z_0, \ldots, z_{p-1}$, by calculations which are reminiscent of Lagrange's formula (p. 138). Grouping the terms which contain a given factor $v_j u^j$ in the sum of $\zeta^{-ik} z_i$, we have

$$\sum_{i=0}^{p-1} \zeta^{-ik} z_i = \sum_{j=0}^{p-1} \left(\sum_{i=0}^{p-1} \zeta^{(j-k)i} \right) v_j u^j \qquad \text{for } k = 0, \ldots, p-1, \tag{13.4}$$

where we have let $v_1 = 1$. If $j \neq k$, then ζ^{j-k} is a p-th root of unity other than 1, whence a root of

$$\Phi_p(X) = \sum_{i=0}^{p-1} X^i.$$

Therefore,

$$\sum_{i=0}^{p-1} \zeta^{(j-k)i} = 0.$$

Hence, all the terms with index $j \neq k$ vanish in the right-hand side of (13.4), and it remains

$$\sum_{i=0}^{p-1} \zeta^{-ik} z_i = p v_k u^k \qquad \text{for } k = 0, \ldots, p-1.$$

This proves that $v_k u^k$ is a rational expression (indeed a linear expression) of $z_0, \ldots, z_{p-1}$ with coefficients in $\mathbb{Q}(\zeta_p)$. In particular, for $k = 1$, we see that u is such an expression, and since $v_k = (v_k u^k) u^{-k}$, it follows that $v_0, v_2, \ldots, v_{p-1}$ also are rational expressions of $z_0, \ldots, z_{p-1}$ with coefficients in $\mathbb{Q}(\zeta_p)$. $\qquad \square$

Now, we let

$$K = \mathbb{C}(x_1, \ldots, x_n),$$

where $x_1, \ldots, x_n$ are independent indeterminates over $\mathbb{C}$, and we denote by F the subfield of symmetric fractions. By Theorem 8.3 (p. 99), we have

$$F = \mathbb{C}(s_1, \ldots, s_n),$$

where $s_1, \ldots, s_n$ are the elementary symmetric polynomials in $x_1, \ldots, x_n$.

13.13. THEOREM (OF NATURAL IRRATIONALITIES). *If an element $v \in K$ lies in a radical extension of F, then there is inside K a radical extension of F containing v.*

Proof. We argue by induction on the height of the radical extension R of F containing v, which is assumed to exist. There is nothing to prove if the height of R is 0 (i.e. if $R = F$) since in this case R lies inside K. We may thus assume the height of R is $h \geq 1$ and consider R as a radical extension of height 1 of some subfield R_1, which is a radical extension of F of height $h - 1$.

If $v \in R_1$, then we are done by the induction hypothesis. For the rest of the proof, we may thus assume that v lies outside R_1. Lemma 13.11 then shows that

$$R = R_1(u)$$

for some element u such that $u^p \in R_1$ (for some prime p) and

$$v = v_0 + u + v_2 u^2 + \cdots + v_{p-1} u^{p-1} \tag{13.5}$$

for some elements $v_0, v_2, \ldots, v_{p-1} \in R_1$. Now, Proposition 10.1 (p. 131) (and its proof) show that every element in K is a root of a polynomial with coefficients in F, which splits into a product of linear factors over K (its roots are the various "values" of the element under the permutations of $x_1, \ldots, x_n$). In particular, v is a root of an equation with coefficients in F (whence in R_1), whose roots all lie in K. Therefore, we can apply Lemma 13.12 to conclude that $u, v_0, v_2, \ldots, v_{p-1} \in K$.

But $u^p, v_0, v_2, \ldots, v_{p-1}$ also lie in R_1, which is a radical extension of height $h - 1$ of F. By the induction hypothesis, $u^p, v_0, v_2, \ldots, v_{p-1}$ all lie in radical extensions of F inside K and, by Corollary 13.4, p. 215, we can find a single radical extension R' of F inside K containing $u^p, v_0, v_2, \ldots, v_{p-1}$. Since $u^p \in R'$, the field $R'(u)$ is a radical extension of R', hence a radical extension of F.

Since moreover we have already observed that $u \in K$, we have $R'(u) \subset K$, and equation (13.5) shows that $v \in R'(u)$. This completes the proof.   □

13.4   Proof of the unsolvability of general equations of degree higher than 4

In order to prove that general equations of degree higher than 4 are not solvable by radicals, we have to show, according to Definitions 13.1 above, that for $n \geq 5$ there is no radical extension of $\mathbb{C}(s_1, \ldots, s_n)$ containing a root x_i of the general equation of degree n

$$(X - x_1) \cdots (X - x_n) = X^n - s_1 X^{n-1} + \cdots + (-1)^n s_n = 0.$$

The proof we give below is based upon Ruffini's last proof (1813) [51, vol. 2, pp. 162–170]. It is sometimes called the Wantzel modification of Abel's proof (see [51, vol. 2, p. 505] and Serret [53, n° 516]), although Wantzel was relying on Ruffini's papers (see Ayoub [4, p. 270]).

13.14. LEMMA. *Let u and a be elements of $\mathbb{C}(x_1, \ldots, x_n)$ such that*

$$u^p = a$$

for some prime number p, and assume $n \geq 5$. If a is invariant under the permutations

$$\sigma: x_1 \mapsto x_2 \mapsto x_3 \mapsto x_1; \quad x_i \mapsto x_i \quad for\, i > 3$$

and

$$\tau: x_3 \mapsto x_4 \mapsto x_5 \mapsto x_3; \quad x_i \mapsto x_i \quad for\, i = 1,\, 2\, and\, i > 5,$$

then so is u.

Proof. Applying σ to both sides of the equation $u^p = a$, we get $\sigma(u)^p = a$, hence

$$\sigma(u)^p = u^p.$$

Since the lemma is trivial if $u = 0$, we may assume $u \neq 0$ and divide both sides of the preceding equation by u^p. We thus obtain

$$\left(\frac{\sigma(u)}{u}\right)^p = 1,$$

whence

$$\sigma(u) = \omega_\sigma u$$

for some p-th root of unity ω_σ. Applying σ to both sides of this last equation, we get $\sigma^2(u) = \omega_\sigma^2 u$, next $\sigma^3(u) = \omega_\sigma^3 u$. Since σ^3 is the identity map, we have $\sigma^3(u) = u$, whence

$$\omega_\sigma^3 = 1. \tag{13.6}$$

Arguing similarly with τ instead of σ, we find

$$\tau(u) = \omega_\tau u$$

with

$$\omega_\tau^3 = 1. \tag{13.7}$$

From these equations, we also deduce

$$\sigma \circ \tau(u) = \omega_\sigma \omega_\tau u \quad \text{and} \quad \sigma^2 \circ \tau(u) = \omega_\sigma^2 \omega_\tau u.$$

However, since

$$\sigma \circ \tau: \ x_1 \mapsto x_2 \mapsto x_3 \mapsto x_4 \mapsto x_5 \mapsto x_1; \quad x_i \mapsto x_i \quad \text{for } i > 5$$

and

$$\sigma^2 \circ \tau: \ x_1 \mapsto x_3 \mapsto x_4 \mapsto x_5 \mapsto x_2 \mapsto x_1; \quad x_i \mapsto x_i \quad \text{for } i > 5,$$

we have $(\sigma \circ \tau)^5 = (\sigma^2 \circ \tau)^5 = \text{Id}$ (the identity map), whence the arguments above yield

$$(\omega_\sigma \omega_\tau)^5 = (\omega_\sigma^2 \omega_\tau)^5 = 1. \tag{13.8}$$

Since

$$\omega_\sigma = \omega_\sigma^6 (\omega_\sigma \omega_\tau)^5 (\omega_\sigma^2 \omega_\tau)^{-5},$$

equations (13.6) and (13.8) yield

$$\omega_\sigma = 1.$$

From (13.8), we then deduce $\omega_\tau^5 = 1$, and since

$$\omega_\tau = \omega_\tau^6 \omega_\tau^{-5},$$

it follows from equation (13.7) that $\omega_\tau = 1$. This shows that u is invariant under σ and τ. $\qquad\square$

13.15. COROLLARY. *Let R be a radical extension of $\mathbb{C}(s_1, \ldots, s_n)$ contained in $\mathbb{C}(x_1, \ldots, x_n)$. If $n \geq 5$, then every element of R is invariant under the permutations σ and τ of Lemma 13.14.*

Proof. We argue by induction on the height of R, which we denote by h. If $h = 0$, then $R = \mathbb{C}(s_1, \ldots, s_n)$ and the corollary is obvious. If $h \geq 1$, then there is an element $u \in R$ and a radical extension R_1 of height $h - 1$ of $\mathbb{C}(s_1, \ldots, s_n)$ such that

$$R = R_1(u) \qquad \text{and} \qquad u^p \in R_1$$

for some prime number p. By induction, we may assume that every element of R_1 is invariant under σ and τ. The lemma then shows that u is also invariant under σ and τ, and, since the elements in R are rational expressions of u, it readily follows that every element in R is invariant under σ and τ. $\qquad\square$

We thus reach the conclusion:

13.16. THEOREM. *If $n \geq 5$, the general equation of degree n*

$$P(X) = (X - x_1) \cdots (X - x_n) = X^n - s_1 X^{n-1} + \cdots + (-1)^n s_n = 0$$

is not solvable by radicals over $\mathbb{Q}(s_1, \ldots, s_n)$, nor over $\mathbb{C}(s_1, \ldots, s_n)$.

Proof. According to Corollary 13.8, it suffices to show that $P(X) = 0$ is not solvable by radicals over $\mathbb{C}(s_1, \ldots, s_n)$. Assume on the contrary that there is a radical extension R of $\mathbb{C}(s_1, \ldots, s_n)$ containing a root x_i of P. Changing the numbering of $x_1, \ldots, x_n$ if necessary, we may assume that $i = 1$. Moreover, by the theorem of natural irrationalities (Theorem 13.13), this radical extension R may be assumed to lie within $\mathbb{C}(x_1, \ldots, x_n)$. Then, Corollary 13.15 shows that every element of R is invariant under σ and τ. But $x_1 \in R$ and x_1 is not invariant under σ. This is a contradiction. $\qquad\square$

Exercises

1. Show that over $\mathbb{R}$ and over $\mathbb{C}$, every equation of any degree is solvable by radicals.

2. Show that the general cubic equation

$$(X - x_1)(X - x_2)(X - x_3) = X^3 - s_1 X^2 + s_2 X - s_3 = 0$$

is solvable by radicals over $\mathbb{Q}(s_1, s_2, s_3)$. Construct explicitly a radical extension of $\mathbb{Q}(s_1, s_2, s_3)$ containing one of the roots of this cubic and show that this radical extension is not contained in $\mathbb{Q}(x_1, x_2, x_3)$. Thus, the solution of the general cubic equation by radicals over $\mathbb{Q}(s_1, s_2, s_3)$ involves accessory irrationalities.

Same questions for the general equation of degree four.

3. Let ζ_7 (resp. ζ_3) be a primitive 7-th (resp. cube) root of unity. Show that $\mathbb{Q}(\zeta_7)$ is not a radical extension of $\mathbb{Q}$, but that $\mathbb{Q}(\zeta_7, \zeta_3)$ is a radical extension of $\mathbb{Q}$.

4. Let R be a radical extension of a field F, of the form $R = F(a^{1/p})$, for some $a \in F$ which is not a p-th power in F. Find an isomorphism which is the identity on F

$$F[X]/(X^p - a) \xrightarrow{\sim} R.$$

Conclude that all the fields of the form $F(a^{1/p})$ are isomorphic, under isomorphisms leaving F invariant.

5. Show that there are three different subfields of $\mathbb{C}$ of the form $\mathbb{Q}(2^{1/3})$. Show that if F is a subfield of $\mathbb{C}$ containing a primitive p-th root of unity, then for any $a \in F$ which is not a p-th power in F there is only one subfield of $\mathbb{C}$ of the form $F(a^{1/p})$.

6. *To make up partially for the lack of details on the early stages of the theory of groups in Ruffini's and Cauchy's works, the following exercise presents a result of Cauchy on the number of values of rational fractions under permutations of the indeterminates, which was used in Abel's proof that general equations are not solvable by radicals.*

Let n be an integer, $n \geq 3$, let $\Delta = \Delta(x_1, \dots, x_n)$ be the polynomial defined in §8.3 and let $I(\Delta) \subset S_n$ be the isotropy group of Δ, i.e.

$$I(\Delta) = \{\sigma \in S_n \mid \sigma(\Delta) = \Delta\}.$$

(This subgroup of S_n is called the *alternating group* on $\{1, \dots, n\}$, and denoted A_n.)

    (a) Show that any permutation of n elements is a composition of permutations which interchange two elements and leave the other elements invariant. (Permutations of this type are called *transpositions*.)

(b) Show that a permutation leaves Δ invariant if and only if it is a composition of an even number of transpositions.

(c) Let p be an odd prime, $p \leq n$. Show that the cyclic permutations of length p

$$i_1 \mapsto i_2 \mapsto \cdots \mapsto i_p \mapsto i_1$$

(where $i_1, \ldots, i_p \in \{1, \ldots, n\}$) generate $I(\Delta)$.

[Hint: By (b), it suffices to show that the composition of any two transpositions is a composition of cycles of length p.]

(d) Let again p be an odd prime, $p \leq n$, and let V be a rational fraction in x_1, $\ldots$, x_n which takes strictly less than p values under the permutations of $x_1, \ldots, x_n$. Show that V has the form $V = R + \Delta S$ where R and S are symmetric rational fractions, hence that the number of values of V is 1 or 2.

[Hint: Show that V is invariant under the cyclic permutations of length p.]

(e) Translate the result above in the following purely group-theoretical terms: if $G \subset S_p$ is a subgroup of index $< p$ (with p prime), then G contains the alternating group A_p.

[Hint: Use Proposition 10.5, p. 146.]

Chapter 14

Galois

14.1 Introduction

After Gauss, Ruffini and Abel, the two major classes of equations have been treated thoroughly, with divergent results: the cyclotomic equations of any degree are solvable by radicals, while the general equations of degree at least five are not. Thus, the next obvious question arises: which are the equations that are solvable by radicals?

Abel himself addressed this question and returned several times to the theory of equations, which he called his "thème favori" [1, t. II, p. 260]. Following a clue from Gauss, he discovered a large class of solvable equations, which contains in particular the cyclotomic equations. In the introduction to the seventh chapter of "Disquisitiones Arithmeticae," in which he discusses cyclotomic equations, Gauss had written [24, Art. 335]:

> Moreover, the principles of the theory that we are about to explain extend much farther than we let it see here. Indeed, they apply not only to circular functions, but also with the same success to numerous other transcendental functions, e.g. to those which depend on the integral $\int \frac{dx}{\sqrt{1-x^4}}$ and also to various kinds of congruences.

The integral $\int \frac{dx}{\sqrt{1-x^4}}$ occurs in the calculation of the length of an arc of the lemniscate, as the integral $\int \frac{dx}{\sqrt{1-x^2}} = \sin^{-1} x$ occurs for the arc of the circle.

Following this clue, Abel realized that Gauss' method for cyclotomic equations could also be applied to the equations which arise from the division of the lemniscate. In complete analogy with Gauss' results on the constructibility of regular polygons with ruler and compass, Abel even proved that the lemniscate can be divided into $2^n + 1$ equal parts by ruler and compass, whenever $2^n + 1$ is a prime number (see [1, t. II, p. 261], Rosen [49]). Pushing his investigations further, Abel eventually came to the following grand generalization (published in 1829) [1, t. I, p. 479]:

THEOREM (ABEL). *Let P be a polynomial with roots $r_1, \dots, r_n$. If the roots $r_2, \dots, r_n$ can be rationally expressed in terms of r_1, i.e. if there exist rational fractions $\theta_2, \dots, \theta_n$ such that*

$$r_i = \theta_i(r_1) \qquad for\ i = 2, \dots, n$$

and if moreover

$$\theta_i \theta_j(r_1) = \theta_j \theta_i(r_1) \qquad for\ all\ i,\ j$$

then the equation $P(X) = 0$ is solvable by radicals.

This theorem applies in particular to cyclotomic equations

$$\Phi_p(X) = X^{p-1} + X^{p-2} + \cdots + X + 1 = 0,$$

for p prime. Indeed, the roots of Φ_p are the primitive p-th roots of unity and, denoting one of the roots by ζ, the other roots are powers of ζ. Rational fractions $\theta_2, \dots, \theta_{p-1}$ as above can thus be chosen to be

$$\theta_i(X) = X^i \qquad for\ i = 2, \dots, p-1.$$

The above condition obviously holds, since

$$\theta_i \theta_j(\zeta) = \zeta^{ij} = \theta_j \theta_i(\zeta).$$

Elaborating again on these results, Abel was closing in on general necessary and sufficient conditions for an equation to be solvable by radicals. He was working on a comprehensive memoir on this subject [1, t. II, n° 18], when he was prematurely carried off by tuberculosis in 1829.

The honor of finding a complete solution to the problem eventually fell to another young genius, Évariste Galois (1811–1832), who was only 18 in 1830 when he submitted to the Paris Academy of Sciences a memoir on the theory of equations. In this memoir, he described what is now known as the Galois group of an equation, and applied this new tool to derive conditions for an equation to be solvable by radicals. The referee was Jean-Baptiste Joseph Fourier (1768–1830), who died a few weeks later. Galois' memoir was then lost in Fourier's papers (see however Galois' collected papers [21, pp. 103–109]). The next year, a second memoir was submitted by Galois, but rejected by the Academy because it was not sufficiently developed. Galois died in a duel the following year, without having had the occasion to submit a more thorough (or rather, a less sketchy) exposition of his ideas. His "Mémoire sur les conditions de résolubilité des équations par radicaux" [21, pp. 43 ff] (see also Edwards [20, App. 1]) is indeed very terse and makes rather difficult reading. Fortunately, Joseph Liouville (1809–1882) generously took the trouble to decipher Galois' memoir, and he published it in 1846 with some explanations of his own, thus rescuing Galois theory from complete oblivion.

The basic idea of Galois is to associate to any* equation a group of permutations of the roots. This group consists of all the permutations which preserve the relations among the roots; it thus shows to what extent the roots are interchangeable. Galois' brilliant insight was that this group provides an effective measure of the difficulty of an equation. In particular, the solvability of the equation by radicals can be translated in terms of the associated group. This is achieved by describing the behavior of the group under extension of the base field. Of these fertile new ideas, Galois offers a single application, proving that irreducible equations of prime degree are solvable by radicals if and only if any of the roots can be rationally expressed in terms of two of them.

This brief summary of Galois' memoir does not do justice to the novelty of the ideas it contains. Indeed, it is not clear at all how to characterize the permutations which preserve the relations among the roots in this general context. (For the particular case of cyclotomic equations, see §§11.3 and 12.4.) This difficulty seems especially overwhelming if one avoids making use of the notion of field, which is the central notion in Galois theory, but which was not available at the time when Galois wrote his memoir. Galois solves the problem by using the irreducibility of polynomials with awesome virtuosity. The concept of field (and of extension of fields) becomes transparent in the first few lines of his memoir, where he emphasizes that in his discussion of irreducibility, the base field can be arbitrary:

> Definitions. An equation is said to be reducible if it admits
> rational divisors; otherwise it is irreducible.
>
> It is necessary to explain what is meant by the word rational,
> because it will appear frequently.

---

*almost any, in fact: see the beginning of §14.2.

When the equation has coefficients that are all numeric and rational, this means simply that the equation can be decomposed into factors which have coefficients that are numeric and rational.

But when the coefficients of an equation are not *all* numeric and rational, one must mean by a rational divisor a divisor whose coefficients can be expressed as rational functions of the coefficients of the proposed equation, and, more generally, by a rational quantity a quantity that can be expressed as a rational function of the coefficients of the proposed equation.

More than this: one can agree to regard as rational all rational functions of a certain number of determined quantities, supposed to be known *a priori*. For example, one can choose a particular root of a whole number and regard as rational every rational function of this radical.

When we agree to regard certain quantities as known in this manner, we shall say that we *adjoin* them to the equation to be resolved. We shall say that these quantities are *adjoined* to the equation.

With these conventions, we shall call *rational* any quantity which can be expressed as a rational function of the coefficients of the equation and of a certain number of *adjoined* quantities arbitrarily agreed upon. [...]

One sees, moreover, that the properties and the difficulties of an equation can be altogether different, depending on what quantities are adjoined to it. [20, pp. 101–102]

Our discussion of Galois' memoir follows Galois' own order of propositions. We thus begin with the definition of the Galois group of an equation, next investigate the behavior of the Galois group under extension of the base field, and deduce a necessary and sufficient condition for an equation to be solvable by radicals, in terms of its Galois group. In the final section, the application of this condition to irreducible equations of prime degree will be described. In an appendix, we review Galois' notation for groups of permutations, which deviates from the modern notation.

14.2   The Galois group of an equation

In this chapter, as in the preceding one, we consider only fields of characteristic zero, so that we are allowed to divide by non-zero integers unconcernedly. Another word of caution: we shall often have to substitute in a rational fraction $f \in F(x_1, \ldots, x_n)$ elements $a_1, \ldots, a_n$ in F for the indeterminates $x_1, \ldots, x_n$, yielding an element $f(a_1, \ldots, a_n) \in F$. Whenever this is done, it is implicitly assumed that the rational fraction f can be represented in the form $f = P/Q$ where P and Q are polynomials in $F[x_1, \ldots, x_n]$ such that $Q(a_1, \ldots, a_n) \neq 0$. We then set

$$f(a_1, \ldots, a_n) = \frac{P(a_1, \ldots, a_n)}{Q(a_1, \ldots, a_n)}.$$

For technical reasons which will be pointed out below (see Lemma 14.6), Galois associates a group to equations with simple roots only. This is not a serious restriction, since Hudde's method transforms any equation into an equation with simple roots, see Theorem 5.21 p. 54 and Remark 5.23, p. 55. For the rest of this section, we shall thus consider a monic polynomial $P(X)$ of degree n over a field F, which has n distinct roots in some field containing F (see Girard's theorem (Theorem 9.3, p. 116)),

$$P(X) = X^n - a_1 X^{n-1} + a_2 X^{n-2} - \cdots + (-1)^n a_n = (X - r_1) \cdots (X - r_n)$$

with $a_1, \ldots, a_n$ in F and $r_1, \ldots, r_n$ in some field containing F. Extending slightly the notation of Proposition 12.15, p. 179, we denote by $F(r_1, \ldots, r_n)$ the field of rational fractions in $r_1, \ldots, r_n$ with coefficients in F. Thus

$$F(r_1, \ldots, r_n) = \{ f(r_1, \ldots, r_n) \mid f \in F(x_1, \ldots, x_n) \}.$$

It is worth emphasizing that, since $r_1, \ldots, r_n$ are not independent indeterminates over F, an element in $F(r_1, \ldots, r_n)$ can be written in the form $f(r_1, \ldots, r_n)$ in more than one way. For instance, 0 can be written as $P(r_i)$ for any $i = 1, \ldots,$ n. This is very important in view of the fact that we shall consider permutations σ of $r_1, \ldots, r_n$; although $f(\sigma(r_1), \ldots, \sigma(r_n))$ is well-defined for any rational fraction $f \in F(x_1, \ldots, x_n)$ whose denominator does not vanish for $x_i = \sigma(r_i)$, defining $\sigma(f(r_1, \ldots, r_n))$ by

$$\sigma(f(r_1, \ldots, r_n)) = f(\sigma(r_1), \ldots, \sigma(r_n)) \tag{14.1}$$

requires caution, since it is not clear that the right-hand side depends on the value $f(r_1, \ldots, r_n)$ only, and not on the rational fraction $f(x_1, \ldots, x_n)$. More precisely, we have to check that if g is another rational fraction such that

$$g(r_1, \ldots, r_n) = f(r_1, \ldots, r_n),$$

then

$$g(\sigma(r_1), \ldots, \sigma(r_n)) = f(\sigma(r_1), \ldots, \sigma(r_n)).$$

If this is not the case, then equation (14.1) does not make sense.

The distinction between the *form* of an element $f(r_1, \ldots, r_n)$ (i.e. the rational fraction $f(x_1, \ldots, x_n)$) and its *value* (in $F(r_1, \ldots, r_n)$) is emphasized by Galois himself [21, p. 50], [20, p. 104]:

> Here we call a function invariant not only if its form is un-
> changed by the substitutions of the roots, but also if its numerical
> value does not vary when these substitutions are applied.

In order to define the Galois group of the equation $P(X) = 0$, some prelim-
inary results are needed. The proofs of these results will be given later, to avoid
interrupting by lengthy proofs the course of reasoning.

RESULT 1. *There is an element* $V \in F(r_1, \ldots, r_n)$ *such that*

$$r_i \in F(V) \qquad for\ i = 1, \ldots, n.$$

The proof will be given below, see Proposition 14.7, p. 245.

The elements V for which this condition holds are called *Galois resolvents* of
the equation $P(X) = 0$ over the field F. This terminology (which is of course
not due to Galois) stems from the observation that in order to solve the equation
$P(X) = 0$ it suffices to determine V, since the roots $r_1, \ldots, r_n$ of P are rational
fractions in V.

RESULT 2. *For every element* $u \in F(r_1, \ldots, r_n)$, *there is a unique monic irreducible polynomial* $\pi \in F[X]$ *such that* $\pi(u) = 0$. *This polynomial* π *splits into
a product of linear factors over* $F(r_1, \ldots, r_n)$.

The proof will be given in Proposition 14.8 below (p. 247).

The polynomial π is called the *minimum polynomial of* u *over* F. (Compare
Remark 12.16, p. 180.)

The Galois group of the equation $P(X) = 0$ over F can now be described as
follows: let V be a Galois resolvent, so that for $i = 1, \ldots, n$,

$$r_i = f_i(V)$$

for some rational fraction $f_i(X) \in F(X)$. Let $V_1, \ldots, V_m \in F(r_1, \ldots, r_n)$ be the roots of the minimum polynomial of V over F (with $V = V_1$, say).

RESULT 3. *For any $j = 1, \ldots, m$, the elements $f_1(V_j), f_2(V_j), \ldots, f_n(V_j)$ are the roots $r_1, \ldots, r_n$ of P, in some order.*

The proof will be given in Proposition 14.10 below (p. 249).

From this result, it follows that for $j = 1, \ldots, m$, the map

$$\sigma_j \colon r_i \mapsto f_i(V_j) \qquad \text{for } i = 1, \ldots, n$$

is a permutation of $r_1, \ldots, r_n$. The set $\{\sigma_1, \ldots, \sigma_m\}$ is called the *Galois group* of $P(X) = 0$ over F, and denoted by $\mathrm{Gal}(P/F)$. To justify this terminology, we shall prove

RESULT 4. *The set $\mathrm{Gal}(P/F)$ is a subgroup of the group of all permutations of $r_1, \ldots, r_n$. It does not depend on the choice of the Galois resolvent V.*

The proof will be given in Corollaries 14.13 and 14.14 below (pp. 252 and 253).

It is noteworthy that the order of the Galois group $\mathrm{Gal}(P/F)$, which is denoted above by m, is equal to the degree of the minimum polynomial π of a Galois resolvent V. Without further assumption on P, this order is not related in any way whatsoever to the degree n of P (see however Exercise 1).

In the course of proving Result 4 (which is not to be found explicitly in Galois' memoir), we shall establish the following major property of the Galois group, which is Proposition 1 in Galois' memoir [21, p. 51], [20, p. 104]:

RESULT 5. *Let $f(x_1, \ldots, x_n)$ be a rational fraction in n indeterminates $x_1, \ldots, x_n$ with coefficients in F. Then*

$$f(r_1, \ldots, r_n) \in F$$

if and only if for all $\sigma \in \mathrm{Gal}(P/F)$,

$$f(r_1, \ldots, r_n) = f\bigl(\sigma(r_1), \ldots, \sigma(r_n)\bigr).$$

The proof will be given in Theorem 14.11 below (p. 250).

This result will enable us to prove moreover that the equation

$$\sigma\bigl(f(r_1, \ldots, r_n)\bigr) = f\bigl(\sigma(r_1), \ldots, \sigma(r_n)\bigr)$$

(see equation (14.1)) makes sense for $\sigma \in \text{Gal}(P/F)$ and defines an extension of σ to an automorphism of $F(r_1, \ldots, r_n)$ which leaves every element in F invariant.

To illustrate the steps which lead to the construction of the Galois group of an equation, we consider the following easy example: let

$$P(X) = (X - 1)(X^2 - 2)(X^2 - 3)$$
$$= (X - 1)(X - \sqrt{2})(X + \sqrt{2})(X - \sqrt{3})(X + \sqrt{3}).$$

We denote the roots of P as follows:

$$r_1 = 1, \quad r_2 = \sqrt{2}, \quad r_3 = -\sqrt{2}, \quad r_4 = \sqrt{3}, \quad r_5 = -\sqrt{3}.$$

In order to determine the Galois group of $P(X) = 0$ over the field $\mathbb{Q}$ of rational numbers, we first choose a Galois resolvent. We claim that the element

$$V = \sqrt{2} + \sqrt{3} = r_2 + r_4$$

satisfies the required conditions. To prove it, we square both sides of the equation $V - r_2 = r_4$, which yields

$$V^2 - 2r_2 V + 2 = 3. \tag{14.2}$$

We then obtain r_2, whence also r_3 since $r_3 = -r_2$, as rational expressions of V, namely

$$r_2 = \frac{V^2 - 1}{2V}, \qquad r_3 = \frac{1 - V^2}{2V}.$$

Similarly, from $V - r_4 = r_2$ we get

$$r_4 = \frac{V^2 + 1}{2V}, \qquad r_5 = -\frac{V^2 + 1}{2V}.$$

Since $r_1 = 1$ is a rational expression of r_1 in V (trivially), this shows that every root of P has a rational expression in V, hence V is a Galois resolvent of $P(X) = 0$ over $\mathbb{Q}$, as claimed.

The next step is to find the minimum polynomial of V over $\mathbb{Q}$. From equation (14.2), a rational equation in V can be obtained by isolating on one side the term containing r_2 and squaring both sides. We thus get

$$V^4 - 10V^2 + 1 = 0,$$

hence V is a root of the polynomial $X^4 - 10X^2 + 1 \in \mathbb{Q}[X]$. It is easy to check that this polynomial factors as

$$X^4 - 10X^2 + 1 =$$
$$\left(X - (\sqrt{2} + \sqrt{3})\right)\left(X - (\sqrt{2} - \sqrt{3})\right)\left(X - (-\sqrt{2} + \sqrt{3})\right)\left(X - (-\sqrt{2} - \sqrt{3})\right)$$

and that the factors on the right-hand side cannot be combined to yield a non-trivial divisor of $X^4 - 10X^2 + 1$ with rational coefficients. Therefore, $X^4 - 10X^2 + 1$ is irreducible, hence it is the minimum polynomial of V over $\mathbb{Q}$.

At the same time, we have found the roots of this polynomial, namely

$$V_1 = V = \sqrt{2} + \sqrt{3}, \qquad V_2 = \sqrt{2} - \sqrt{3},$$
$$V_3 = -\sqrt{2} + \sqrt{3}, \qquad V_4 = -\sqrt{2} - \sqrt{3}.$$

The determination of the Galois group of $P(X) = 0$ is now only a matter of straightforward calculations: in the rational fractions $f_i(X)$ which are such that $r_i = f_i(V)$ for $i = 1, \dots, 5$, i.e.

$$f_1(X) = 1, \qquad f_2(X) = \frac{X^2 - 1}{2X}, \qquad f_3(X) = -f_2(X),$$
$$f_4(X) = \frac{X^2 + 1}{2X}, \qquad f_5(X) = -f_4(X),$$

we substitute successively V_1, V_2, V_3, V_4 for X and we obtain the elements of $\mathrm{Gal}(P/\mathbb{Q})$ as

$$\sigma_j \colon r_i \mapsto f_i(V_j) \qquad \text{for } j = 1, \dots, 4.$$

Explicitly,

$$\sigma_1 = \mathrm{Id} \quad \text{(the identity)}$$

$$\sigma_2 \colon \begin{cases} r_1 \mapsto r_1 \\ r_2 \mapsto r_2 \\ r_3 \mapsto r_3 \\ r_4 \mapsto r_5 \\ r_5 \mapsto r_4 \end{cases} \quad \sigma_3 \colon \begin{cases} r_1 \mapsto r_1 \\ r_2 \mapsto r_3 \\ r_3 \mapsto r_2 \\ r_4 \mapsto r_4 \\ r_5 \mapsto r_5 \end{cases} \quad \sigma_4 \colon \begin{cases} r_1 \mapsto r_1 \\ r_2 \mapsto r_3 \\ r_3 \mapsto r_2 \\ r_4 \mapsto r_5 \\ r_5 \mapsto r_4. \end{cases}$$

Thus, the Galois group of $P(X) = 0$ over $\mathbb{Q}$ consists of the permutations of r_1, $\dots$, r_5 which leave r_1 invariant and which either leave invariant or interchange r_2 and r_3 on one side and r_4 and r_5 on the other side.

This was predictable from the heuristic point of view that the permutations in the Galois group are the permutations which preserve the relations among the roots. Indeed, the roots $\sqrt{2}$ and $-\sqrt{2}$ play exactly the same role with respect to rational numbers, there is no way to distinguish one from the other with the aid of rational numbers. They can therefore be interchanged by the Galois group. Similar arguments hold for $\sqrt{3}$ and $-\sqrt{3}$, but the roots of the various factors $X - 1$, $X^2 - 2$ and $X^2 - 3$ of P cannot be interchanged, since for instance r_2 satisfies $r_2^2 - 2 = 0$ whereas r_4 does not. The permutations σ_1, σ_2, σ_3, σ_4 above are therefore the only permutations which preserve the relations among the roots.

With hindsight, it appears that the most tricky points in the determination of the Galois group of an equation are

    (a) to find the roots of the given equation,

    (b) to find a Galois resolvent,

    (c) to determine its minimum polynomial,

    (d) to find the roots of the minimum polynomial.

In fact, point (b) is not too much of a problem, since the proof of the existence of a Galois resolvent (which will be given in Proposition 14.7 below, p. 245) is sufficiently explicit to provide a method to find one. Likewise, the proof of the existence of the minimum polynomial (see Proposition 14.8 below, p. 247) yields a polynomial of which the Galois resolvent is a root. It thus "suffices" to find an irreducible factor of this polynomial which has the Galois resolvent as a root. This could be a formidable task, however. Similarly, to find the roots of the given equation and of the minimum polynomial explicitly enough so that the subsequent calculations could be performed can prove to be a daunting problem.

Of course, Galois was well aware of these problems:

> If you now give me an equation that you have chosen at your pleasure, and if you want to know if it is or is not solvable by radicals, I could do no more than to indicate to you the means of answering your question, without wanting to give myself or anyone else the task of doing it. In a word, the calculations are impracticable.
>
> From that, it would seem that there is no fruit to derive from the solution that we propose. Indeed, it would be so if the question usually arose from this point of view. But, most of the time, in the applications of the Algebraic Analysis, one is led to equations of which one knows beforehand all the properties: prop-

erties by means of which it will always be easy to answer the question by the rules we are going to explain. [...] All, that makes this theory beautiful and at the same time difficult, is that one has always to indicate the course of analysis and to foresee its results without ever being able to perform [the calculations]. [21, pp. 39–40]

These last remarks will be clear from the following examples.

14.1. Example. The Galois group of the general equation of degree n

$$P(X) = X^n - s_1 X^{n-1} + \cdots + (-1)^n s_n = (X - x_1) \cdots (X - x_n) = 0$$

over the field F of rational fractions in $s_1, \ldots, s_n$ (over some field of constants k) is the group of all permutations of $x_1, \ldots, x_n$. It can thus be identified with the full symmetric group S_n.

Indeed, if we assume by way of contradiction that $\mathrm{Gal}(P/F)$ is not the group of all permutations of $x_1, \ldots, x_n$, then by Proposition 10.5 (p. 146) we can find a rational fraction $f(x_1, \ldots, x_n) \in k(x_1, \ldots, x_n) \ (= F(x_1, \ldots, x_n))$ which is not symmetric (i.e. not in F) but such that

$$f(\sigma(x_1), \ldots, \sigma(x_n)) = f(x_1, \ldots, x_n) \quad \text{for all } \sigma \in \mathrm{Gal}(P/F).$$

This contradicts Result 5 above.

14.2. Example. The Galois group over $\mathbb{Q}$ of the cyclotomic equation of prime index

$$\Phi_p(X) = X^{p-1} + X^{p-2} + \cdots + X + 1 = 0$$

(with p prime) is a cyclic group of order $p - 1$.

To prove this, we retrace the steps in the determination of the Galois group. Let ζ be any primitive p-th root of unity, i.e. any root of $\Phi_p(X)$. Since the other roots of $\Phi_p(X)$ are powers of ζ, we can choose ζ itself as a Galois resolvent of $\Phi_p(X)$. Since $\Phi_p(X)$ is irreducible, the minimum polynomial of ζ is $\Phi_p(X)$. Choosing a primitive root g of p, we denote

$$\zeta_i = \zeta^{g^i} \qquad \text{for } i = 0, \ldots, p - 2,$$

as in §12.4. Thus, the roots of $\Phi_p(X)$ are $\zeta_0, \ldots, \zeta_{p-2}$, and the rational fractions f_i are now

$$f_i(X) = X^{g^i}.$$

According to the definition (p. 237), the elements of $\text{Gal}(\Phi_p/\mathbb{Q})$ are

$$\sigma_j\colon \zeta_i \mapsto f_i(\zeta_j) \qquad \text{for } j = 0, \ldots, p-2.$$

Since

$$f_i(\zeta_j) = (\zeta^{g^j})^{g^i} = \zeta^{g^{i+j}},$$

and since, by Fermat's theorem (Theorem 12.5, p. 171), $g^{p-1} \equiv g^0 \bmod p$, it follows that the above description of σ_j can be simplified to

$$\sigma_j\colon \zeta_i \mapsto \zeta_{i+j},$$

where the subscript $i+j$ is taken modulo $p-1$ (i.e. replaced by the integer between 0 and $p-2$ congruent to $i+j$, if $i+j \geq p-1$).

Therefore,

$$\sigma_j = \sigma_1^j \qquad \text{for } j = 0, \ldots, p-2,$$

hence $\text{Gal}(\Phi_p/\mathbb{Q})$ is generated by the single element σ_1 (which was denoted σ in §12.4). It is thus a cyclic group of order $p - 1$.

14.3. Example. Let P be a polynomial with simple roots $r_1, \ldots, r_n$ for which Abel's condition (in the theorem quoted in §14.1) holds, i.e. there are rational fractions $\theta_i(X) \in F(X)$ such that

$$r_i = \theta_i(r_1) \qquad \text{for } i = 2, \ldots, n$$

and

$$\theta_i \theta_j(r_1) = \theta_j \theta_i(r_1) \qquad \text{for all } i, j.$$

Then the Galois group of the equation $P(X) = 0$ over F is commutative. (This is why commutative groups are often called *abelian* groups.)

Indeed, in this case one can choose r_1 as Galois resolvent. Since r_1 is a root of P, its minimum polynomial divides P, by Lemma 12.14, p. 178, hence the roots of the minimum polynomial of r_1 are among $r_1, \ldots, r_n$. Changing the numbering if necessary, we may assume that these roots are $r_1, \ldots, r_m$ (with $m \leq n$). According to the definition of the Galois group (p. 237), the elements of $\text{Gal}(P/F)$ are $\sigma_1, \ldots, \sigma_m$ where

$$\sigma_j\colon r_i \mapsto \theta_i(r_j) \qquad \text{for } i = 1, \ldots, n \text{ and } j = 1, \ldots, m.$$

Since

$$\theta_i(r_j) = \theta_i\theta_j(r_1)$$

and since Abel's condition holds, we have

$$\theta_i(r_j) = \theta_j\theta_i(r_1) = \theta_j(r_i),$$

hence

$$\sigma_j \colon r_i \mapsto \theta_j(r_i) \qquad \text{for } i = 1, \ldots, n \text{ and } j = 1, \ldots, m.$$

It then follows that, for all j, k between 1 and m,

$$\sigma_j \circ \sigma_k \colon r_i \mapsto \theta_j\theta_k(r_i) = \theta_j\theta_k\theta_i(r_1)$$

and

$$\sigma_k \circ \sigma_j \colon r_i \mapsto \theta_k\theta_j(r_i) = \theta_k\theta_j\theta_i(r_1).$$

Therefore, commutativity of $\mathrm{Gal}(P/F)$ readily follows from Abel's condition.

We now turn to the proofs of the results quoted above. First, we prove the following easy elaboration on Lemma 12.14 (p. 178), which will be repeatedly used in the sequel:

14.4. LEMMA. *Let $f \in F(X)$ be a rational fraction in one indeterminate over a field F and let V be a root of some irreducible polynomial $\pi \in F[X]$ (in some field containing F). If $f(V) = 0$, then $f(W) = 0$ for every root W of π.*

Proof. Let $f = P/Q$ for some polynomials P, $Q \in F[X]$ such that $Q(V) \neq 0$ and $P(V) = 0$. By Lemma 12.14 (p. 178), this last equation implies that π divides P, hence $P(W) = 0$ for every root W of π. On the other hand, if $Q(W) = 0$ for any root W of π, then the same argument shows that $Q(V) = 0$, a contradiction. Therefore, $P(W) = 0$ and $Q(W) \neq 0$, hence $f(W) = 0$. $\square$

(Compare Lemma 1 in Galois' memoir [21, p. 47], [20, p. 102].)

14.5. LEMMA. *Let g be a polynomial in n indeterminates $x_1, \ldots, x_n$ over some field K. If g is invariant under every permutation of $x_2, \ldots, x_n$, then it can be written as a polynomial in x_1 and the elementary symmetric polynomials $s_1, \ldots, s_{n-1}$ in $x_1, \ldots, x_n$.*

Proof. We consider g as a polynomial in $x_2, \ldots, x_n$ with coefficients in $K[x_1]$. From Theorem 8.4 (p. 99) and Remark 8.8(b) (p. 105) it follows that g can be written as a polynomial in the elementary symmetric polynomials $s_1', \ldots, s_{n-1}'$ in $x_2, \ldots, x_n$, with coefficients in $K[x_1]$. Therefore, there exists a polynomial g' such that

$$g(x_1, \ldots, x_n) = g'(x_1, s_1', \ldots, s_{n-1}'), \tag{14.3}$$

where

$$s_1' = x_2 + \cdots + x_n, \quad s_2' = x_2 x_3 + \cdots + x_{n-1} x_n, \quad \cdots,$$
$$s_{n-1}' = x_2 x_3 \cdots x_n.$$

To complete the proof, it now suffices to observe that one can substitute for s_1', $\ldots, s_{n-1}'$ polynomials in x_1 and $s_1, \ldots, s_{n-1}$. A simple way to obtain explicit formulas for $s_1', \ldots, s_{n-1}'$ is to divide by $X - x_1$ the general polynomial

$$(X - x_1) \cdots (X - x_n) = X^n - s_1 X^{n-1} + \cdots + (-1)^n s_n$$

and to identify the result with

$$(X - x_2) \cdots (X - x_n) = X^{n-1} - s_1' X^{n-2} + \cdots + (-1)^{n-1} s_{n-1}'.$$

We thus get

$$s_1' = s_1 - x_1,$$
$$s_2' = s_2 - s_1 x_1 + x_1^2,$$
$$s_3' = s_3 - s_2 x_1 + s_1 x_1^2 - x_1^3,$$
$$\cdots$$
$$s_{n-1}' = s_{n-1} - s_{n-2} x_1 + \cdots + (-1)^{n-1} x_1^{n-1},$$

hence, substituting for $s_1', \ldots, s_{n-1}'$ in equation (14.3) we obtain

$$g(x_1, \ldots, x_n) = g'(x_1, s_1 - x_1, \ldots, s_{n-1} - s_{n-2} x_1 + \cdots + (-1)^{n-1} x_1^{n-1}),$$

and the right-hand side is a polynomial in x_1 and $s_1, s_2, \ldots, s_{n-1}$. $\square$

From this point on, we make use of the notation set at the beginning of this section. Thus, P is a polynomial of degree n over some field F, with distinct roots $r_1, \ldots, r_n$ in some field containing F,

$$P(X) = X^n - a_1 X^{n-1} + \cdots + (-1)^n a_n = (X - r_1) \cdots (X - r_n).$$

14.6 LEMMA. *There is a polynomial* $f \in F[x_1, \ldots, x_n]$ *such that the various elements in* $F(r_1, \ldots, r_n)$ *obtained from* f *by substituting* $r_1, \ldots, r_n$ *for the indeterminates* $x_1, \ldots, x_n$ *in all* $n!$ *possible ways are all pairwise distinct.*

Proof. Let $L(x_1, \ldots, x_n) = A_1 x_1 + \cdots + A_n x_n$, where $A_1, \ldots, A_n$ are indeterminates. The equality between two values of L obtained by substituting $r_1, \ldots, r_n$ for $x_1, \ldots, x_n$ in some ways is a linear equation in $A_1, \ldots, A_n$ (with coefficients in $F(r_1, \ldots, r_n)$). Writing down all the possible equalities yields a finite number $(\binom{n!}{2}$, in fact) of homogeneous linear equations in $A_1, \ldots, A_n$, none of which is trivial, since $r_1, \ldots, r_n$ are pairwise distinct. The solutions of these equations in F^n form a finite union of proper (vector-) subspaces of F^n. Now, since F is infinite (as its characteristic is assumed to be zero), F^n is not a union of a finite number of proper subspaces, hence we can find a n-tuple $(\alpha_1, \ldots, \alpha_n)$ in F^n for which none of the equations in $A_1, \ldots, A_n$ holds. The resulting polynomial

$$f(x_1, \ldots, x_n) = \alpha_1 x_1 + \cdots + \alpha_n x_n$$

satisfies the condition of the lemma. $\square$

This lemma is Lemma 2 in Galois' memoir [21, p. 47], [20, p. 102]. It obviously does not hold if multiple roots are allowed, i.e. if $r_1, \ldots, r_n$ are not pairwise distinct. We can now prove Result 1 (p. 236), which asserts the existence of Galois resolvents:

14.7 PROPOSITION. *There is an element* $V \in F(r_1, \ldots, r_n)$ *such that*

$$r_i \in F(V) \quad \text{for } i = 1, \ldots, n.$$

Proof. Let $f \in F[x_1, \ldots, x_n]$ be a polynomial as in Lemma 14.6, and let

$$V = f(r_1, \ldots, r_n) \in F(r_1, \ldots, r_n).$$

We are going to show that $r_1, \ldots, r_n$ are in $F(V)$. It is of course sufficient to spell out the arguments for one of the roots $r_1, \ldots, r_n$, say for r_1, since the same proof applies to any of them, by a simple change of numbering.

We consider the polynomial

$$g(x_1, \ldots, x_n) = \prod_{\sigma} \Big(V - f\big(x_1, \sigma(x_2), \ldots, \sigma(x_n)\big) \Big) \in F(V)[x_1, \ldots, x_n],$$

where σ runs over all the permutations of $x_2, \ldots, x_n$. Since g is symmetric in $x_2, \ldots, x_n$, Lemma 14.5 shows that g can be written as a polynomial in x_1 and the elementary symmetric polynomials $s_1, \ldots, s_{n-1}$ in $x_1, \ldots, x_n$. Let

$$g(x_1, x_2, \ldots, x_n) = h(x_1, s_1, \ldots, s_{n-1})$$

for some polynomial h with coefficients in $F(V)$. Therefore, substituting in various ways the roots $r_1, \ldots, r_n$ of P for the indeterminates $x_1, \ldots, x_n$, which has the effect of substituting $a_1, \ldots, a_{n-1} \in F$ for $s_1, \ldots, s_{n-1}$, we obtain

$$g(r_1, r_2, \ldots, r_n) = h(r_1, a_1, \ldots, a_{n-1}) \qquad (14.4)$$

and

$$g(r_i, r_1, r_2, \ldots, r_{i-1}, r_{i+1}, \ldots, r_n) = h(r_i, a_1, \ldots, a_{n-1}). \qquad (14.5)$$

Now, since f satisfies the property in Lemma 14.6 and $V = f(r_1, \ldots, r_n)$, we have

$$V \neq f(r_i, \sigma(r_1), \sigma(r_2), \ldots, \sigma(r_{i-1}), \sigma(r_{i+1}), \ldots, \sigma(r_n))$$

for $i \neq 1$ and for any permutation σ of $\{r_1, \ldots, r_{i-1}, r_{i+1}, \ldots, r_n\}$. Therefore,

$$g(r_i, r_1, r_2, \ldots, r_{i-1}, r_{i+1}, \ldots, r_n) \neq 0 \qquad \text{for } i \neq 1.$$

On the other hand, the definitions of g and V readily show that

$$g(r_1, \ldots, r_n) = 0.$$

In view of equations (14.4) and (14.5), these last relations show that the polynomial

$$h(X, a_1, \ldots, a_{n-1}) \in F(V)[X]$$

vanishes for $X = r_1$ but not for $X = r_i$ with $i \neq 1$. Therefore, it is divisible by $X - r_1$ but not by $X - r_i$ for $i \neq 1$.

We then consider the monic greatest common divisor $D(X)$ of $P(X)$ and $h(X, a_1, \ldots, a_{n-1})$ in $F(V)[X]$. Since in $F(r_1, \ldots, r_n)[X]$,

$$P(X) = (X - r_1) \cdots (X - r_n),$$

it follows that D splits over $F(r_1, \ldots, r_n)$ in a product of factors $X - r_i$. Since $X - r_1$ divides both $P(X)$ and $h(X, a_1, \ldots, a_{n-1})$, it divides D. On the other hand, $h(X, a_1, \ldots, a_{n-1})$ is not divisible by $X - r_i$ for $i \neq 1$, hence D has no other factor than $X - r_1$. Thus, $D = X - r_1$ whence $r_1 \in F(V)$ since $D \in F(V)[X]$. □

This proposition is Lemma 3 in Galois' memoir [21, p. 49], [20, p. 103]. We now turn to the proof of Result 2 (p. 236), about the existence of minimum polynomials:

14.8. PROPOSITION.   (a) *Every element* $u \in F(r_1, \ldots, r_n)$ *has a polynomial expression in* $r_1, \ldots, r_n$, *namely*

$$u = \varphi(r_1, \ldots, r_n)$$

for some polynomial $\varphi \in F[x_1, \ldots, x_n]$.
(b) *For every element* $u \in F(r_1, \ldots, r_n)$, *there is a unique monic irreducible polynomial* $\pi \in F[X]$ *such that* $\pi(u) = 0$. *This polynomial* π *splits into a product of linear factors over* $F(r_1, \ldots, r_n)$.

Proof. The proofs of these two results are intertwined: we first establish *(b)* for those elements u which have a polynomial expression in $r_1, \ldots, r_n$ and then deduce *(a)*. The proof of *(b)* will then be complete.

Step 1: proof of (b) for the elements which have a polynomial expression in r_1, $\ldots, r_n$. Let $u \in F(r_1, \ldots, r_n)$ be such that

$$u = \varphi(r_1, \ldots, r_n)$$

for some *polynomial* $\varphi \in F[x_1, \ldots, x_n]$. According to the observations before Proposition 12.15 (p. 179) and Remark 12.16 (p. 180), it suffices to show that u is a root of some polynomial with coefficients in F which splits into a product of linear factors over $F(r_1, \ldots, r_n)$. Let

$$\Theta(X, x_1, \ldots, x_n) = \prod_{\sigma} \left(X - \varphi(\sigma(x_1), \ldots, \sigma(x_n)) \right)$$

where σ runs over the set of all permutations of $x_1, \ldots, x_n$. Since Θ is symmetric in $x_1, \ldots, x_n$, we can write Θ as a polynomial in X and the elementary symmetric polynomials $s_1, \ldots, s_n$ in $x_1, \ldots, x_n$, by Theorem 8.4 (p. 99) and Remark 8.8(b) (p. 105). Let

$$\Theta(X, x_1, \ldots, x_n) = \Psi(X, s_1, \ldots, s_n)$$

for some polynomial Ψ with coefficients in F. Substituting $r_1, \ldots, r_n$ for the indeterminates $x_1, \ldots, x_n$, we obtain

$$\Theta(X, r_1, \ldots, r_n) = \Psi(X, a_1, \ldots, a_n) \in F[X].$$

Since, by definition of Θ,

$$\Theta(u, r_1, \ldots, r_n) = 0,$$

it follows that $\Psi(X, a_1, \ldots, a_n)$ is a polynomial in $F[X]$ which has u as a root. Moreover, since $\Theta(X, r_1, \ldots, r_n)$ is a product of linear factors, it follows that $\Psi(X, a_1, \ldots, a_n)$ splits into a product of linear factors over $F(r_1, \ldots, r_n)$.

(The point of taking for φ a polynomial is that $\Theta(X, x_1, \ldots, x_n)$ is then a polynomial, hence $\Theta(X, r_1, \ldots, r_n)$ is defined. Otherwise, it would not be clear that no denominator vanish when $r_1, \ldots, r_n$ are substituted for $x_1, \ldots, x_n$.)

Step 2: proof of (a). Let $V \in F(r_1, \ldots, r_n)$ be defined as in the proof of Proposition 14.7. Since $r_1, \ldots, r_n$ have been shown to be rational fractions in V, it follows that u also is a rational fraction in V, so $u \in F(V)$. Since V has a polynomial expression in $r_1, \ldots, r_n$, we can apply step 1 and Proposition 12.15 (p. 179) to derive that u can be expressed as a polynomial in V. Let

$$u = Q(V)$$

for some polynomial $Q \in F[X]$. Substituting $f(r_1, \ldots, r_n)$ for V, we obtain

$$u = Q(f(r_1, \ldots, r_n)).$$

This is a polynomial expression in $r_1, \ldots, r_n$, since Q and f are polynomials. $\square$

14.9. COROLLARY. *Let V be any Galois resolvent of $P(X) = 0$ over F and let $V_1, \ldots, V_m$ be the roots of its minimum polynomial over F (among which V lies). Then*

$$F(r_1, \ldots, r_n) = F(V) = F(V_1, \ldots, V_m).$$

Proof. Since $r_1, \ldots, r_n$ are rational fractions in V, we have

$$F(r_1, \ldots, r_n) \subset F(V).$$

On the other hand, by the preceding proposition, the roots $V_1, \ldots, V_m$ of the minimum polynomial of V are in $F(r_1, \ldots, r_n)$, hence

$$F(V_1, \ldots, V_m) \subset F(r_1, \ldots, r_n).$$

Since the inclusion

$$F(V) \subset F(V_1, \ldots, V_m)$$

is obvious (as V lies among $V_1, \ldots, V_m$), the three inclusions above yield

$$F(r_1, \ldots, r_n) = F(V) = F(V_1, \ldots, V_m).$$

$\square$

We now come to the proof of Result 3 (p. 237), which is Lemma 4 in Galois' memoir [21, p. 49], [20, p. 104]. We let V be a Galois resolvent of $P(X) = 0$ and we let

$$r_i = f_i(V) \qquad \text{for } i = 1, \ldots, n,$$

for some rational fraction $f_i(X) \in F(X)$. We denote by $V_1 = V, V_2, \ldots, V_m$ the roots of the minimum polynomial of V over F, which are in $F(r_1, \ldots, r_n)$, by Result 2.

14.10. PROPOSITION. *For $i = 1, \ldots, n$ and $j = 1, \ldots, m$, the element $f_i(V_j)$ is a root of P. Moreover, for any given $j = 1, \ldots, m$, the roots $f_1(V_j), \ldots, f_n(V_j)$ are pairwise distinct, so that*

$$\{f_1(V_j), \ldots, f_n(V_j)\} = \{r_1, \ldots, r_n\}.$$

Proof. Since $f_i(V_1) = r_i$ for $i = 1, \ldots, n$, we have $P\big(f_i(V_1)\big) = 0$. Therefore, by Lemma 14.4, applied to the rational fraction $P\big(f_i(X)\big) \in F(X)$, it follows that

$$P(f_i(V_j)) = 0 \qquad \text{for } j = 1, \ldots, m.$$

(From this argument, it follows at the same time that $f_i(V_j)$ is defined.)
Moreover, if for some $i, k = 1, \ldots, n$ and some $j = 1, \ldots, m$,

$$f_i(V_j) = f_k(V_j),$$

then V_j is a root of the rational fraction $f_i - f_k$, hence by Lemma 14.4 again,

$$f_i(V_1) = f_k(V_1).$$

This shows that $r_i = r_k$, whence $i = k$ since the roots $r_1, \ldots, r_n$ are assumed to be pairwise distinct. $\square$

This proposition shows that for all $j = 1, \ldots, m$, the maps

$$\sigma_j \colon r_i = f_i(V_1) \mapsto f_i(V_j) \qquad \text{for } i = 1, \ldots, n$$

are permutations of $r_1, \ldots, r_n$. We set

$$\text{Gal}(P/F) = \{\sigma_1, \ldots, \sigma_m\}$$

(although it is not yet clear at this stage that this set does not depend on the choice of the Galois resolvent V), and we prove the following major property of $\text{Gal}(P/F)$, which was announced as Result 5 on p. 237:

14.11. THEOREM. *Let $f(x_1, \ldots, x_n)$ be a rational fraction in n indeterminates $x_1, \ldots, x_n$, with coefficients in F. For $\sigma \in \text{Gal}(P/F)$, the element*

$$f(\sigma(r_1), \ldots, \sigma(r_n)) \in F(r_1, \ldots, r_n)$$

is defined whenever $f(r_1, \ldots, r_n)$ is defined. Moreover,

$$f(r_1, \ldots, r_n) \in F$$

if and only if

$$f(\sigma(r_1), \ldots, \sigma(r_n)) = f(r_1, \ldots, r_n)$$

for all $\sigma \in \text{Gal}(P/F)$.

Proof. Let $f = \varphi/\psi$, where $\varphi, \psi \in F[x_1, \ldots, x_n]$. We have to prove first that if $\psi(r_1, \ldots, r_n) \neq 0$, then $\psi(\sigma(r_1), \ldots, \sigma(r_n)) \neq 0$ for every $\sigma \in \text{Gal}(P/F)$. Substituting for $r_1, \ldots, r_n$ their rational expression in V, we get

$$\psi(r_1, \ldots, r_n) = \psi(f_1(V), \ldots, f_n(V)) = g(V)$$

for some rational fraction $g \in F(X)$.

Let now σ be a permutation in $\text{Gal}(P/F)$. If V' is the root of π such that

$$\sigma: r_i \mapsto f_i(V'),$$

then

$$\psi(\sigma(r_1), \ldots, \sigma(r_n)) = \psi(f_1(V'), \ldots, f_n(V')) = g(V').$$

Therefore, if $\psi(\sigma(r_1), \ldots, \sigma(r_n)) = 0$, then V' is a root of g and by Lemma 14.4 it follows that $g(V) = 0$, whence $\psi(r_1, \ldots, r_n) = 0$.

This shows that $f(\sigma(r_1), \ldots, \sigma(r_n))$ is defined when $f(r_1, \ldots, r_n)$ is defined. To prove the rest, we substitute for $r_1, \ldots, r_n$ their rational expression in V in $f(r_1, \ldots, r_n)$, obtaining

$$f(r_1, \ldots, r_n) = f(f_1(V), \ldots, f_n(V)) = h(V)$$

where

$$h(X) = f\big(f_1(X), \ldots, f_n(X)\big) \in F(X).$$

If $f(r_1, \ldots, r_n) \in F$, then

$$h(X) - f(r_1, \ldots, r_n) \in F(X).$$

Since this rational fraction vanishes for $X = V$, it also vanishes for $X = V_1, \ldots, V_m$, by Lemma 14.4. Therefore,

$$h(V_j) = f\big(f_1(V_j), \ldots, f_n(V_j)\big) = f(r_1, \ldots, r_n) \qquad \text{for } j = 1, \ldots, m$$

whence, using the definition of σ_j,

$$f\big(\sigma_j(r_1), \ldots, \sigma_j(r_n)\big) = f(r_1, \ldots, r_n) \qquad \text{for } j = 1, \ldots, m.$$

Conversely, if this last equation holds, then

$$f(r_1, \ldots, r_n) = h(V_j), \qquad \text{for } j = 1, \ldots, m$$

hence

$$f(r_1, \ldots, r_n) = \frac{1}{m}\big(h(V_1) + \cdots + h(V_m)\big). \qquad (14.6)$$

Since the rational fraction $h(x_1) + \cdots + h(x_m)$ is clearly symmetric in the indeterminates $x_1, \ldots, x_m$, it can be expressed as a rational fraction in the elementary symmetric polynomials $s_1, \ldots, s_m$. Therefore, substituting $V_1, \ldots, V_m$ for $x_1, \ldots, x_m$, it follows that the right-hand side of (14.6) can be rationally calculated from the coefficients of the polynomial π which has as roots $V_1, \ldots, V_m$, and is thus an element in F. Hence, equation (14.6) shows that

$$f(r_1, \ldots, r_n) \in F.$$

$\square$

14.12. COROLLARY. *Each permutation $\sigma \in \mathrm{Gal}(P/F)$ can be extended to an automorphism of $F(r_1, \ldots, r_n)$ which leaves every element in F invariant, by setting*

$$\sigma\big(f(r_1, \ldots, r_n)\big) = f\big(\sigma(r_1), \ldots, \sigma(r_n)\big)$$

for any rational fraction $f(x_1, \ldots, x_n)$ for which $f(r_1, \ldots, r_n)$ is defined.

Proof. As pointed out at the beginning of this section, we first have to prove that $\sigma\big(f(r_1,\ldots,r_n)\big)$ is well-defined by the equation above, i.e. that it does not really depend on the rational fraction $f \in F(x_1,\ldots,x_n)$, but only on $f(r_1,\ldots,r_n)$. Assume thus

$$f(r_1,\ldots,r_n) = g(r_1,\ldots,r_n)$$

for some rational fractions $f, g \in F(x_1,\ldots,x_n)$. Then the rational fraction $f-g$ vanishes for $x_i = r_i$ ($i = 1,\ldots,n$), i.e.

$$(f-g)(r_1,\ldots,r_n) = 0 \in F.$$

The preceding theorem shows that for all $\sigma \in \mathrm{Gal}(P/F)$,

$$(f-g)\big(\sigma(r_1),\ldots,\sigma(r_n)\big) = (f-g)(r_1,\ldots,r_n) = 0,$$

hence

$$f\big(\sigma(r_1),\ldots,\sigma(r_n)\big) = g\big(\sigma(r_1),\ldots,\sigma(r_n)\big).$$

This shows that $f\big(\sigma(r_1),\ldots,\sigma(r_n)\big)$ depends only on the value of the element $f(r_1,\ldots,r_n) \in F(r_1,\ldots,r_n)$, and not on the choice of the rational fraction $f(x_1,\ldots,x_n)$ which represents it.

Since σ is clearly bijective on $F(r_1,\ldots,r_n)$, the fact that it is an automorphism of $F(r_1,\ldots,r_n)$ readily follows from its definition, since

$$f\big(\sigma(r_1),\ldots,\sigma(r_n)\big) + g\big(\sigma(r_1),\ldots,\sigma(r_n)\big) = (f+g)\big(\sigma(r_1),\ldots,\sigma(r_n)\big)$$

and

$$f\big(\sigma(r_1),\ldots,\sigma(r_n)\big) \cdot g\big(\sigma(r_1),\ldots,\sigma(r_n)\big) = (fg)\big(\sigma(r_1),\ldots,\sigma(r_n)\big).$$

$\square$

14.13. COROLLARY. *The set* $\mathrm{Gal}(P/F)$ *does not depend on the choice of the Galois resolvent* V.

Proof. Let $V' \in F(r_1,\ldots,r_n)$ be another Galois resolvent of $P(X) = 0$ and let π' be its minimum polynomial over F. Let also $f_i' \in F(X)$, for $i = 1,\ldots,n$, be a rational fraction such that

$$r_i = f_i'(V') \qquad \text{for } i = 1,\ldots,n. \tag{14.7}$$

We have to show that every element of $\text{Gal}(P/F)$, as defined above with the aid of V, is also an element of $\text{Gal}(P/F)$ as defined with respect to V'. The converse will then be clear, by interchanging V and V'.

Let thus $\sigma \in \text{Gal}(P/F)$. We have to show

$$\sigma: r_i \mapsto f_i'(W')$$

for some root W' of π'. In order to find a suitable W', we use the extension of σ to $F(r_1, \ldots, r_n)$. From equation (14.7), it follows by applying σ to both sides

$$\sigma(r_i) = f_i'\big(\sigma(V')\big),$$

since σ leaves every element in F invariant. Similarly, since V' is a root of π', it follows that

$$\pi'\big(\sigma(V')\big) = 0,$$

i.e. $\sigma(V')$ is a root of π'. The element $W' = \sigma(V')$ thus satisfies the required conditions. $\qquad\square$

We now complete the proof of Result 4 (p. 237), and thus finish proving the results which were announced at the beginning of this section.

14.14. COROLLARY. $\text{Gal}(P/F)$ *is a subgroup of the group of all permutations of* $r_1, \ldots, r_n$.

Proof. That the identity map is in $\text{Gal}(P/F)$ is clear, since this map is σ_1 in the definition of the Galois group, p. 237. It thus remains to show that $\text{Gal}(P/F)$ is stable under composition of maps and under inversion.

Let $\sigma \in \text{Gal}(P/F)$. By definition of $\text{Gal}(P/F)$,

$$\sigma(r_i) = f_i(V_j)$$

for some $j = 1, \ldots, m$. Proposition 14.10 shows that V_j is also a Galois resolvent of $P(X) = 0$, hence we can define $\text{Gal}(P/F)$ with the aid of V_j instead of $V = V_1$. Since V_1 and V_j are roots of the same minimum polynomial π, it then follows that the map

$$f_i(V_j) \mapsto f_i(V_1)$$

is also an element of $\text{Gal}(P/F)$. This map is the inverse of σ, hence we have shown that $\text{Gal}(P/F)$ is stable under inversion.

In order to prove that for any $\tau \in \mathrm{Gal}(P/F)$ the composition $\tau \circ \sigma$ also is in $\mathrm{Gal}(P/F)$, we consider again the definition of $\mathrm{Gal}(P/F)$ with respect to V_j. Thus,

$$\tau \colon f_i(V_j) \mapsto f_i(V_k) \qquad \text{for some } k = 1, \ldots, m$$

and it follows that

$$\tau \circ \sigma \colon r_i \mapsto f_i(V_k).$$

Therefore, $\tau \circ \sigma \in \mathrm{Gal}(P/F)$. □

14.3   The Galois group under field extension

In the definition of the Galois group of an equation, the base field F plays a rather inconspicuous, yet important, role. It is the purpose of this section to bring it into focus and to investigate what happens to the Galois group when the base field is enlarged by the adjunction of roots of auxiliary polynomials. In view of applications to solvability by radicals, the crucial case is the adjunction of p-th roots of elements, i.e. roots of auxiliary equations of the type $X^p - a$ (where p can be chosen to be prime: see the beginning of §13.2).

As in the preceding section, we denote by P a monic polynomial of degree n over some field F, which has n distinct roots $r_1, \ldots, r_n$ in some field S containing F,

$$P(X) = X^n - a_1 X^{n-1} + \cdots + (-1)^n a_n = (X - r_1) \cdots (X - r_n).$$

The existence of such a field S follows from Girard's theorem (Theorem 9.3, p. 116). In fact, the field S can be chosen arbitrarily large, since only the subfield $F(r_1, \ldots, r_n)$ matters for the determination of the Galois group of $P(X) = 0$ over F. Therefore, if some field K containing F is given, we can assume that S contains K. Indeed, it suffices to apply Girard's theorem with base field K instead of F. This allows us to mix elements in K and elements in $F(r_1, \ldots, r_n)$ in calculations and, in particular, to consider the field $K(r_1, \ldots, r_n)$ of rational fractions in $r_1, \ldots, r_n$ with coefficients in K. We can then determine the Galois group of $P(X) = 0$ over K as well as over F, by the method of the preceding section. Here is how these Galois groups compare:

14.15. PROPOSITION. *If K is a field containing F, then* $\mathrm{Gal}(P/K)$ *is a subgroup of* $\mathrm{Gal}(P/F)$.

Proof. Let V be a Galois resolvent of $P(X) = 0$ over F. For $i = 1, \ldots, n$,

$$r_i \in F(V)$$

and since $F(V) \subset K(V)$, every root r_i of P is a rational fraction of V with coefficients in K, hence V is also a Galois resolvent of $P(X) = 0$ over K. If, for $i = 1, \ldots, n$,

$$r_i = f_i(V)$$

for some rational fraction $f_i \in F(X)$, then the same fraction f_i can be used to determine $\mathrm{Gal}(P/K)$ and $\mathrm{Gal}(P/F)$. The only difference is that the minimum polynomial π of V over F may not be irreducible over K. The minimum polynomial of V over K, which we denote by θ, is then different from π, but in any case θ divides π, by Lemma 12.14, p. 178. Therefore, the roots of θ are among those of π.

Since the permutations in $\mathrm{Gal}(P/K)$ are of the form

$$\sigma \colon r_i \mapsto f_i(V') \qquad \text{for } i = 1, \ldots, n$$

where V' is a root of θ, while the permutations in $\mathrm{Gal}(P/F)$ have the same form, but with V' a root of π, it follows that $\mathrm{Gal}(P/K) \subset \mathrm{Gal}(P/F)$. $\qquad \square$

Our aim in the rest of this section is to obtain additional information on the relations between $\mathrm{Gal}(P/K)$ and $\mathrm{Gal}(P/F)$, under certain assumptions on K. More precisely, we shall show that if K is obtained by adjoining a root of an irreducible auxiliary equation $T(X) = 0$, then the quotient

$$\frac{|\mathrm{Gal}(P/F)|}{|\mathrm{Gal}(P/K)|},$$

i.e. the index[†] of $\mathrm{Gal}(P/K)$ in $\mathrm{Gal}(P/F)$, divides the degree of T. If on the other hand the field K is obtained by adjoining *all* the roots of the equation $T(X) = 0$, then the following property holds:

$$\sigma \circ \tau \circ \sigma^{-1} \in \mathrm{Gal}(P/K) \qquad \text{for } \sigma \in \mathrm{Gal}(P/F) \text{ and } \tau \in \mathrm{Gal}(P/K).$$

This property is expressed by saying that $\mathrm{Gal}(P/K)$ is a *normal* subgroup of $\mathrm{Gal}(P/F)$.

[†]Recall from the proof of Theorem 10.3 (p. 141), that the *index* of a subgroup H in a group G, denoted by $(G : H)$, is the number of (left) cosets of H in G. If G is finite, the proof of Lagrange's theorem (Theorem 10.3) shows that equivalently $(G : H) = |G|/|H|$.

14.16. LEMMA. *Let π be an irreducible polynomial over a field F, and let K be a field containing F and such that π splits into a product of linear factors over K. Let also f, g, $h \in F[X, Y]$. If, for some root V of π in K,*

$$f(X, V) = g(X, V)h(X, V) \qquad in\ K[X]$$

then

$$f(X, W) = g(X, W)h(X, W) \qquad in\ K[X]$$

for every root W of π.

Proof. Regarding f, g and h as polynomials in one indeterminate X over $F[Y]$, we can write

$$f(X, Y) - g(X, Y)h(X, Y) = c_r(Y)X^r + \cdots + c_0(Y)$$

for some polynomials $c_r(Y), \ldots, c_0(Y) \in F[Y]$. The hypothesis that $f(X, V) = g(X, V)h(X, V)$ implies that

$$c_i(V) = 0 \qquad \text{for } i = 0, \ldots, r.$$

Therefore, Lemma 14.4 (p. 243) implies

$$c_i(W) = 0 \qquad \text{for } i = 0, \ldots, r,$$

for every root W of π, hence

$$f(X, W) = g(X, W)h(X, W).$$

$\square$

Henceforth, we denote by T an irreducible polynomial of degree t over F. From Corollary 5.22 (p. 55) it follows that T has only simple roots in any field containing F. Thus, over a suitable field,

$$T(X) = (X - u_1) \cdots (X - u_t)$$

where $u_1, \ldots, u_t$ are pairwise distinct.

14.17. THEOREM. *The index of* $\mathrm{Gal}(P/F(u_1))$ *in* $\mathrm{Gal}(P/F)$ *divides t.*

Proof. Let V be a Galois resolvent of $P(X) = 0$ over F. As we have seen in the proof of Proposition 14.15, V is also a Galois resolvent of $P(X) = 0$ over $F(u_1)$. We let θ (resp. π) denote its minimum polynomial over $F(u_1)$ (resp. F). Since the permutations in $\text{Gal}(P/F)$ are in 1–1 correspondence with the roots of π, and those in $\text{Gal}\big(P/F(u_1)\big)$ with the roots of θ, we have to prove

$$\frac{\deg \pi}{\deg \theta} \text{ divides } t.$$

From Lemma 12.14, p. 178, we know that θ divides π. Let then

$$\pi = \theta\lambda \tag{14.8}$$

for some polynomial $\lambda \in F(u_1)[X]$. Let also

$$\theta(X) = X^r + b_{r-1}X^{r-1} + \cdots + b_1X + b_0.$$

Since $b_0, \ldots, b_{r-1} \in F(u_1)$, these elements have polynomial expressions in u_1, by Proposition 12.15, p. 179. Let

$$b_i = \theta_i(u_1) \qquad \text{for } i = 0, \ldots, r-1,$$

for some polynomial $\theta_i \in F[Y]$. Let then

$$\Theta(X, Y) = X^r + \theta_{r-1}(Y)X^{r-1} + \cdots + \theta_1(Y)X + \theta_0(Y) \in F[X, Y],$$

so that

$$\Theta(X, u_1) = \theta(X).$$

Acting similarly with λ, we construct a polynomial $\Lambda(X, Y) \in F[X, Y]$ such that

$$\Lambda(X, u_1) = \lambda(X).$$

Equation (14.8) can then be rewritten

$$\pi(X) = \Theta(X, u_1)\Lambda(X, u_1)$$

and the preceding lemma yields

$$\pi(X) = \Theta(X, u_i)\Lambda(X, u_i) \qquad \text{for } i = 1, \ldots, t.$$

Multiplying these equations, we get

$$\pi(X)^t = \Theta(X, u_1) \cdots \Theta(X, u_t)\Lambda(X, u_1) \cdots \Lambda(X, u_t) \tag{14.9}$$

in $F(u_1, \ldots, u_t)[X]$.

We claim that in fact the product $\Theta(X, u_1) \cdots \Theta(X, u_t)$ is a polynomial with coefficients in F. Indeed, since the polynomial

$$\Theta(X, Y_1) \cdots \Theta(X, Y_t)$$

is clearly symmetric in the indeterminates $Y_1, \ldots, Y_t$, it can be expressed as a polynomial in X and the elementary symmetric polynomials in $Y_1, \ldots, Y_t$. Therefore, substituting $u_1, \ldots, u_t$ for $Y_1, \ldots, Y_t$ yields a polynomial in X whose coefficients can be calculated from the coefficients of the equation $T(Y) = 0$ which has $u_1, \ldots, u_t$ as roots. Since the coefficients of T are in F, it follows that

$$\Theta(X, u_1) \cdots \Theta(X, u_t) \in F[X],$$

as claimed.

Equation (14.9) shows that this product divides $\pi(X)^t$. Since π is irreducible over F, it follows that

$$\Theta(X, u_1) \cdots \Theta(X, u_t) = \pi(X)^k \qquad (14.10)$$

for some integer k between 1 and t. Comparing the degrees of both sides, we get

$$tr = k \deg \pi$$

and since $r = \deg \theta$, it follows that

$$\frac{\deg \pi}{\deg \theta} \text{ divides } t.$$

$\square$

With the same notation, we now prove the other property announced at the beginning of this section:

14.18. THEOREM. $\mathrm{Gal}\big(P/F(u_1, \ldots, u_t)\big)$ *is a normal subgroup in* $\mathrm{Gal}(P/F)$, *i.e. if* $\sigma \in \mathrm{Gal}(P/F)$ *and* $\tau \in \mathrm{Gal}\big(P/F(u_1, \ldots, u_t)\big)$, *then*

$$\sigma \circ \tau \circ \sigma^{-1} \in \mathrm{Gal}\big(P/F(u_1, \ldots, u_t)\big).$$

Proof. Let V be a Galois resolvent of $P(X) = 0$ over F (whence also over $F(u_1, \ldots, u_t)$). We let φ (resp. π) denote the minimum polynomial of V over $F(u_1, \ldots, u_t)$ (resp. over F) and let $f_1, \ldots, f_n \in F(X)$ be rational fractions such that

$$r_i = f_i(V) \qquad \text{for } i = 1, \ldots, n.$$

Any permutation $\tau \in \mathrm{Gal}\big(P/F(u_1, \ldots, u_t)\big)$ then has the form

$$\tau \colon r_i = f_i(V) \mapsto f_i(V') \qquad \text{for } i = 1, \ldots, n \qquad (14.11)$$

where V' is a root of φ.

If $\sigma \in \mathrm{Gal}(P/F)$, then σ extends to an automorphism of $F(r_1, \ldots, r_n)$ which leaves every element of F invariant, by Corollary 14.12. Therefore, applying σ to both sides of the equation

$$\pi(V) = 0,$$

we get

$$\pi\big(\sigma(V)\big) = 0.$$

This shows that $\sigma(V)$ is a root of π, hence, by Proposition 14.10, every root r_1, $\ldots$, r_n of P is a rational fraction in $\sigma(V)$. In other words, $\sigma(V)$ is a Galois resolvent of $P(X) = 0$ over F, hence also over $F(u_1, \ldots, u_t)$.

Since $\mathrm{Gal}\big(P/F(u_1, \ldots, u_t)\big)$ does not depend on the choice of a Galois resolvent (Corollary 14.13), to describe its elements we can choose any Galois resolvent we find convenient. It turns out that $\sigma(V)$ is quite suitable to describe $\sigma \circ \tau \circ \sigma^{-1}$. Indeed, (14.11) readily yields

$$\sigma \circ \tau \circ \sigma^{-1} \colon f_i\big(\sigma(V)\big) \mapsto f_i\big(\sigma(V')\big).$$

Therefore, in order to prove that $\sigma \circ \tau \circ \sigma^{-1} \in \mathrm{Gal}\big(P/F(u_1, \ldots, u_t)\big)$, it suffices to prove that $\sigma(V')$ is a root of the minimum polynomial of $\sigma(V)$ over $F(u_1, \ldots, u_t)$.

Let W be a Galois resolvent of $T(X) = 0$ and let $W_1, \ldots, W_s$ be the roots of its minimum polynomial over F (among which W lies). By Corollary 14.9, we have

$$F(u_1, \ldots, u_t) = F(W) = F(W_1, \ldots, W_s).$$

In fact, since W can be any of $W_1, \ldots, W_s$, we also have

$$F(u_1, \ldots, u_t) = F(W_i) \qquad \text{for any } i = 1, \ldots, s.$$

The extension $F(u_1, \ldots, u_t)$ can thus be regarded as an extension of F by a single element W_1. Duplicating the arguments in the proof of Theorem 14.17 above, we produce a polynomial $\Phi(X, Y) \in F[X, Y]$ such that

$$\varphi(X) = \Phi(X, W_1) \in F(u_1, \ldots, u_t)[X]$$

and we obtain as in this proof an equation similar to (14.10), namely

$$\Phi(X, W_1)\cdots\Phi(X, W_s) = \pi(X)^\ell$$

for some integer ℓ between 1 and s. Since $\sigma(V)$ is a root of π, this equation shows that $\sigma(V)$ is a root of some factor $\Phi(X, W_k)$.

In order to show that $\Phi(X, W_k)$ is the minimum polynomial of $\sigma(V)$ over $F(u_1, \ldots, u_t)$, it suffices to prove that this polynomial is irreducible. If it factors over $F(u_1, \ldots, u_t)$, then since $F(u_1, \ldots, u_t) = F(W_k)$, the factorization can be written in the form

$$\Phi(X, W_k) = \Gamma(X, W_k)\Delta(X, W_k)$$

for some polynomials Γ, Δ with coefficients in F. Hence, by Lemma 14.16,

$$\Phi(X, W_1) = \Gamma(X, W_1)\Delta(X, W_1).$$

Since $\Phi(X, W_1) = \varphi(X)$ is irreducible, it follows that the above factorization is trivial, whence also the factorization of $\Phi(X, W_k)$. Therefore, $\Phi(X, W_k)$ is the minimum polynomial of $\sigma(V)$ over $F(u_1, \ldots, u_t)$.

Thus, what we have to prove is that $\sigma(V')$ is a root of $\Phi(X, W_k)$, as $\sigma(V)$, assuming that V' is a root of $\Phi(X, W_1)$ ($= \varphi(X)$), as V.

Since by Corollary 14.9 (p. 248), $F(r_1, \ldots, r_n) = F(V)$, we have

$$V' = g(V) \tag{14.12}$$

for some rational fraction $g(X) \in F(X)$. In fact, by Proposition 12.15, p. 179, we can choose g to be a polynomial in $F[X]$. Since $\Phi(V', W_1) = 0$, we have

$$\Phi\big(g(V), W_1\big) = 0$$

hence V is a root of the polynomial $\Phi\big(g(X), W_1\big) \in F(u_1, \ldots, u_t)[X]$ and, by Lemma 12.14, p. 178, $\Phi(X, W_1)$ divides $\Phi\big(g(X), W_1\big)$. Let

$$\Phi\big(g(X), W_1\big) = \Phi(X, W_1)\Psi(X, W_1)$$

for some polynomial $\Psi \in F[X, Y]$. Lemma 14.16 (p. 256) then shows that

$$\Phi\big(g(X), W_k\big) = \Phi(X, W_k)\Psi(X, W_k)$$

and since $\sigma(V)$ is a root of $\Phi(X, W_k)$ it follows that

$$\Phi\big(g(\sigma(V)), W_k\big) = 0.$$

Now, applying σ to both sides of (14.12) yields

$$\sigma(V') = g(\sigma(V)),$$

hence the preceding equation shows that $\sigma(V')$ is a root of $\Phi(X, W_k)$, as was to be shown. □

14.19. Remark. This theorem does not yield the index of $\mathrm{Gal}(P/F(u_1, \ldots, u_t))$ in $\mathrm{Gal}(P/F)$, but some information can be obtained from Theorem 14.17. Indeed, since u_2 is a root of the polynomial

$$\frac{T(X)}{X - u_1} \in F(u_1)[X],$$

which has degree $t - 1$, it follows that the degree of the minimum polynomial of u_2 over $F(u_1)$ is at most $t - 1$, hence by Theorem 14.17,

$$\Big(\mathrm{Gal}(P/F(u_1)) : \mathrm{Gal}(P/F(u_1, u_2))\Big) \le t - 1.$$

Likewise, since u_3 is a root of

$$\frac{T(X)}{(X - u_1)(X - u_2)} \in F(u_1, u_2)[X],$$

which has degree $t - 2$, we have

$$\Big(\mathrm{Gal}(P/F(u_1, u_2)) : \mathrm{Gal}(P/F(u_1, u_2, u_3))\Big) \le t - 2,$$

and so on.

Now, the index of $\mathrm{Gal}(P/F(u_1, \ldots, u_t))$ in $\mathrm{Gal}(P/F)$ can be calculated as

$$\frac{|\mathrm{Gal}(P/F)|}{|\mathrm{Gal}(P/F(u_1, \ldots, u_t))|} =$$

$$\frac{|\mathrm{Gal}(P/F)|}{|\mathrm{Gal}(P/F(u_1))|} \cdot \frac{|\mathrm{Gal}(P/F(u_1))|}{|\mathrm{Gal}(P/F(u_1, u_2))|} \cdots \frac{|\mathrm{Gal}(P/F(u_1, \ldots, u_{t-1}))|}{|\mathrm{Gal}(P/F(u_1, \ldots, u_t))|},$$

hence the above bounds and Theorem 14.17 yield

$$\Big(\mathrm{Gal}(P/F) : \mathrm{Gal}(P/F(u_1, \ldots, u_t))\Big) \le t!.$$

14.20. Example. As an illustration of Theorems 14.17 and 14.18, we now show how the Galois group of the general equation of degree 4 is affected by the adjunction of one or all of the roots of a resolvent cubic. Let thus

$$P(X) = (X - x_1)(X - x_2)(X - x_3)(X - x_4) =$$
$$X^4 - s_1X^3 + s_2X^2 - s_3X + s_4 = 0$$

be the general equation of degree 4, with base field the field of rational fractions in s_1, s_2, s_3, s_4,

$$F = k(s_1, s_2, s_3, s_4)$$

(for some field of constants k; for instance $k = \mathbb{Q}$ or $\mathbb{C}$). By Example 14.1 (p. 241), the Galois group $\mathrm{Gal}(P/F)$ is the group of all permutations of x_1, x_2, x_3, x_4, which can be identified with the symmetric group S_4,

$$\mathrm{Gal}(P/F) = S_4.$$

From Lagrange's discussion of the solution of equations of degree 4 (see §10.2), it follows that the roots of Ferrari's resolvent cubic equation are

$$u_1 = -\tfrac{1}{2}(x_1 + x_2)(x_3 + x_4),$$
$$u_2 = -\tfrac{1}{2}(x_1 + x_3)(x_2 + x_4),$$
$$u_3 = -\tfrac{1}{2}(x_1 + x_4)(x_2 + x_3).$$

Since none of these roots is in F, the resolvent cubic is irreducible over F, by Corollary 5.13, p. 51. We can therefore apply Theorems 14.17 and 14.18 (and Remark 14.19) to conclude that $\mathrm{Gal}(P/F(u_1))$ is a subgroup of index 3 (whence of order 8) in $\mathrm{Gal}(P/F)$, and that $\mathrm{Gal}(P/F(u_1, u_2, u_3))$ is a normal subgroup in $\mathrm{Gal}(P/F)$, of index at most 6.

In fact, it is not difficult to determine these groups explicitly, as we now show. By Theorem 14.11, the permutations in $\mathrm{Gal}(P/F(u_1))$ leave u_1 invariant. Therefore, denoting by $I(u_1)$ as in §10.3 the subgroup of S_4 consisting of all the permutations which leave u_1 invariant, we have

$$\mathrm{Gal}(P/F(u_1)) \subset I(u_1).$$

Since, by Lagrange's theorem (Theorem 10.2, p. 139), the index of $I(u_1)$ in S_4 is 3 (i.e. the number of values of u_1 under the permutations in S_4), it follows that

$$|I(u_1)| = |\mathrm{Gal}(P/F(u_1))|,$$

hence

$$\mathrm{Gal}\big(P/F(u_1)\big) = I(u_1).$$

More explicitly, the permutations in $I(u_1)$ are those which are induced on the vertices of the square

by the isometries which leave the square globally invariant. Such a group is called a *dihedral group* of order 8. The subgroups $I(u_1)$, $I(u_2)$ and $I(u_3)$ of S_4 correspond to the three inequivalent numberings of the vertices of a square.

The group $I(u_1)$ is not normal in S_4. Indeed, if $\sigma \in S_4$ is a permutation which transforms u_1 into u_2, then

$$\sigma \circ I(u_1) \circ \sigma^{-1} = I(u_2) \quad (\neq I(u_1)).$$

By contrast, the subgroup $\mathrm{Gal}\big(P/F(u_1, u_2, u_3)\big)$ is normal in S_4, and Theorem 14.11 shows that u_1, u_2 and u_3 are all invariant under the permutations in this group. Therefore,

$$\mathrm{Gal}\big(P/F(u_1, u_2, u_3)\big) \subset I(u_1) \cap I(u_2) \cap I(u_3).$$

The permutations in $I(u_1) \cap I(u_2) \cap I(u_3)$ are easy to find: they are Id (the identity), and the permutations which interchange 1, 2, 3, 4 by pairs,

$$\sigma_1 \colon \begin{cases} 1 \leftrightarrow 2, \\ 3 \leftrightarrow 4, \end{cases} \qquad \sigma_2 \colon \begin{cases} 1 \leftrightarrow 3, \\ 2 \leftrightarrow 4, \end{cases} \qquad \sigma_3 \colon \begin{cases} 1 \leftrightarrow 4, \\ 2 \leftrightarrow 3. \end{cases}$$

Thus,

$$\big|\mathrm{Gal}\big(P/F(u_1, u_2, u_3)\big)\big| \le |I(u_1) \cap I(u_2) \cap I(u_3)| = 4.$$

On the other hand, since the index of $\mathrm{Gal}\big(P/F(u_1, u_2, u_3)\big)$ in S_4 is at most 6, we have

$$\big|\mathrm{Gal}\big(P/F(u_1, u_2, u_3)\big)\big| \ge 4.$$

Therefore,

$$\mathrm{Gal}\big(P/F(u_1, u_2, u_3)\big) = \{\mathrm{Id}, \sigma_1, \sigma_2, \sigma_3\}.$$

To finish this section, we record the following straightforward consequence of Theorems 14.17 and 14.18:

14.21 COROLLARY. *Let K be a radical extension of height 1 of F,*

$$K = F(u) \qquad with \qquad u^p = a$$

for some prime number p and some $a \in F$ which is not a p-th power in F. If F contains a primitive p-th root of unity, then $\mathrm{Gal}(P/K)$ is a normal subgroup of index 1 or p in $\mathrm{Gal}(P/F)$.

Proof. We apply the results above to the polynomial

$$T(X) = X^p - a,$$

which is irreducible, by Lemma 13.9, p. 219. Since F contains a primitive p-th root of unity ζ, it follows that adjoining u, which is one of the roots of T, amounts to adjoining all the roots of T, since the other roots are $\zeta u, \zeta^2 u, \ldots, \zeta^{p-1} u$. Thus,

$$K = F(u) = F(u, \zeta u, \zeta^2 u, \ldots, \zeta^{p-1} u)$$

and therefore $\mathrm{Gal}(P/K)$ is a subgroup of $\mathrm{Gal}(P/F)$ which is of index 1 or p by Theorem 14.17 and is normal by Theorem 14.18.   □

Remark. The applications to the solvability of equations by radicals use Theorems 14.17 and 14.18 only through the above corollary. In fact, only the special case of the corollary was stated instead of Theorem 14.18 in the original version of Galois' memoir, with a sketch of proof. It was replaced by the general statement of Theorem 14.18 at a later stage, presumably on the eve of the duel, with the comment "one will find the proof." The above proof is taken from Edwards [20].

14.4 Solvability by radicals

The solvability of an equation by radicals can now be translated into a condition on the Galois group of the equation. However, the notion of solvability by radicals in Galois' memoir is slightly different from that of §13.2, in that Galois requires *all* the roots of the equation (instead of one of them) to have an expression by radicals. To distinguish this condition from that of §13.2, we say that a polynomial equation with coefficients in a field F is *completely solvable (by radicals) over F* if there is a radical extension of F containing all the roots of the equation.

This distinction is significant when dealing with arbitrary equations, and more specifically with equations $P(X) = 0$ in which the left-hand side is a reducible polynomial. In this case, solving the equation amounts to finding a root of one of the factors of P, and the difficulty of finding such a root can be completely different from factor to factor. For instance, over $\mathbb{C}(s_1, \ldots, s_n)$ the equation

$$(X - 1)(X^n - s_1 X^{n-1} + s_2 X^{n-2} - \cdots + (-1)^n s_n) = 0$$

is solvable by radicals, since $X - 1 = 0$ is solvable, but it is not completely solvable by radicals if $n \geq 5$, since general equations of degree at least 5 are not solvable by radicals (Theorem 13.16, p. 227).

We shall prove however that if the polynomial P is irreducible over F, then the equation $P(X) = 0$ is solvable by radicals over F if and only if it is completely solvable by radicals over F.

In his memoir, Galois only considers the complete solvability of equations without multiple roots. Since the crucial case is the solution of irreducible equations (to which one is led by factoring the given polynomial) and since in this case both notions are equivalent, it turns out that Galois' results are actually sufficient to investigate the more general notion of solvability of equations by radicals.

The central result of this section is the following:

14.22. THEOREM. *Let P be a polynomial over a field F, and assume P has only simple roots in any field containing F. The equation $P(X) = 0$ is completely solvable over F if and only if its Galois group* $\mathrm{Gal}(P/F)$ *contains a sequence of subgroups*

$$\mathrm{Gal}(P/F) = G_0 \supset G_1 \supset G_2 \supset \cdots \supset G_t = \{\mathrm{Id}\}$$

such that, for $i = 1, \ldots, t$, the subgroup G_i is normal of prime index in G_{i-1}. (Possibly, $t = 0$, i.e. $\mathrm{Gal}(P/F) = \{\mathrm{Id}\}$.)

Accordingly, a finite group G is said to be *solvable* if it satisfies the condition of the theorem, i.e. if it contains a sequence of subgroups starting with G and ending with $\{\mathrm{Id}\}$, such that each subgroup is normal of prime index in the preceding one.

The result above is Proposition 5 in Galois' memoir [21, pp. 57 ff], [20, pp. 108–109]. Its proof can actually be adapted to yield a necessary and sufficient condition for one of the roots r of P to have an expression by radicals. The condition is the same except that G_t is not required to be reduced to the identity alone, but is instead required to contain only permutations which leave r invariant.

Although the "only if" part follows relatively easily from the preceding results, the "if" part requires some preparation. More specifically, we need some results on group theory. First, we recall from the proof of Theorem 10.3 (p. 141) that (left) cosets of a subgroup H in a group G are the subsets of G of the form

$$\sigma H = \{\sigma\xi \mid \xi \in H\}, \qquad \text{for } \sigma \in G,$$

and that the number of distinct cosets of H in G is the index $(G : H)$, which is equal to the quotient $|G|/|H|$ if G is finite, by Lagrange's theorem (Theorem 10.3, p. 141).

14.23. LEMMA. $\sigma H = \tau H$ *if and only if* $\sigma^{-1}\tau \in H$.

Proof. If $\sigma H = \tau H$, then, in particular, $\tau \cdot 1 \in \sigma H$, hence $\tau = \sigma\xi$ for some $\xi \in H$, and

$$\sigma^{-1}\tau = \xi \in H.$$

Conversely, if $\sigma^{-1}\tau \in H$, then the equation

$$\sigma\xi = \tau\big((\sigma^{-1}\tau)^{-1}\xi\big) \qquad \text{for } \xi \in H$$

shows that $\sigma H \subset \tau H$, while the equation

$$\tau\xi = \sigma\big((\sigma^{-1}\tau)\xi\big) \qquad \text{for } \xi \in H$$

shows that $\tau H \subset \sigma H$. □

14.24. LEMMA. *Let* $G_1 \supset G_2 \supset G_3$ *be a chain of subgroups. If* G_1 *is finite, then*

$$(G_1 : G_3) = (G_1 : G_2)(G_2 : G_3).$$

In particular, if G_3 *is a subgroup of prime index in* G_1, *then either* $G_2 = G_1$ *or* $G_2 = G_3$.

Proof. This is clear if the index of G_3 in G_1 is calculated as

$$\frac{|G_1|}{|G_3|} = \frac{|G_1|}{|G_2|} \cdot \frac{|G_2|}{|G_3|}.$$

□

Remark. The lemma also holds without the hypothesis that G_1 is finite, but the proof is more delicate. This more general case will not be needed.

14.25. PROPOSITION. *Let H and N be subgroups of a group G, and define a subset $H \cdot N$ of G by*

$$H \cdot N = \{\xi\nu \mid \xi \in H, \ \nu \in N\}.$$

If N is normal in G, then $H \cdot N$ is a subgroup of G and $H \cap N$ is a normal subgroup of H. If moreover N has prime index in G, and G is finite, then either $H \cdot N = N$ or else $H \cdot N = G$.

If $H \cdot N = N$, then $H \subset N$, hence $H \cap N = H$.

If $H \cdot N = G$, then the index of $H \cap N$ in H is equal to the index of N in G,

$$(H : H \cap N) = (G : N).$$

Moreover, in this case, every coset of N in G has the form ξN for some $\xi \in H$.

Proof. The normality of $H \cap N$ in H readily follows from that of N in G, and showing that $H \cdot N$ is a subgroup of G is a straightforward verification. First, the unit element 1 in G is in $H \cdot N$ since it can be written as the product of the element $1 \in H$ and the element $1 \in N$.

Next, $H \cdot N$ is stable under products, since for $\xi_1, \xi_2 \in H$ and $\nu_1, \nu_2 \in N$,

$$(\xi_1\nu_1)(\xi_2\nu_2) = (\xi_1\xi_2)\big((\xi_2^{-1}\nu_1\xi_2)\nu_2\big),$$

where $\xi_2^{-1}\nu_1\xi_2 \in N$ since N is normal.

Finally, $H \cdot N$ contains the inverse of each of its elements, since for $\xi \in H$ and $\nu \in N$,

$$(\xi\nu)^{-1} = \xi^{-1}(\xi\nu^{-1}\xi^{-1}).$$

We thus have inclusions of subgroups

$$G \supset H \cdot N \supset N.$$

If the index of N in G is prime, then it follows from Lemma 14.24 that either $H \cdot N = N$ or $H \cdot N = G$. In the first case, we have $H \subset N$ since H is obviously contained in $H \cdot N$. In the latter case, we can find, for every element $\sigma \in G$, elements $\xi \in H$ and $\nu \in N$ such that

$$\sigma = \xi\nu.$$

From Lemma 14.23, it then follows that

$$\sigma N = \xi N,$$

hence every coset of N in G has the form ξN for some $\xi \in H$.

To prove the rest, we define a bijection between the set of cosets of $H \cap N$ in H and the set of cosets of N in G, by

$$\xi(H \cap N) \mapsto \xi N \qquad \text{for } \xi \in H.$$

That this map is onto follows from the last observation, that every coset of N has the form ξN for some $\xi \in H$. To prove the injectivity, we assume ξ_1 and $\xi_2 \in H$ are such that $\xi_1 N = \xi_2 N$. Lemma 14.23 then shows that

$$\xi_1^{-1}\xi_2 \in N,$$

hence

$$\xi_1^{-1}\xi_2 \in H \cap N$$

since ξ_1 and ξ_2 are both in H. Applying Lemma 14.23 again, we obtain

$$\xi_1(H \cap N) = \xi_2(H \cap N).$$

We have thus proved that

$$(G : N) = (H : H \cap N).$$

$\square$

Remark. The results in this proposition can be put in a somewhat better perspective by making use of the notion of *factor group* of a group by a normal subgroup. Essentially, the proposition asserts that the inclusion of H in $H \cdot N$ induces an isomorphism of factor groups

$$\frac{H}{H \cap N} \xrightarrow{\sim} \frac{H \cdot N}{N}.$$

This result is valid even when G is infinite. We have avoided this presentation, however, since the notion of factor group does not appear in Galois' papers and may have been unknown to Galois.

14.26. COROLLARY. *Let N be a normal subgroup of prime index p in a finite group G. If σ is an element of G outside N, then $\sigma^p \in N$.*

Proof. Consider the $p + 1$ cosets

$$N, \quad \sigma N, \quad \ldots, \quad \sigma^p N.$$

Since the index of N in G is p, these cosets cannot be pairwise distinct, hence we can find integers m, n between 0 and p, with $m < n$, such that

$$\sigma^m N = \sigma^n N.$$

Lemma 14.23 then yields

$$\sigma^{n-m} \in N.$$

We may thus consider the smallest integer $k > 0$ such that $\sigma^k \in N$. The preceding argument shows that $k \leq p$. To complete the proof, it thus suffices to show that $k \geq p$. In order to do this, we are going to show that every coset of N in G is one of the following:

$$N, \quad \sigma N, \quad \ldots, \quad \sigma^{k-1} N.$$

It then readily follows that the index p of N in G is at most equal to k. Let $H = \{\sigma^i \mid i \in \mathbb{Z}\}$. This is clearly a subgroup of G and since $\sigma \notin N$, it follows that $H \cdot N \neq N$. Therefore, the preceding proposition shows that $H \cdot N = G$ and that any coset of N in G has the form $\sigma^i N$ for some $i \in \mathbb{Z}$. Dividing i by k, we get $i = kq + r$ for some integers q and r, with $0 \leq r < k$. Then

$$\sigma^{i-r} = (\sigma^k)^q \in N,$$

hence, by Lemma 14.23,

$$\sigma^i N = \sigma^r N.$$

This proves the claim, since r is between 0 and $k - 1$. $\qquad\square$

As a further consequence of Proposition 14.25, we record the following result, which will be quite useful in proofs by induction:

14.27 COROLLARY. *Every subgroup H of a (finite) solvable group G is solvable.*

Proof. Let

$$G = G_0 \supset G_1 \supset \cdots \supset G_t = \{1\}$$

be a sequence of subgroups in G, each of which is normal of prime index in the preceding one. We have to find a similar sequence in H. Consider

$$H = H \cap G_0 \supseteq H \cap G_1 \supseteq \cdots \supseteq H \cap G_t = \{1\}. \qquad (14.13)$$

Applying Proposition 14.25 with G_i instead of G, with G_{i+1} instead of N and $H \cap G_i$ instead of H, we deduce that $H \cap G_{i+1}$ is a normal subgroup of $H \cap G_i$, and that either

$$H \cap G_{i+1} = H \cap G_i \quad \text{or} \quad (H \cap G_i : H \cap G_{i+1}) = (G_i : G_{i+1}).$$

Therefore, after deleting repetitions in the sequence (14.13), we get a sequence of subgroups in H with the required properties. $\qquad\qquad\square$

After all these preliminaries on group theory, we now come back to the solution of equations by radicals. We use the same notation as in the preceding sections. Thus, we consider a polynomial

$$P(X) = X^n - a_1 X^{n-1} + a_2 X^{n-2} - \cdots + (-1)^n a_n = (X - r_1) \cdots (X - r_n)$$

with coefficients $a_1, \ldots, a_n$ in a field F and pairwise distinct roots $r_1, \ldots, r_n$ in some field containing F. Our next result is a kind of converse of Corollary 14.21.

14.28. LEMMA. *Let N be a normal subgroup of prime index p in $\mathrm{Gal}(P/F)$. If F contains a primitive p-th root of unity, then there exists a radical extension K of F in $F(r_1, \ldots, r_n)$, of the form*

$$K = F(a^{1/p})$$

for some $a \in F$, such that

$$\mathrm{Gal}(P/K) = N.$$

Proof. We proceed by several steps. First, we pick a permutation σ in $\mathrm{Gal}(P/F)$ but not in N.

Step 1. Let $x \in F(r_1, \ldots, r_n)$ be such that $\nu(x) = x$ for all $\nu \in N$. We claim:

- If $\sigma(x) = x$, then $x \in F$.
- If $\sigma(x) \neq x$, and if $\tau \in \mathrm{Gal}(P/F)$ is such that $\tau(x) = x$, then $\tau \in N$.

Let X be the set of permutations in $\mathrm{Gal}(P/F)$ which leave x invariant, i.e.

$$X = \{\tau \in \mathrm{Gal}(P/F) \mid \tau(x) = x\}.$$

This set is obviously a group, which contains N, by hypothesis. From the inclusions

$$\mathrm{Gal}(P/F) \supset X \supset N$$

and from the hypothesis that the index of N in $\mathrm{Gal}(P/F)$ is prime, we deduce by Lemma 14.24

$$X = N \qquad \text{or} \qquad X = \mathrm{Gal}(P/F).$$

If $\sigma(x) = x$, then $\sigma \in X$ and therefore $X \neq N$ since $\sigma \notin N$. It then follows that $X = \mathrm{Gal}(P/F)$, hence Theorem 14.11 (p. 250) shows that $x \in F$.

If $\sigma(x) \neq x$, then $\sigma \notin X$, hence $X \neq \mathrm{Gal}(P/F)$. Therefore, $X = N$, which means that every permutation in $\mathrm{Gal}(P/F)$ which leaves x invariant is in N.

Step 2. There is an element $v \in F(r_1, \ldots, r_n)$ which is invariant under every permutation in N but is not in F.

Let $f(x_1, \ldots, x_n)$ be a polynomial in $F[x_1, \ldots, x_n]$ which has the property of Lemma 14.6 (p. 245), i.e. that the $n!$ elements of $F(r_1, \ldots, r_n)$ obtained by substituting $r_1, \ldots, r_n$ for the indeterminates in all possible ways are pairwise different, and let $V = f(r_1, \ldots, r_n)$. The proof of Proposition 14.7 (p. 245) shows that V is a Galois resolvent of $P(X) = 0$ over F, hence the degree of its minimum polynomial over F is equal to $|\mathrm{Gal}(P/F)|$. Consider then the polynomial

$$\prod_{\nu \in N} \left(X - \nu(V) \right).$$

The coefficients of this polynomial are clearly invariant under N. If they were all in F, then V would be a root of a polynomial of degree $|N|$ over F. This is impossible since the minimum polynomial of V over F has degree larger than $|N|$. Therefore, at least one of the coefficients is invariant under N but is not in F. This coefficient can be chosen for v.

For every p-th root of unity $\omega \in F$, we define a kind of Lagrange resolvent

$$t(\omega) = v + \omega\sigma(v) + \cdots + \omega^{p-1}\sigma^{p-1}(v).$$

Step 3. $\sigma\big(t(\omega)\big) = \omega^{-1}t(\omega)$, and $\nu\big(t(\omega)\big) = t(\omega)$ for all $\nu \in N$.

The powers of ω are in F and are therefore invariant under every permutation in $\mathrm{Gal}(P/F)$, by Theorem 14.11 (p. 250). Thus,

$$\sigma\big(t(\omega)\big) = \sigma(v) + \omega\sigma^2(v) + \cdots + \omega^{p-1}\sigma^p(v)$$

or, equivalently,

$$\sigma\big(t(\omega)\big) = \omega^{-1}\big(\sigma^p(v) + \omega\sigma(v) + \cdots + \omega^{p-1}\sigma^{p-1}(v)\big),$$

and

$$\nu\big(t(\omega)\big) = \nu(v) + \omega\nu\sigma(v) + \cdots + \omega^{p-1}\nu\sigma^{p-1}(v),$$

for every $\nu \in N$.

Corollary 14.26 shows that $\sigma^p \in N$, hence $\sigma^p(v) = v$ and it readily follows that

$$\sigma\big(t(\omega)\big) = \omega^{-1}t(\omega).$$

On the other hand, since N is a normal subgroup of $\mathrm{Gal}(P/F)$, we have

$$\sigma^{-i} \circ \nu \circ \sigma^i \in N \qquad \text{for every } \nu \in N \text{ and every } i = 0, \ldots, p-1$$

whence

$$\sigma^{-i} \circ \nu \circ \sigma^i(v) = v \qquad \text{for every } \nu \in N \text{ and every } i = 0, \ldots, p-1.$$

Applying σ^i to both sides of this equation, we get

$$\nu \circ \sigma^i(v) = \sigma^i(v) \qquad \text{for every } \nu \in N \text{ and every } i = 0, \ldots, p-1,$$

hence

$$\nu\big(t(\omega)\big) = t(\omega) \quad \text{for every } \nu \in N.$$

Step 4. $t(\omega)^p \in F$ for every p-th root of unity ω, and there is a p-th root of unity $\omega \neq 1$ such that $t(\omega) \neq 0$.

From step 3 it follows that $t(\omega)^p$ is invariant under σ and under every permutation in N. Step 1 shows that $t(\omega)^p$ therefore lies in F.

If we assume $t(\omega) = 0$ for every p-th root of unity $\omega \neq 1$, then Lagrange's formula (p. 138)

$$v = \frac{1}{p}\Big(\sum_{\omega} t(\omega)\Big)$$

yields

$$v = \frac{1}{p}t(1)$$

and this equality shows, by Step 3, that v is invariant under σ. Since it is also invariant under N, it follows by Step 1 that v is in F; this is a contradiction, since v has been chosen outside F in Step 2.

Let thus ω be a p-th root of unity such that $\omega \neq 1$ and $t(\omega) \neq 0$, and let

$$K = F\big(t(\omega)\big).$$

Step 4 and Proposition 13.2 (p. 214) show that K is a radical extension of F, of the form $F(a^{1/p})$. To complete the proof, it now suffices to show:

Step 5. $\mathrm{Gal}(P/K) = N$.

Since $t(\omega) \neq 0$ and $\omega \neq 1$, Step 3 shows that $t(\omega)$ is not invariant under σ. Therefore, $K \neq F$, hence K is a radical extension of height 1 of F. From Corollary 14.21 (p. 264), it then follows that $\mathrm{Gal}(P/K)$ is a subgroup of index p in $\mathrm{Gal}(P/F)$, whence

$$|\mathrm{Gal}(P/K)| = |N|.$$

Moreover, as $\sigma\big(t(\omega)\big) \neq t(\omega)$, Step 1 shows that every permutation in $\mathrm{Gal}(P/F)$ which leaves $t(\omega)$ invariant is in N. As $t(\omega) \in K$, the permutations in $\mathrm{Gal}(P/K)$ leave $t(\omega)$ invariant, by Theorem 14.11 (p. 250), hence

$$\mathrm{Gal}(P/K) \subseteq N.$$

Since these groups have the same order, $\mathrm{Gal}(P/K)$ cannot be strictly smaller than N, so

$$\mathrm{Gal}(P/K) = N.$$

$\square$

Proof of Theorem 14.22. We first prove, by induction on $|\mathrm{Gal}(P/F)|$, that if the equation $P(X) = 0$ is completely solvable over F, then $\mathrm{Gal}(P/F)$ is solvable. If $|\mathrm{Gal}(P/F)| = 1$, then $\mathrm{Gal}(P/F) = \{\mathrm{Id}\}$, and this group is trivially solvable. We may thus assume that completely solvable equations with Galois group of order less than that of $P(X) = 0$ over F have solvable Galois groups. Let R be a radical extension of F which contains all the roots of P. From Theorem 14.11 (p. 250) it follows that every element in R is invariant under $\mathrm{Gal}(P/R)$. Therefore, every root of P is invariant under the permutations in $\mathrm{Gal}(P/R)$, which means that $\mathrm{Gal}(P/R) = \{\mathrm{Id}\}$. This shows that there exist radical extensions K of F such that

$$|\mathrm{Gal}(P/K)| < |\mathrm{Gal}(P/F)|.$$

We may thus consider the smallest prime p for which the extraction of a p-th root decreases the order of the Galois group of P. Explicitly, we let p be the smallest prime number for which there exists a radical extension L of F such that

$$\mathrm{Gal}(P/L) = \mathrm{Gal}(P/F)$$

and

$$\left|\mathrm{Gal}\big(P/L(a^{1/p})\big)\right| < \left|\mathrm{Gal}(P/F)\right|$$

for some $a \in L$ which is not a p-th power in L.

By Proposition 13.5 (p. 216), there is a radical extension of L which contains a primitive p-th root of unity. Moreover, inspection of the proof of this proposition shows that there is such an extension R' which is obtained from L by extractions of q-th roots for prime numbers $q < p$. Therefore, by definition of p, we have

$$\mathrm{Gal}(P/R') = \mathrm{Gal}(P/L) = \mathrm{Gal}(P/F).$$

Moreover, by Proposition 14.15 (p. 254),

$$\mathrm{Gal}\big(P/R'(a^{1/p})\big) \subset \mathrm{Gal}\big(P/L(a^{1/p})\big),$$

hence

$$\left|\mathrm{Gal}\big(P/R'(a^{1/p})\big)\right| < \left|\mathrm{Gal}(P/F)\right|.$$

Since R' contains a primitive p-th root of unity, $\mathrm{Gal}\big(P/R'(a^{1/p})\big)$ is a normal subgroup of index p in $\mathrm{Gal}(P/R') = \mathrm{Gal}(P/F)$, by Corollary 14.21 (p. 264). Since $P(X) = 0$ is completely solvable over F, it is also completely solvable over $R'(a^{1/p})$, by Theorem 13.7 (p. 218). (Note that the proof of that theorem holds without change for complete solvability instead of solvability.) Therefore, by the induction hypothesis, we can find a sequence of subgroups

$$\mathrm{Gal}\big(P/R'(a^{1/p})\big) \supset G_2 \supset \cdots \supset G_t = \{\mathrm{Id}\}$$

such that each subgroup is normal of prime index in the preceding one. The sequence

$$\mathrm{Gal}(P/F) \supset \mathrm{Gal}\big(P/R'(a^{1/p})\big) \supset G_2 \supset \cdots \supset G_t = \{\mathrm{Id}\}$$

then shows that $\mathrm{Gal}(P/F)$ is solvable.

We now prove that, conversely, solvability of $\mathrm{Gal}(P/F)$ implies complete solvability of $P(X) = 0$ by radicals over F. We argue again by induction on $|\mathrm{Gal}(P/F)|$.

If $|\mathrm{Gal}(P/F)| = 1$, then the only permutation in $\mathrm{Gal}(P/F)$ is the identity, which leaves every root invariant. Therefore, by Theorem 14.11 (p. 250), all the roots of P are in F, which is a radical extension of height 0 of itself.

We may thus assume, by induction, that equations with solvable Galois group of order less than that of P over F are completely solvable. Since $\mathrm{Gal}(P/F)$ is solvable, it contains a normal subgroup N of prime index. Let

$$p = (\mathrm{Gal}(P/F) : N).$$

By Proposition 13.5 (p. 216), there exists a radical extension R of F which contains all the p-th roots of unity. If

$$|\mathrm{Gal}(P/R)| < |\mathrm{Gal}(P/F)|,$$

then we can resort to the induction hypothesis, since by Corollary 14.27 (p. 269), $\mathrm{Gal}(P/R)$ is a solvable group. The equation $P(X) = 0$ is thus completely solvable over R, hence there exists a radical extension R' of R which contains all the roots of P. Since the field R' is also a radical extension of F, the proof is complete in this case. If on the contrary

$$\mathrm{Gal}(P/R) = \mathrm{Gal}(P/F),$$

then we resort to Lemma 14.28, which shows that there is a radical extension R'' of R such that $\mathrm{Gal}(P/R'') = N$. Since then

$$|\mathrm{Gal}(P/R'')| < |\mathrm{Gal}(P/F)|,$$

we conclude as above by the induction hypothesis. $\square$

14.29. Remark. Assume all the roots of unity are in the base field F. The last part of the proof above then shows that if the Galois group $\mathrm{Gal}(P/F)$ is solvable, i.e. if there exists a sequence of subgroups

$$\mathrm{Gal}(P/F) = G_0 \supset G_1 \supset \cdots \supset G_t = \{\mathrm{Id}\},$$

with G_i normal of prime index in G_{i-1} for $i = 1, \ldots, t$, then a radical extension of F containing all the roots of P can be obtained by t extractions of roots. First, the extraction of a $(G_0 : G_1)$-th root, which reduces the Galois group to G_1, next the extraction of a $(G_1 : G_2)$-th root, which reduces the Galois group to G_2, and so on.

14.30. Example. Theorem 14.22 (p. 265) illuminates the solution of equations of degree 3 and 4 by radicals, as we now show. First, we define for any integer $n \geq 2$

a subgroup A_n of the symmetric group S_n: it is the group of all permutations in S_n which leave invariant the polynomial $\Delta(x_1, \ldots, x_n)$ used in the definition of the discriminant (see §8.3),

$$\Delta(x_1, \ldots, x_n) = \prod_{1 \leq i < j \leq n} (x_i - x_j).$$

Thus, with the notation of §10.2,

$$A_n = I(\Delta).$$

The group A_n is called the *alternating group* on $\{1, \ldots, n\}$. (See Exercise 6 of Chapter 13.)

As noted in §8.3, any permutation of the indeterminates either leaves Δ invariant or transforms it into its opposite $-\Delta$. Therefore, by Lagrange's theorem (Theorem 10.2, p. 139), the subgroup A_n has index 2 in S_n,

$$|A_n| = \frac{n!}{2}.$$

Moreover, it is easily seen that A_n is normal in S_n. One has to see that for all $\sigma \in S_n$ and for all $\tau \in A_n$, $\sigma \circ \tau \circ \sigma^{-1} \in A_n$. This is clear if $\sigma \in A_n$, since A_n is stable under products; if on the contrary $\sigma \notin A_n$, then $\sigma^{-1} \notin A_n$, whence

$$\sigma(\Delta) = \sigma^{-1}(\Delta) = -\Delta.$$

Therefore,

$$\sigma \circ \tau \circ \sigma^{-1}(\Delta) = -\sigma \circ \tau(\Delta) = -\sigma(\Delta) = \Delta,$$

which shows that $\sigma \circ \tau \circ \sigma^{-1} \in A_n$, as claimed.

Consider now the general equation of degree n,

$$P(X) = (X - x_1) \cdots (X - x_n) = X^n - s_1 X^{n-1} + \cdots + (-1)^n s_n = 0$$

over $F = \mathbb{C}(s_1, \ldots, s_n)$. (The field of constants is chosen to be $\mathbb{C}$ so that all the roots of unity are in F.) By Example 14.1 (p. 241), the Galois group $\mathrm{Gal}(P/F)$ can be identified with S_n. Since Δ^2 is the discriminant,

$$\Delta^2 = D(s_1, \ldots, s_n) \in F,$$

it follows that Δ is a root of the polynomial

$$X^2 - D(s_1, \ldots, s_n) \in F[X].$$

Hence, by Theorem 14.17 (p. 256), $\text{Gal}\big(P/F(\Delta)\big)$ is a subgroup of index 2 in $\text{Gal}(P/F) = S_n$. Since the elements in $\text{Gal}\big(P/F(\Delta)\big)$ leave Δ invariant (Theorem 14.11, p. 250), we have

$$\text{Gal}\big(P/F(\Delta)\big) \subseteq A_n$$

whence

$$\text{Gal}\big(P/F(\Delta)\big) = A_n$$

since these groups both have index 2 in S_n and have therefore the same number of elements. (The fact that A_n is normal in S_n then also follows from Theorem 14.18 (p. 258), since adjoining one of the root of a quadratic polynomial amounts to adjoining both roots.)

Let now $n = 3$. The sequence of subgroups

$$S_3 \supset A_3 \supset \{\text{Id}\}$$

shows that S_3 is solvable, since the index of A_3 in S_3 is 2 and that of $\{\text{Id}\}$ in A_3 is $|A_3| = 3$. Therefore, the general equation of degree 3 is completely solvable by radicals. Moreover, a radical extension of F containing all the roots can be obtained by two extractions of roots: first, the extraction of a square root, which reduces the Galois group from S_3 to A_3, and then the extraction of a cube root, which reduces the Galois group to $\{\text{Id}\}$. More precisely, the discussion above shows that the square root which has to be extracted first is that of the discriminant $D(s_1, s_2, s_3)$, since indeed $\text{Gal}\big(P/F(\Delta)\big) = A_3$. (In Cardano's formula, the expression under the square root is indeed the discriminant, see §8.3.)

Now, consider $n = 4$. In Example 14.20 (p. 262), we have seen that the adjunction to F of all the roots of Ferrari's resolvent cubic reduces the Galois group to

$$V = \{\text{Id}, \sigma_1, \sigma_2, \sigma_3\},$$

where σ_1, σ_2 and σ_3 commute and satisfy $\sigma_1^2 = \sigma_2^2 = \sigma_3^2 = \text{Id}$. Therefore, $\{\text{Id}, \sigma_1\}$, $\{\text{Id}, \sigma_2\}$ and $\{\text{Id}, \sigma_3\}$ are normal subgroups of index 2 in V. Moreover, a direct verification shows that σ_1, σ_2 and σ_3 leave Δ invariant, whence

$$V \subset A_4.$$

Counting elements, we get

$$(A_4 : V) = 3.$$

Moreover, since V is normal in S_4 (see Example 14.20, p. 262), it is *a fortiori* normal in A_4. The sequence

$$S_4 \supset A_4 \supset V \supset \{\mathrm{Id}, \sigma_1\} \supset \{\mathrm{Id}\}$$

then shows that S_4 is solvable. Consequently, the general equation of degree 4 is completely solvable by radicals over F. The above sequence of subgroups shows moreover that a radical extension of F containing all the roots is obtained by the following operations:

(1) the extraction of a square root of the discriminant $D(s_1, s_2, s_3, s_4)$, which reduces the Galois group to A_4;

(2) the extraction of a cube root, which reduces the Galois group to V;

(3) and (4) the successive extraction of two square roots, which reduce the Galois group to $\{\mathrm{Id}, \sigma_1\}$ and then to $\{\mathrm{Id}\}$.

In fact, the first two steps are achieved by the solution of Ferrari's resolvent cubic, which requires first the extraction of a square root of its discriminant. Therefore, extracting a square root of $D(s_1, s_2, s_3, s_4)$ amounts to extracting a square root of the discriminant of the resolvent cubic. Indeed, it is readily verified by a direct computation that these two discriminants are equal. (See Exercise 3 of Chapter 8.)

From the preceding discussion, it follows at the same time that every equation of degree 3 or 4 is completely solvable by radicals, since the Galois group of such an equation is a group of permutation of the roots, which can be identified, via a numbering of the roots, to a subgroup of S_3 or S_4 and is therefore solvable, by Corollary 14.27 (p. 269).

To finish this section, we turn to the (not necessarily complete) solvability of equations by radicals.

14.31. THEOREM. *Let r be a root of a polynomial P over some field F, and assume P has only simple roots in any field containing F. The root r has an expression by radicals over F if and only if $\mathrm{Gal}(P/F)$ contains a sequence of subgroups*

$$\mathrm{Gal}(P/F) = G_0 \supset G_1 \supset G_2 \supset \cdots \supset G_t$$

such that G_i is normal of prime index in G_{i-1} for $i = 1, \ldots, t$, and r is invariant under every permutation in G_t. (Possibly, $t = 0$.)

The proof is the same as for Theorem 14.22 (p. 265), except that one has to use induction on other integers than $|\mathrm{Gal}(P/F)|$. For the "only if" part, use induction on the number of elements in the set

$$\{\sigma(r) \mid \sigma \in \mathrm{Gal}(P/F)\}$$

(which is called the *orbit* of r under $\mathrm{Gal}(P/F)$); for the "if" part, use induction on the length t of the sequence of subgroups of $\mathrm{Gal}(P/F)$. Details are left to the reader.

Using this theorem and Theorem 14.22 (p. 265), we are now able to show that, as claimed in the introduction to this section, the two notions of solvability by radicals are equivalent for irreducible polynomials. We shall need the following characterization of irreducible equations through their Galois group:

14.32. PROPOSITION. *A non-constant polynomial* $P \in F[X]$ *without multiple root in any extension of* F *is irreducible over* F *if and only if the Galois group* $\mathrm{Gal}(P/F)$ *is transitive on the roots of* P, *i.e., for any two roots* r_i, r_j *of* P, *there is a permutation* $\sigma \in \mathrm{Gal}(P/F)$ *such that*

$$\sigma(r_i) = r_j.$$

Proof. If P is not irreducible, let

$$P = P_1 P_2$$

for some non-constant polynomials P_1, $P_2 \in F[X]$. If r_1 is a root of P_1, then

$$P_1(r_1) = 0,$$

whence, applying $\sigma \in \mathrm{Gal}(P/F)$ to both sides of this equation,

$$P_1\big(\sigma(r_1)\big) = 0 \qquad \text{for all } \sigma \in \mathrm{Gal}(P/F).$$

Thus, no permutation in $\mathrm{Gal}(P/F)$ carries r_1 to a root of P_2. Therefore, this group is not transitive on the roots of P. Conversely, assume $\mathrm{Gal}(P/F)$ is not transitive on the roots of P. Then, there are roots r_i, r_j of P such that r_i is not mapped onto r_j by any permutation in $\mathrm{Gal}(P/F)$. Let

$$R = \{\sigma(r_i) \mid \sigma \in \mathrm{Gal}(P/F)\}$$

and let

$$P_1 = \prod_{r \in R} (X - r).$$

The coefficients of P_1 are in F, by Theorem 14.11 (p. 250), since they are invariant under $\mathrm{Gal}(P/F)$. Moreover, the elements in R are roots of P, hence P_1 divides P. On the other hand, the hypothesis on r_i, r_j implies that $r_j \notin R$. There is thus a root of P which is not a root of P_1. Therefore, the degree of P_1 is strictly smaller than that of P, and it follows that P is not irreducible in $F[X]$. $\qquad\square$

14.33. PROPOSITION. *Let P be an irreducible polynomial over some field F. The equation $P(X) = 0$ is completely solvable by radicals over F if and only if it is solvable by radicals over F.*

Proof. It obviously suffices to prove the "if" part. We first note that, by Theorem 5.21 (p. 54), irreducible polynomials have no multiple root in any extension of the base field. We may thus use Theorems 14.22 (p. 265) and 14.31, together with Proposition 14.32, to translate the proposition into the following purely group-theoretical statement:

Claim: Let G be a transitive group of permutations of a set $\{r_1, \dots, r_n\}$. If G contains a sequence of subgroups

$$G = G_0 \supset G_1 \supset \cdots \supset G_t \qquad (S)$$

such that G_i is normal of prime index in G_{i-1} for $i = 1, \dots, t$, and such that every permutation in the last subgroup G_t leaves one of the elements (r_1, say) invariant, then G is solvable.

Since G is transitive, we can find, for $i = 2, \dots, n$, a permutation $\sigma_i \in G$ such that

$$\sigma_i(r_1) = r_i.$$

We then use the inner automorphism $\tau \mapsto \sigma_i \circ \tau \circ \sigma_i^{-1}$ of G, which transforms the given sequence of subgroups into

$$G = G_0 \supset (\sigma_i \circ G_1 \circ \sigma_i^{-1}) \supset \cdots \supset (\sigma_i \circ G_t \circ \sigma_i^{-1}), \qquad (S_i)$$

a sequence in which each subgroup is normal of prime index in the preceding one, and the last subgroup $\sigma_i \circ G_t \circ \sigma_i^{-1}$ leaves r_i invariant.

Intersecting with G_t all the subgroups in the sequence (S_2), we get

$$G_t \supseteq (G_t \cap (\sigma_2 \circ G_1 \circ \sigma_2^{-1})) \supseteq (G_t \cap (\sigma_2 \circ G_2 \circ \sigma_2^{-1})) \supseteq \cdots$$
$$\supseteq (G_t \cap (\sigma_2 \circ G_t \circ \sigma_2^{-1})). \qquad (S_2')$$

By Proposition 14.25 (p. 267) (see also the proof of Corollary 14.27, p. 269), each subgroup is normal of index 1 or a prime number in the preceding subgroup. The given sequence (S) can thus be continued down to $G_t \cap (\sigma_2 \circ G_t \circ \sigma_2^{-1})$, which leaves invariant both r_1 and r_2.

Intersecting with $G_t \cap (\sigma_2 \circ G_t \circ \sigma_2^{-1})$ all the subgroups in sequence (S_3), we get a sequence similar to (S_2'), beginning with $G_t \cap (\sigma_2 \circ G_t \circ \sigma_2^{-1})$ and ending with

$$G_t \cap (\sigma_2 \circ G_t \circ \sigma_2^{-1}) \cap (\sigma_3 \circ G_t \circ \sigma_3^{-1}),$$

a subgroup which leaves invariant r_1, r_2 and r_3. This sequence can be used to extend the sequence previously constructed. Continuing in the same way (and deleting the possible repetitions), we construct a sequence of subgroups of G, each of which is normal of prime index in the preceding one, which ends with

$$G_t \cap (\sigma_2 \circ G_t \circ \sigma_2^{-1}) \cap \cdots \cap (\sigma_n \circ G_t \circ \sigma_n^{-1}).$$

Since this group leaves invariant all of $r_1, \ldots, r_n$, it is reduced to $\{\mathrm{Id}\}$, and the sequence of subgroups thus constructed shows that G is solvable. $\qquad\square$

14.5 Applications

We present two applications of Galois' theory: the first one is due to Galois himself, and deals with irreducible equations of prime degree; the second one proves the theorem of Abel stated in §14.1, namely that equations which satisfy Abel's condition are solvable by radicals.

In the last part of his memoir, Galois determines the Galois groups of irreducible equations of prime degree which are solvable by radicals (completely or not; these properties are equivalent, by Proposition 14.33). In view of Theorem 14.22 (p. 265) and Proposition 14.32, this amounts to the determination of the solvable transitive groups of permutations of p elements, for p prime, i.e. of the solvable transitive subgroups of S_p.

Before discussing Galois' result, we review some basic observations about groups of permutations, which will play a crucial role in the sequel.

14.34 DEFINITIONS. Let G be a group of permutations of a set E. For any $a \in E$, we define the *orbit* of a under G as the set of elements in E where a can be mapped by a permutation in G, i.e.

$$G(a) = \{\sigma(a) \mid \sigma \in G\}.$$

We also define the *isotropy subgroup* of a in G as the set of permutations in G which leave a invariant,

$$I_G(a) = \{\sigma \in G \mid \sigma(a) = a\}$$

(compare §10.3).

The same arguments as in the proof of Lagrange's theorem (Theorem 10.2, p. 139) yield the following result:

14.35. THEOREM. *If G is finite, then for any $a \in E$,*

$$|G| = |G(a)| \cdot |I_G(a)|.$$

Proof. For any $b \in E$, let

$$I_G(a \mapsto b) = \{\sigma \in G \mid \sigma(a) = b\}.$$

Arranging the elements of G according to where they map a, we obtain a decomposition of G into disjoint subsets

$$G = \bigcup_{b \in G(a)} I_G(a \mapsto b),$$

whence

$$|G| = \sum_{b \in G(a)} |I_G(a \mapsto b)|. \tag{14.14}$$

Now, if $b \in G(a)$ then $b = \sigma(a)$ for some $\sigma \in G$ and it is readily checked that

$$I_G(a \mapsto b) = \sigma \circ I_G(a).$$

Therefore, all the sets $I_G(a \mapsto b)$ for $b \in G(a)$ have the same number of elements as $I_G(a)$, and (14.14) yields

$$|G| = |G(a)| \cdot |I_G(a)|.$$

$\square$

The following observations on the orbits under G will often be useful in the sequel:

14.36. PROPOSITION. *With the same notation as in Definitions 14.34,*

(a) *for any $x \in G(a)$, we have $G(x) = G(a)$;*

(b) any two orbits under G are either disjoint or identical;

(c) the set E decomposes into a union of disjoint orbits under G.

Proof. *(a)* Let $x = \sigma(a)$ for some $\sigma \in G$. Then

$$G(x) = G(\sigma(a)) = (G \circ \sigma)(a).$$

Since composition on the right with σ is a bijection from G onto G, it follows that $G \circ \sigma = G$, hence

$$G(x) = G(a).$$

(b) Let $a, b \in E$ be such that $G(a) \cap G(b)$ is not empty. There is then an element $x \in G(a) \cap G(b)$. By *(a)*, we have $G(x) = G(a)$ and $G(x) = G(b)$, hence $G(a)$ and $G(b)$ coincide.

(c) This readily follows from *(b)*. Indeed, if A is a subset of E containing one element from each orbit under G, then

$$E = \bigcup_{a \in A} G(a),$$

and the orbits $G(a)$ are pairwise disjoint. $\square$

We thus obtain a first result on transitive subgroups of S_p:

14.37. COROLLARY. *Let G be a transitive group of permutations of a set E with p elements (p prime), and let N be a normal subgroup of G. If $N \neq \{\mathrm{Id}\}$, then N is transitive on E.*

Proof. Decompose E into a union of disjoint orbits under N,

$$E = \bigcup_{a \in A} N(a).$$

Counting elements, it follows that

$$p = \sum_{a \in A} |N(a)|.$$

To complete the proof, it now suffices to show that any two orbits have the same number of elements. Indeed, denoting this number by n, the equality above yields

$$p = n \cdot |A|$$

and since p is prime, there are only two possibilities: either $n = 1$, which means that N leaves every element of E invariant, hence that $N = \{\mathrm{Id}\}$; or $n = p$ and $|A| = 1$, which means that N is transitive on E. Thus, it only remains to prove:

Claim: $|N(a)| = |N(b)|$ for any $a, b \in E$.

Since G is transitive on E, there is a permutation $\sigma \in G$ such that $\sigma(a) = b$. Then

$$\sigma \circ N \circ \sigma^{-1}(b) = \sigma \circ N(a).$$

On the other hand $\sigma \circ N \circ \sigma^{-1} = N$ since N is normal in G, hence

$$N(b) = \sigma \circ N(a).$$

Therefore, composition with the permutation σ induces a bijection from $N(a)$ onto $N(b)$. This proves the claim. $\qquad\qquad\qquad\qquad\qquad\qquad\qquad\square$

In order to state Galois' classification of the solvable transitive subgroups of S_p, we recall the group $\mathrm{GA}(p)$ defined before Theorem 10.7, p. 148. For notational convenience, we consider S_p as the group of permutations of the set $\{0, 1, \ldots, p - 1\}$ (instead of $\{1, \ldots, p\}$), and we define

$$\tau \colon 0 \mapsto 1 \mapsto \cdots \mapsto p - 1 \mapsto 0.$$

Using the congruence relation modulo p, we can recast the definition of τ as

$$\tau \colon x \mapsto x + 1 \qquad \mathrm{mod}\ p$$

for $x \in \{0, \ldots, p - 1\}$, since $(p - 1) + 1 \equiv 0 \bmod p$.

For $i = 1, \ldots, p - 1$, we also define permutations σ_i by

$$\sigma_i \colon x \mapsto ix \qquad \mathrm{mod}\ p.$$

(Compare the definition before Theorem 10.7, p. 148.) The group $\mathrm{GA}(p)$ is the subgroup of S_p generated by $\sigma_1, \ldots, \sigma_{p-1}$ and τ. It is readily checked that the elements of $\mathrm{GA}(p)$ are the permutations of the form

$$x \mapsto ax + b \qquad \mathrm{mod}\ p$$

where $a \in \{1, \ldots, p - 1\}$ and $b \in \{0, \ldots, p - 1\}$. (This shows that $|\mathrm{GA}(p)| = p(p - 1)$, as claimed in §10.3.)

Galois' result (in Proposition 7 of his memoir [21, pp. 65 ff], [20, pp. 111–112]) is that $\mathrm{GA}(p)$ is essentially the only solvable transitive subgroup of S_p:

14.38. THEOREM. *Every solvable transitive group of S_p is conjugate to a subgroup of* GA(p), *i.e. is of the form* $\alpha \circ H \circ \alpha^{-1}$ *for some* $\alpha \in S_p$ *and some subgroup H of* GA(p). *In particular, the order of such a group divides* $p(p-1)$. *Conversely, every subgroup of S_p which is conjugate to a subgroup of* GA(p) *is solvable.*

An interpretation of this theorem, in the light of Lagrange's investigations, is that no reduction can be carried out beyond the equation of degree $(p-2)!$ of Theorem 10.7, p. 148. The following property of GA(p) is crucial for the proof of the theorem:

14.39. LEMMA. *Let* $\tau \in$ GA(p) *be defined as above by*

$$\tau(x) = x + 1 \mod p \qquad \text{for } x \in \{0, \ldots, p-1\}$$

and let $\theta \in S_p$. *If* $\theta \circ \tau \circ \theta^{-1} \in$ GA(p), *then* $\theta \in$ GA(p).

Proof. We first show that $\theta \circ \tau \circ \theta^{-1}$ is a power of τ. Assume on the contrary

$$\theta \circ \tau \circ \theta^{-1} : x \mapsto ax + b \mod p$$

for some $a \not\equiv 0, 1 \mod p$. Then it is easily seen by induction on i that

$$(\theta \circ \tau \circ \theta^{-1})^i : x \mapsto a^i x + (a^{i-1} + \cdots + a + 1)b \mod p.$$

In particular,

$$(\theta \circ \tau \circ \theta^{-1})^{p-1} : x \mapsto a^{p-1} x + (a^{p-2} + \cdots + a + 1)b \mod p.$$

Now, since $a \not\equiv 0 \mod p$, we have $a^{p-1} \equiv 1 \mod p$ by Fermat's theorem (Theorem 12.5, p. 171). Moreover, multiplying the coefficient of b by $a - 1$ we get

$$(a-1)(a^{p-2} + \cdots + a + 1) = a^{p-1} - 1 \equiv 0 \mod p.$$

Since $a - 1 \not\equiv 0 \mod p$, it follows that the coefficient of b is $0 \mod p$, hence

$$(\theta \circ \tau \circ \theta^{-1})^{p-1} = \text{Id}.$$

Since

$$(\theta \circ \tau \circ \theta^{-1})^{p-1} = \theta \circ \tau^{p-1} \circ \theta^{-1},$$

this last equality yields $\tau^{p-1} = \text{Id}$, a contradiction. Therefore, as claimed,

$$\theta \circ \tau \circ \theta^{-1} = \tau^i$$

for some integer i between 1 and $p - 1$.

Composing both sides by θ on the right, we get

$$\theta \circ \tau = \tau^i \circ \theta$$

which means that for all $x = 0, \ldots, p - 1$,

$$\theta(x + 1) = \theta(x) + i \mod p.$$

Arguing by induction on x, it follows that

$$\theta(x) = ix + \theta(0) \mod p$$

for all $x = 0, \ldots, p - 1$. This shows that $\theta = \tau^{\theta(0)} \circ \sigma_i$, whence $\theta \in \mathrm{GA}(p)$. $\square$

Proof of Theorem 14.38. Let G be a solvable transitive subgroup of S_p and let

$$G = G_0 \supset G_1 \supset \cdots \supset G_t = \{\mathrm{Id}\} \tag{14.15}$$

be a sequence of subgroups, each of which is normal of prime index in the preceding one. Since G is transitive, Corollary 14.37 (p. 283) implies that G_1 is transitive (unless $G_1 = \{\mathrm{Id}\}$, i.e. $t = 1$), whence also that G_2 is transitive (unless $G_2 = \{\mathrm{Id}\}$, i.e. $t = 2$), and so on up to G_{t-1}.

By Theorem 14.35 (p. 282), the number of elements in the orbit of any element in $\{0, \ldots, p - 1\}$ divides $|G_{t-1}|$. Since G_{t-1} is transitive on $\{0, \ldots, p - 1\}$, it follows that p divides $|G_{t-1}|$; but the order of G_{t-1}, which is the index of G_t in G_{t-1}, is prime, hence

$$|G_{t-1}| = p.$$

The subgroup of G_{t-1} generated by any element other than Id is then equal to G_{t-1}, since its order is a divisor of p. It follows that G_{t-1} is generated by a single permutation, which must be a cycle of length p, i.e. a permutation

$$\gamma \colon i_1 \mapsto i_2 \mapsto \cdots \mapsto i_p \mapsto i_1$$

where $i_1, i_2, \ldots, i_p$ are $0, 1, \ldots, p - 1$ in some order. Let $\alpha \in S_p$ be the permutation

$$\alpha \colon \begin{cases} 0 \mapsto i_1 \\ 1 \mapsto i_2 \\ \quad \vdots \\ p - 1 \mapsto i_p. \end{cases}$$

Then

$$\alpha^{-1} \circ \gamma \circ \alpha: \ 0 \mapsto 1 \mapsto 2 \mapsto \cdots \mapsto p - 1 \mapsto 0,$$

i.e. with the same notation as above,

$$\alpha^{-1} \circ \gamma \circ \alpha = \tau.$$

For $i = 0, 1, \ldots, t$, let G_i' be the image of G_i under the inner automorphism $\xi \mapsto \alpha^{-1} \circ \xi \circ \alpha$ of S_p,

$$G_i' = \alpha^{-1} \circ G_i \circ \alpha.$$

Transforming the given sequence (14.15) by this inner automorphism, we get a similar sequence

$$G_0' \supset G_1' \supset \cdots \supset G_t' = \{\mathrm{Id}\}$$

in which each subgroup is normal of prime index in the preceding subgroup. Moreover, since G_{t-1} is generated by γ, the next-to-last subgroup G_{t-1}' is generated by τ. Thus, $G_{t-1}' \subset \mathrm{GA}(p)$.

The subgroup G_{t-1}' is normal in G_{t-2}', hence for any $\theta \in G_{t-2}'$,

$$\theta \circ \tau \circ \theta^{-1} \in G_{t-1}'.$$

Consequently,

$$\theta \circ \tau \circ \theta^{-1} \in \mathrm{GA}(p) \qquad \text{for all } \theta \in G_{t-2}',$$

since $G_{t-1}' \subset \mathrm{GA}(p)$. Lemma 14.39 then shows that $G_{t-2}' \subset \mathrm{GA}(p)$. We can now repeat the same arguments with G_{t-2}' and G_{t-3}' instead of G_{t-1}' and G_{t-2}', to conclude that $G_{t-3}' \subset \mathrm{GA}(p)$. Repeating the same arguments as many times as needed, we eventually obtain $G_0' \subset \mathrm{GA}(p)$. Since $G = \alpha \circ G_0' \circ \alpha^{-1}$, it follows that G is conjugate to the subgroup G_0' of $\mathrm{GA}(p)$.

In order to prove that, conversely, every subgroup of S_p which is conjugate to a subgroup of $\mathrm{GA}(p)$ is solvable, it suffices, by Corollary 14.27 (p. 269), to prove that $\mathrm{GA}(p)$ itself is solvable. To this end, we choose a primitive root g of p (see Theorem 12.1, p. 169) and, for any factor e of $p - 1$, we define a subset H_e of $\mathrm{GA}(p)$ by

$$H_e =$$
$$\{x \mapsto g^{ei}x + c \bmod p \mid c = 0, \ldots, p - 1 \text{ and } i = 0, \ldots, (p-1)e^{-1} - 1\}.$$

A straightforward verification shows that this set is a normal subgroup of $GA(p)$. Moreover, we clearly have

$$|H_e| = \frac{p(p-1)}{e}.$$

If e, e' are factors of $p-1$ such that e divides e', then

$$H_e \supset H_{e'},$$

and by comparing the orders of these groups we see that the index of $H_{e'}$ in H_e is e'/e. Let now

$$p - 1 = q_1 \cdots q_r$$

be the decomposition of $p-1$ into a product of (not necessarily distinct) primes, and let

$$e_0 = 1, \quad e_1 = q_1, \quad e_2 = q_1 q_2, \quad \ldots, \quad e_{r-1} = q_1 \cdots q_{r-1}, \quad e_r = p - 1,$$

so that e_{i-1} divides e_i for $i = 1, \ldots, r$, with quotient $e_i/e_{i-1} = q_i$, a prime number. The sequence of subgroups

$$GA(p) = H_{e_0} \supset H_{e_1} \supset \cdots \supset H_{e_r} \supset \{\mathrm{Id}\}$$

then shows that $GA(p)$ is solvable. $\square$

Another characterization of the solvable transitive subgroups of S_p can be derived from the preceding theorem:

14.40. THEOREM. *A transitive subgroup of S_p is solvable if and only if no permutation in G leaves two elements of $\{0, \ldots, p-1\}$ invariant, except the identity.*

Proof. Assume first that G is solvable. By Theorem 14.38, there is a permutation $\alpha \in S_p$ such that

$$\alpha^{-1} \circ G \circ \alpha \subset GA(p).$$

If $\theta \in G$ leaves two elements u, v invariant, then $\alpha^{-1} \circ \theta \circ \alpha \in GA(p)$ leaves $\alpha^{-1}(u)$ and $\alpha^{-1}(v)$ invariant. But it is readily verified that no permutation of the form

$$x \mapsto ax + b \quad \text{with } a \in \{1, \ldots, p-1\} \text{ and } b \in \{0, \ldots, p-1\}$$

(i.e. no permutation in $GA(p)$) except the identity, leaves two elements invariant. Therefore, $\alpha^{-1} \circ \theta \circ \alpha = \mathrm{Id}$, hence $\theta = \mathrm{Id}$.

Conversely, if no permutation in G, besides the identity, leaves two elements invariant, then for $u, v \in \{0, \ldots, p-1\}$ with $u \neq v$,

$$I_G(u) \cap I_G(v) = \{\text{Id}\}.$$

Therefore, the set of permutations in G, other than Id, which leave an element invariant decomposes into the following union of disjoint subsets of G:

$$\bigcup_{u \in \{0, \ldots, p-1\}} (I_G(u) - \{\text{Id}\}). \tag{14.16}$$

In order to calculate the number of permutations in this set, we observe that, since G is transitive,

$$G(u) = \{0, \ldots, p-1\} \qquad \text{for all } u \in \{0, \ldots, p-1\}.$$

Therefore, by Theorem 14.35,

$$p \cdot |I_G(u)| = |G| \qquad \text{for all } u \in \{0, \ldots, p-1\}.$$

Letting $q = |I_G(u)|$ for $u \in \{0, \ldots, p-1\}$, we have

$$|G| = pq$$

and, using the decomposition (14.16), it follows that the number of elements in G, other than Id, which leave an element invariant is $p(q-1)$. There are therefore $p-1$ permutations in G which leave no element invariant. Let θ be such a permutation.

Claim: θ is a cycle of length p,

$$\theta: i_1 \mapsto i_2 \mapsto \cdots \mapsto i_p \mapsto i_1$$

where $i_1, \ldots, i_p$ are $0, 1, \ldots, p-1$ in some order, and the elements of G which leave no element invariant are $\theta, \theta^2, \ldots, \theta^{p-1}$.

Let T be the subgroup of G generated by θ,

$$T = \{\theta^k \mid k \in \mathbb{Z}\}.$$

We first show that $I_T(u) = \{\text{Id}\}$ for every $u \in \{0, \ldots, p-1\}$. Indeed, if

$$\theta^k(u) = u,$$

then, applying θ to both sides, we get

$$\theta^k(\theta(u)) = \theta(u),$$

hence every permutation in $I_T(u)$ leaves invariant the two elements u, $\theta(u)$. From the hypothesis on G, it follows that $I_T(u)$ is reduced to $\{\mathrm{Id}\}$.

By Theorem 14.35, this result implies that

$$|T(u)| = |T| \qquad \text{for all } u \in \{0, \dots, p-1\}.$$

Considering then the decomposition of $\{0, \dots, p-1\}$ into a union of disjoint orbits under T,

$$\{0, \dots, p-1\} = \bigcup_{u \in U} T(u).$$

By counting elements, we get

$$p = n \cdot |T|$$

where n is the number of distinct orbits. Since p is prime and $|T| > 1$, it follows that $|T| = p$ and $n = 1$, hence θ is a cycle of length p. Then, θ, θ^2, $\dots$, θ^{p-1} leave no element invariant and lie in G. There is no other permutation in G with this property, since their total number is $p - 1$, as previously noted. This proves the claim.

Let then $\alpha \in S_p$ be defined by

$$\alpha: \begin{cases} 0 \mapsto i_1 \\ 1 \mapsto i_2 \\ \quad \vdots \\ p-1 \mapsto i_p \end{cases}$$

so that, with the same notation as above,

$$\alpha^{-1} \circ \theta \circ \alpha = \tau.$$

Let ρ be any element in G. Since $\rho \circ \theta \circ \rho^{-1}$ is an element of G which leaves no element of E invariant, it is some power of θ. Let

$$\rho \circ \theta \circ \rho^{-1} = \theta^k$$

for some k between 1 and $p - 1$. Transforming this equation by the inner automorphism $\xi \mapsto \alpha^{-1} \circ \xi \circ \alpha$ of S_p, we get

$$(\alpha^{-1} \circ \rho \circ \alpha) \circ \tau \circ (\alpha^{-1} \circ \rho \circ \alpha)^{-1} = \tau^k.$$

Lemma 14.39 then shows that $\alpha^{-1} \circ \rho \circ \alpha \in \mathrm{GA}(p)$.

We have thus proved

$$\alpha^{-1} \circ G \circ \alpha \subset \text{GA}(p),$$

hence G is conjugate to a subgroup of $\text{GA}(p)$, and is therefore solvable. □

Of course, in order to justify the introduction of groups and to demonstrate the power and usefulness of this new tool, one has to come up with some new results which do not refer to groups in their statement but require some group theory in their proof. Only a couple of such results are quoted by Galois. The following is Proposition 8 in his memoir [21, p. 69], [20, p. 113]:

14.41. COROLLARY. *Let P be an irreducible polynomial of prime degree over a field F. The equation $P(X) = 0$ is solvable by radicals over F if and only if all the roots of P can be rationally expressed over F from any two of them.*

Proof. Denoting by $r_1, \ldots, r_p$ the roots of P (in some extension of F) and transforming the condition that $P(X) = 0$ is solvable by radicals into a condition on groups by Theorem 14.22, p. 265 (and Proposition 14.33, p. 280), we have to prove that $\text{Gal}(P/F)$ is solvable if and only if

$$r_1, \ldots, r_p \in F(r_i, r_j) \qquad \text{for any } i, j = 1, \ldots, p \text{ with } i \neq j.$$

First, we note that the irreducibility hypothesis on P implies that $\text{Gal}(P/F)$ is transitive on $r_1, \ldots, r_p$, by Proposition 14.32 (p. 279). We may thus apply the preceding results on transitive subgroups of S_p.

If $r_1, \ldots, r_p$ have rational expressions in r_i, r_j over F, then every permutation in $\text{Gal}(P/F)$ which leaves invariant r_i and r_j must leave invariant $r_1, \ldots, r_p$. It is thus the identity. From the characterization of solvable transitive groups of permutations of p elements in Theorem 14.40, it then follows that $\text{Gal}(P/F)$ is solvable.

Conversely, if $\text{Gal}(P/F)$ is solvable, then by the same characterization,

$$\text{Gal}(P/F(r_i, r_j)) = \{\text{Id}\} \qquad \text{for any } i, j = 1, \ldots, p \text{ with } i \neq j,$$

since Theorem 14.11 (p. 250) shows that $\text{Gal}(P/F(r_i, r_j))$ only contains permutations which leave r_i and r_j invariant. Therefore, this group leaves $r_1, \ldots, r_p$ invariant, whence

$$r_1, \ldots, r_p \in F(r_i, r_j),$$

by Theorem 14.11. □

This corollary can be effectively used to produce examples of non-solvable equations over the field $\mathbb{Q}$ of rational numbers, as we now show.

14.42. COROLLARY. *Let P be an irreducible polynomial of prime degree over $\mathbb{Q}$. If at least two roots of P, but not all, are real, then $P(X) = 0$ is not solvable by radicals over $\mathbb{Q}$.*

Proof. Let $r_1, \ldots, r_p$ be the roots of P, and assume $r_1, r_2 \in \mathbb{R}$ and $r_p \notin \mathbb{R}$. Then $\mathbb{Q}(r_1, r_2) \subset \mathbb{R}$, hence $r_p \notin \mathbb{Q}(r_1, r_2)$ and the preceding corollary shows that $P(X) = 0$ is not solvable over $\mathbb{Q}$. $\square$

The above condition on the roots of P is not hard to check in specific examples. If the degree of P is a prime congruent to 1 mod 4, it can even be done purely arithmetically with the aid of the discriminant, as the next corollary shows:

14.43. COROLLARY. *Let P be a monic irreducible polynomial of prime degree p over $\mathbb{Q}$. Assume $p \equiv 1$ mod 4. If the discriminant of P is negative, then $P(X) = 0$ is not solvable by radicals over $\mathbb{Q}$.*

Proof. By Exercise 4 of Chapter 8, the condition on the discriminant readily implies that the number of real roots of P is not 1 nor p. $\square$

As a specific example, equations

$$X^5 - pqX + p = 0$$

where p is prime and q is an integer, $q \geq 2$ (or $q \geq 1$ and $p \geq 13$) are not solvable by radicals over $\mathbb{Q}$, since $X^5 - pqX + p$ is irreducible over $\mathbb{Q}$, as Eisenstein's criterion (Proposition 12.12, p. 176) readily shows, and its discriminant is negative (see Exercise 1 of Chapter 8 for the calculation of this discriminant).

As a last application of Galois' investigations of equations of prime degree, we now give another proof of the Ruffini–Abel Theorem 13.16, p. 227.

14.44. COROLLARY. *For $n \geq 5$, the general equation of degree n is not solvable by radicals (over $k(s_1, \ldots, s_n)$).*

Proof. We have seen in Example 14.1 (p. 241) that the Galois group of the general equation of degree n is the group of all permutations of the roots, which can be identified to S_n. Since this group is obviously transitive on the roots, it follows from Proposition 14.32 (p. 279) that the general equation of degree n is irreducible (over $k(s_1, \ldots, s_n)$), hence that solvability of this equation implies its complete solvability (by Proposition 14.33, p. 280), which implies the solvability of S_n (by Theorem 14.22, p. 265).

Consider first the case $n = 5$. In this case we can apply Theorem 14.38 to conclude that S_5 is not solvable, since $|S_5| > 4 \cdot 5$.

It then follows from Corollary 14.27 (p. 269) that S_n is not solvable for $n \geq 5$, since S_5 can be identified to the subgroup of S_n which leaves all the elements of $\{1, \ldots, n\}$ invariant, except $\{1, \ldots, 5\}$. □

We now turn to the theorem of Abel quoted in §14.1.

14.45. THEOREM. *Let P be a polynomial of degree n over some field F, with roots $r_1, \ldots, r_n$ in some field containing F. If there exist rational fractions $\theta_2, \ldots, \theta_n \in F(X)$ such that*

$$r_i = \theta_i(r_1) \qquad \text{for } i = 2, \ldots, n$$

and

$$\theta_i(\theta_j(r_1)) = \theta_j(\theta_i(r_1)) \qquad \text{for all } i, j,$$

then the equation $P(X) = 0$ is completely solvable by radicals over F.

Proof. Using Hudde's trick in Theorem 5.21 (p. 54), we may assume without loss of generality that the roots $r_1, \ldots, r_n$ are pairwise distinct. By Example 14.3 (p. 242), the Galois group $\mathrm{Gal}(P/F)$ is abelian. Therefore, by Theorem 14.22 (p. 265), it suffices to prove the following group-theoretical statement:

14.46. PROPOSITION. *Every (finite) abelian group is solvable.*

Proof. Since every subgroup of an abelian group is normal, it suffices to prove that every finite abelian group $G \neq \{1\}$ contains a subgroup G_1 of prime index.

Arguing by induction on the order of G, we then construct a sequence of subgroups

$$G \supset G_1 \supset G_2 \supset \cdots \supset G_r = \{1\}$$

each of which is normal of prime index in the preceding one. This sequence shows that G is solvable.

It thus only remains to prove the existence of a subgroup of prime index in each finite non-trivial abelian group. This is a special case ($H = \{1\}$) of the following result:

14.47. LEMMA. *Let H be a subgroup of a finite abelian group G. If $H \neq G$,
then there exists in G a subgroup G_1 of prime index which contains H.*

Proof. We argue by induction on the index $(G : H)$, which is assumed to be at
least 2. If $(G : H) = 2, 3$ or any other prime number, then $G_1 = H$ satisfies the
required conditions. Assume then $(G : H)$ is not prime. Pick σ in G but not in
H and consider the minimal exponent $e > 0$ for which $\sigma^e \in H$. Let also p be a
prime factor of e and

$$\rho = \sigma^{e/p}.$$

Then $\rho \notin H$ (otherwise e would not be minimal), and $\rho^p \in H$. Consider then

$$H' = \{\rho^i \mu \mid i = 0, \ldots, p - 1; \; \mu \in H\}.$$

It is easily checked that H' is a subgroup of G containing H, and that the cosets of
H in H' are $H', \rho H', \rho^2 H', \ldots, \rho^{p-1} H'$ (compare the proof of Corollary 14.26,
p. 268), so that

$$(H' : H) = p.$$

Therefore, $H' \neq G$, since by hypothesis $(G : H)$ is not prime, and

$$(G : H') < (G : H).$$

From the induction hypothesis, it follows that there exists in G a subgroup G_1 of
prime index which contains H', hence also H. $\square$

Remark. Proposition 14.46 can be used to show that the definition of solvability
given after the statement of Theorem 14.22 (p. 265) in the case of finite groups
has other equivalent formulations, which make sense for infinite groups as well.

For each group G, we define a derived subgroup G': it is the subgroup of G
generated by commutators $\sigma \tau \sigma^{-1} \tau^{-1}$, for $\sigma, \tau \in G$. For any positive integer
$n \geq 2$, the n-th derived group $G^{(n)}$ is inductively defined as the derived subgroup
of $G^{(n-1)}$.

PROPOSITION. *The following conditions on a group G are equivalent:*

(a) *there exists an integer n such that $G^{(n)} = \{1\}$;*
(b) *G contains a sequence of subgroups*

$$G = G_0 \supset G_1 \supset \cdots \supset G_t = \{1\}$$

such that each subgroup G_i is normal in the preceding subgroup G_{i-1}, with abelian factor group G_{i-1}/G_i, for $i = 1, \ldots, r$.

Moreover, if G is finite, these conditions are also equivalent to:

(c) G contains a sequence of subgroups

$$G = G_0 \supset G_1 \supset \cdots \supset G_r = \{1\}$$

such that each subgroup is normal of prime index in the preceding one.

Proof. (a) $\Rightarrow$ (b) The sequence defined by $G_i = G^{(i)}$ satisfies the required conditions.

(b) $\Rightarrow$ (a) Since G_{i-1}/G_i is abelian, we have $G'_{i-1} \subset G_i$. By induction, it follows that $G^{(t)} \subset G_t$, hence $G^{(t)} = \{1\}$.

(c) $\Rightarrow$ (b) Each factor group G_{i-1}/G_i for $i = 1, \ldots, r$ has prime order and is therefore abelian, since it is generated by any single element (except the identity).

(b) $\Rightarrow$ (c) *if G is finite:* By Proposition 14.46, each factor group G_{i-1}/G_i contains a sequence of subgroups

$$G_{i-1}/G_i \supset H_{i1} \supset \cdots \supset H_{ir_i} = \{1\}$$

in which each subgroup has prime index in the preceding subgroup. Taking inverse images of $H_{i1}, \ldots, H_{ir_i}$ under the canonical projection $\pi \colon G_{i-1} \to G_{i-1}/G_i$, we obtain a sequence of subgroups

$$G_{i-1} \supset \pi^{-1}(H_{i1}) \supset \cdots \supset \pi^{-1}(H_{ir_i}) = G_i$$

in which each subgroup is normal of prime index in the preceding subgroup. The sequences thus obtained can be joined end to end to produce a sequence of subgroups starting at G and ending with $\{1\}$, in which each subgroup is normal of prime index in the preceding one. This shows (c). $\square$

Appendix: Galois' description of groups of permutations

Although groups of permutations have been widely used in the preceding discussion of Galois' results, it should be observed that the notion of group in Galois' papers is slightly different from the modern one. Indeed, in Galois' approach to groups of permutations of a set E, the central role is played by the *arrangements*[‡]

[‡]Galois uses the term "permutation" for what is called an "arrangement" here, and "substitution" for what is usually called "permutation" nowadays. Because of possible confusions, we avoid using

of the elements of E, which are the various ways of ranging the elements of E in a row, while nowadays the fundamental objects are the *substitutions* (or permutations), i.e. the 1–1 mappings from E onto itself.

The purpose of this appendix is to present Galois' description of groups and to point out how Galois' definitions are related to modern ones.

Let E be a finite set and let Ω be the set of arrangements of the elements of E. Thus, if for instance $E = \{a, b, c\}$, then

$$\Omega = \{abc, acb, bac, bca, cab, cba\}.$$

We denote by $\mathrm{Sym}(E)$ the set of substitutions of E (which have been up to here called permutations of E). The substitutions of E induce substitutions of Ω in an obvious way: a substitution σ transforms an arrangement $\alpha = abc\ldots$ into $\sigma(\alpha) = \sigma(a)\sigma(b)\sigma(c)\ldots$. This action of $\mathrm{Sym}(E)$ on Ω has the following remarkable, yet obvious, property: for any α, $\beta \in \Omega$, there is one and only one substitution $\sigma \in \mathrm{Sym}(E)$ such that $\sigma(\alpha) = \beta$. (This property is sometimes expressed as follows: Ω is a *principal homogeneous set* under $\mathrm{Sym}(E)$.)

DEFINITIONS. A *group of arrangements* of E is a non-empty subset A of Ω which has the following property: for any ξ, η, $\zeta \in A$, the substitution which transforms ξ into η transforms ζ into an arrangement which also belongs to A. In other words, if $\sigma \in \mathrm{Sym}(E)$ is such that $\sigma(\xi) \in A$ for some $\xi \in A$, then $\sigma(\zeta) \in A$ for all $\zeta \in A$, i.e. $\sigma(A) \subset A$ (and in fact $\sigma(A) = A$ since the number of arrangements in $\sigma(A)$ is the same as in A).

A *group of substitutions* of E is (as usual) a subgroup of $\mathrm{Sym}(E)$.

PROPOSITION. *There is a 1–1 correspondence between groups of substitutions of E and groups of arrangements of E which contain a given arrangement α. This correspondence associates to any group of substitutions G the orbit $G(\alpha) \subset \Omega$, and to any group of arrangements A the set $\{\sigma \in \mathrm{Sym}(E) \mid \sigma(\alpha) \in A\}$.*

Proof. First, we show that $G(\alpha)$ is a group of arrangements of E. Let $\sigma \in \mathrm{Sym}(E)$ be such that $\sigma(\xi) \in G(\alpha)$ for some $\xi \in G(\alpha)$; we have to prove $\sigma(G(\alpha)) = G(\alpha)$. The hypotheses that ξ and $\sigma(\xi)$ are in G yield

$$\xi = \tau(\alpha) \quad \text{and} \quad \sigma(\xi) = \theta(\alpha) \quad \text{for some } \tau, \theta \in G.$$

the term "permutation" as far as possible in the appendix, and use "arrangement" and "substitution" instead.

Hence, $\sigma \circ \tau(\alpha) = \theta(\alpha)$ and therefore

$$\sigma \circ \tau = \theta.$$

This equality shows that $\sigma \in G$, whence $\sigma \circ G = G$ and $\sigma(G(\alpha)) = G(\alpha)$.

Next, we prove that if A is a group of arrangements containing α, then the set

$$S_\alpha(A) = \{\sigma \in \mathrm{Sym}(E) \mid \sigma(\alpha) \in A\}$$

is a subgroup of $\mathrm{Sym}(E)$. This set contains the identity, since

$$\mathrm{Id}(\alpha) = \alpha \in A.$$

If $\sigma, \tau \in S_\alpha(A)$, then the property of groups of arrangements, applied with $\xi = \alpha$, $\eta = \sigma(\alpha)$ and $\zeta = \tau(\alpha)$, yields

$$\sigma \circ \tau(\alpha) \in A,$$

hence $\sigma \circ \tau \in S_\alpha(A)$. Likewise, if $\sigma \in S_\alpha(A)$, the same property applied with $\xi = \sigma(\alpha), \eta = \alpha$ and $\zeta = \alpha$ yields

$$\sigma^{-1}(\alpha) \in A,$$

hence $\sigma^{-1} \in S_\alpha(A)$. This shows that $S_\alpha(A)$ is a group of substitutions.

To complete the proof, it remains to see that the maps

$$G \mapsto G(\alpha) \quad \text{and} \quad A \mapsto S_\alpha(A)$$

are reciprocal bijections, i.e. that

$$S_\alpha(G(\alpha)) = G \tag{14.17}$$

and

$$S_\alpha(A)(\alpha) = A. \tag{14.18}$$

These equalities both readily follow from the definitions (and the fact that Ω is a principal homogeneous set under $\mathrm{Sym}(E)$). $\qquad \square$

COROLLARY. *If α and β both belong to a group of arrangements A, then*

$$\{\sigma \in \mathrm{Sym}(E) \mid \sigma(\alpha) \in A\} = \{\sigma \in \mathrm{Sym}(E) \mid \sigma(\beta) \in A\},$$

i.e., with the notation of the preceding proof,

$$S_\alpha(A) = S_\beta(A).$$

Proof. Equation (14.18) yields

$$\beta \in S_\alpha(A)(\alpha),$$

which means that β is in the orbit of α under the group $S_\alpha(A)$. Therefore, by Proposition 14.36(a) (p. 282),

$$S_\alpha(A)(\alpha) = S_\alpha(A)(\beta)$$

whence, by equation (14.18),

$$A = S_\alpha(A)(\beta).$$

Taking the images of both sides under S_β and applying (14.17) (with β instead of α and $S_\alpha(A)$ instead of G), we get

$$S_\beta(A) = S_\alpha(A).$$

$\square$

This corollary shows that the group of substitutions $S_\alpha(A)$ which corresponds to a given group of arrangements A does not depend on the choice of a particular reference arrangement α in A. By contrast, the choice of a reference arrangement α plays an important role for the passage from groups of substitutions to groups of arrangements, since different groups of arrangements may correspond to the same group of substitutions. For instance, if $E = \{a, b, c\}$ as above and if $G = \{\mathrm{Id}, \tau\}$, where τ interchanges a and b and leaves c invariant, then choosing as reference α the arrangement abc we get the group

$$\{abc, bac\}$$

whereas taking bca as reference we get

$$\{bca, acb\}.$$

This shows that groups of substitutions are more natural than groups of arrangements, in that they do not depend on the choice of a reference arrangement. Certain passages in his memoir leave no doubt that Galois was aware of the fact that the basic notion was ultimately that of substitution, instead of arrangement. However, he seems to have settled for arrangements because of their more concrete, tangible nature, as the following quotations suggest.

The following is from the introductory principles [21, p. 47], [20, p. 102]:

The initial permutation one uses to describe substitutions is entirely arbitrary when one is dealing with functions, because there is no reason, in a function of several letters, for a letter to occupy one position rather than another. Nonetheless, since one can hardly comprehend the idea of a substitution without that of a permutation, we shall frequently speak of permutations, and we shall consider substitutions only as the passage from one permutation to another.

After his Proposition 1 [21, p. 53], [20, p. 106] (i.e. Theorem 14.11 above, p. 250), Galois writes:

Scholium. Clearly in the group of permutations under discussion the disposition of the letters is of no importance, but only the *substitutions* of the letters by which one passes from one permutation to the other.

A positive point in Galois' description with groups of arrangements is that the notion of subgroup, and particularly of normal subgroup, arises in a fairly natural way, as we now show.

If H is a subgroup of a group of substitutions G, then the corresponding group of arrangements $H(\alpha)$ (obtained from a reference arrangement α) is clearly a subgroup of $G(\alpha)$. Moreover, the decomposition of G into left cosets of H

$$G = \bigcup_{\sigma \in R} \sigma \circ H$$

where R is a set of representatives of the cosets of H in G, i.e. a subset of G containing one and only one element from each coset, yields a decomposition of the group of arrangements $G(\alpha)$, as follows:

$$G(\alpha) = \bigcup_{\sigma \in R} \sigma(H(\alpha)).$$

The subsets $\sigma(H(\alpha))$, for $\sigma \in R$, are pairwise disjoint and are in fact subgroups of $G(\alpha)$, since the equality

$$\sigma(H(\alpha)) = (\sigma \circ H \circ \sigma^{-1})(\sigma(\alpha))$$

shows that $\sigma(H(\alpha))$ is the orbit of $\sigma(\alpha)$ under the group of substitutions $\sigma \circ H \circ \sigma^{-1}$. The set $\sigma(H(\alpha))$ is therefore the group of arrangements containing $\sigma(\alpha)$ and associated with the group of substitutions $\sigma \circ H \circ \sigma^{-1}$.

The normality of H in G translates as follows: the groups of substitutions of each $\sigma\big(H(\alpha)\big)$ are all equal to H. Indeed, this condition amounts to

$$\sigma \circ H \circ \sigma^{-1} = H \qquad \text{for all } \sigma \in R$$

and since every element in G has the form $\sigma \circ \tau$ for some $\sigma \in R$ and some $\tau \in H$, it follows that

$$\rho \circ H \circ \rho^{-1} = H \qquad \text{for all } \rho \in G.$$

For instance, let $E = \{a, b, c\}$, let $G = \text{Sym}(E)$ and $H = \{\text{Id}, \tau\}$ where τ interchanges a and b and leaves c invariant. Choose $\alpha = abc$ as reference arrangement. Then $G(\alpha)$ is the group of all arrangements of a, b, c, which decomposes into three subgroups: $H(\alpha)$ and two other subgroups of the form $\sigma\big(H(\alpha)\big)$, which are obtained by applying a single substitution to all the arrangements of $H(\alpha)$:

One passes from the first subgroup of arrangements (which is $H(\alpha)$), to the second one by applying on all the arrangements the substitution $b \leftrightarrow c$ (or, equivalently, the substitution $a \mapsto c \mapsto b \mapsto a$), and to the third one by applying $a \mapsto b \mapsto c \mapsto a$ (or, equivalently, $a \leftrightarrow c$).

That H is not normal in G is reflected in the fact that the three groups of arrangements do not have the same group of substitutions. Indeed, the first group of substitutions is H, the second is $\{\text{Id}, a \leftrightarrow c\}$ and the third is $\{\text{Id}, b \leftrightarrow c\}$.

If we choose instead of H the group $N = \{\text{Id}, a \mapsto b \mapsto c \mapsto a, a \mapsto c \mapsto b \mapsto a\}$ (which is the alternating group on E), then the corresponding decomposition of $G(\alpha)$ is

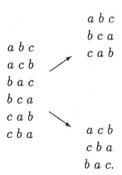

$$
\begin{array}{l}
a\ b\ c \\
a\ c\ b \\
b\ a\ c \\
b\ c\ a \\
c\ a\ b \\
c\ b\ a
\end{array}
\nearrow
\begin{array}{l}
a\ b\ c \\
b\ c\ a \\
c\ a\ b
\end{array}
$$

$$
\searrow
\begin{array}{l}
a\ c\ b \\
c\ b\ a \\
b\ a\ c.
\end{array}
$$

The second subgroup of arrangements is obtained from the first by applying the substitution $b \leftrightarrow c$, and the two subgroups both have N as group of substitutions.

Exercises

1. Let P be a polynomial over some field F. Show that if P is irreducible over F, then $|\mathrm{Gal}(P/F)|$ is divisible by the degree of P.

2. Let V be a Galois resolvent of an equation $P(X) = 0$ over a field F. Show that for any $\sigma \in \mathrm{Gal}(P/F)$, the function $\sigma(V)$ also is a Galois resolvent of $P(X) = 0$.

3. Show that an equation $P(X) = 0$ is completely solvable by radicals over a field F if and only if for each irreducible factor Q of P, the equation $Q(X) = 0$ is solvable by radicals over F.

4. Let $P(X) = (X - x_1) \cdots (X - x_n) = X^n - s_1 X^{n-1} + s_2 X^{n-2} - \cdots + (-1)^n s_n$ be the general polynomial of degree n over some field of constants k, and let $F = k(s_1, \ldots, s_n)$. Let $u \in k(x_1, \ldots, x_n)$ and let $u_1, \ldots, u_r$ be the various (distinct) values of u under the permutations of $x_1, \ldots, x_n$ (with $u = u_1$, say). Show that the polynomial $(X - u_1) \cdots (X - u_r)$ is irreducible over F. Deduce as in Example 14.30 (p. 275) that $\mathrm{Gal}(P/F(u)) = I(u)$.

5. Let G be a group of substitutions of a finite set E, let H be a subgroup of G and let α be an arrangement of the elements of E. Show that the group of arrangements $G(\alpha)$ can be decomposed into subgroups which have H as group of substitutions. Show that this decomposition is identical to $G(\alpha) = \cup_{\sigma \in R} \sigma\big(H(\alpha)\big)$ (where R is a set of representatives of the left cosets of H in G) if and only if H is normal in G.

Chapter 15

Epilogue

Although Galois' memoir is nowadays regarded as the climax of several decades of research on algebraic equations, the first reactions to Galois' theory were negative. It was rejected by the referees, because the arguments were "not clear enough nor developed enough" (Taton [57, p. 121]), but also for another, deeper motive: it did not yield any workable criterion to determine whether an equation is solvable by radicals. In that respect, even the application to equations of prime degree indicated by Galois (see Corollary 14.41, p. 291) is hardly useful, as the referees pointed out:

> However, one should observe that [the memoir] does not contain, as [its] title promised, the condition of solvability of equations by radicals; indeed, assuming as true M. Galois' proposition, one would not derive from it any good way of deciding whether a given equation of prime degree is solvable or not by radicals, since one would have first to verify whether this equation is irreducible and next whether any of its roots can be expressed as a rational fraction of two others. The condition for solvability, if it exists, ought to have an external character which can be verified by inspecting the coefficients of a given equation or, at most, by solving other equations of degrees lower than that of the proposed equation. (Taton [57, p. 121])

Galois' criterion (see Theorem 14.22, p. 265) was very far from being external; indeed, Galois always worked with the roots of the proposed equation, never with its coefficients.* Thus, Galois' theory did not correspond to what was expected, it

---

*It is telling that the proposed equation is nowhere displayed in Galois' memoir.

was too novel to be readily accepted.

After the publication of Galois' memoir by Liouville, its importance dawned upon the mathematical world, and it was eventually realized that Galois had discovered a mathematical gem much more valuable than any hypothetical external characterization of solvable equations. After all, the problem of solving equations by radicals was utterly artificial. It had focused the efforts of several generations of brilliant mathematicians because it displayed some strange, puzzling phenomena. It contained something mysterious, profoundly appealing. Galois had taken the pith out of the problem, by showing that the difficulty of an equation was related to the ambiguity of its roots and pointing out how this ambiguity could be measured by means of a group. He had thus set the theory of equations and, indeed, the whole subject of algebra, on a completely different track.

> Now, I think that the simplifications produced by the elegance of calculations (intellectual simplifications, I mean; there is no material simplification) are limited; I think the moment will come where the algebraic transformations foreseen by the speculations of analysts will not find nor the time nor the place to occur any more; so that one will have to be content with having foreseen them. [...]
> Jump above calculations; group the operations, classify them according to their complexities rather than their appearances; this, I believe, is the mission of future mathematicians; this is the road on which I am embarking in this work. [21, p. 9]

Thereafter, the theory of equations slowly disappeared, while new subjects emerged, such as the theory of groups and of various algebraic structures. This final stage in the evolution of a mathematical theory has been beautifully described by A. Weil [68, p. 52]:

> Nothing is more fruitful, as all mathematicians know, than these dim analogies, these foggy glimpses from one theory to the other, these stealthy caresses, these inexplicable jumbles; nothing also gives more pleasure to the researcher. A day comes when the illusion dissipates; the vagueness changes into certainty; the twin theories disclose their common fount before vanishing; as the Gītā teaches, one reaches knowledge and indifference at the same time. Metaphysics has become mathemat-

ics, ready to make the substance of a treatise whose cold beauty could not move us any more.

The subsequent developments arising from Galois theory do not fall within the scope of these lectures, so we refer to the papers by Kiernan [37] and by Van der Waerden [63] and to the book by Nový [47] for detailed accounts. There is however one major trend in this evolution that we want to point out: the gradual elimination of polynomials and equations from the foundations of Galois theory. Indeed, it is revealing of the profoundness of Galois' ideas to see, through the various textbook expositions, how this theory initially designed to answer a question about equations progressively outgrew its original context.

The first step in this direction is the emergence of the notion of field, through the works of Kronecker and Dedekind. Their approaches were quite different but complementary. Kronecker's point of view was constructivist. To define a field according to this point of view is to describe a process by which the elements of the field can be constructed. By contrast, Dedekind's approach was set-theoretic. He does not hesitate to define the field generated by a set P of complex numbers as the intersection of all the fields which contain P. This definition is hardly useful for determining whether a given complex number belongs to the field thus defined. Although Dedekind's approach has become the usual point of view nowadays, Kronecker's constructivism also led to important results, such as the algebraic construction of fields in which polynomials split into linear factors, see §9.2. The next step is the observation by Dedekind, around the end of the nineteenth century, that the permutations in the Galois group of an equation can be considered as automorphisms of the field of rational fractions of the roots (see Corollary 14.12, p. 251). Moreover, the newly developed linear algebra was brought to bear on the theory of fields, as the larger field in an extension can be regarded as a vector space over the smaller field.

These ideas came to fruition in the first decades of the twentieth century, as witnessed by the famous treatise of B.L. Van der Waerden "Moderne Algebra" (1930) (of which [61] is the seventh edition). The treatment of Galois theory in this book is based on lectures by E. Artin. It states as its "fundamental theorem" a 1–1 correspondence between subfields of certain extensions (those which are obtained by adjoining all the roots of a polynomial without multiple root), nowadays called *Galois* extensions, and the subgroups of the associated Galois group. This correspondence is not quite explicit in Galois' memoir. It can be observed in the dual statements of Corollary 14.21 (p. 264) and Lemma 14.28 (p. 270), and in the proof of the criterion for solvability of an equation by radicals (Theorem 14.22, p. 265), but in this proof it is obscured by the fact that roots of unity are not assumed to be in the base field, while they are needed for the application of Corollary 14.21 or Lemma 14.28.

In Van der Waerden's book, the treatment of Galois theory clearly emphasizes fields and groups, while polynomials and equations play a secondary role. They are used as tools in the proofs, but the main theorems do not involve polynomials in their statement.

A few years later, the exposition of Galois theory further evolved under the influence of Emil Artin, who once wrote [3, p. 380]:

> Since my mathematical youth, I have been under the spell of the classical theory of Galois. This charm has forced me to return to it again and again, and to try to find new ways to prove these fundamental theorems.

In his book "Galois theory" [2] (1942), Artin proposes a new, highly original, definition of Galois extension. The extension is looked at from the point of view of the larger field instead of the smaller. An extension of fields is then called Galois if the smaller field is the field of invariants under a (finite) group of automorphisms of the larger. This definition and some improvements in the proofs enabled Artin to further reduce the role of polynomials in the basic results of Galois theory, so that the fundamental theorem can now be proved without ever mentioning polynomials (see the appendix).

Artin's exposition has nowadays become the classical treatment of Galois theory from an elementary point of view. However, several other expositions have been proposed in more recent times, inspired by the applications of Galois theory in related areas. For instance, the Jacobson–Bourbaki correspondence [33, p. 22] yields a uniform treatment of both the classical Galois theory and the Galois theory for purely inseparable field extensions of height 1, where restricted p-Lie algebras are substituted for groups. In another direction, the Galois theory of commutative rings due to Chase, Harrison and Rosenberg [13], has inspired new expositions which stress the analogy between extensions of fields and coverings of locally compact topological spaces, see Douady [19] (compare also the new version of Bourbaki's treatise [7]).

Through its applications in various areas and as a source of inspiration for new investigations, Galois theory is far from being a closed issue.

Appendix: The fundamental theorem of Galois theory

To conclude these lectures, we now give an account of the 1–1 correspondence which is now regarded as the fundamental theorem of Galois theory, after Artin's classical exposition in [2].

15.1. DEFINITIONS. Let K be a field containing a subfield F. The dimension of K, regarded as a vector space over F, is called the *degree* of K over F, and is denoted by $[K : F]$, so

$$[K : F] = \dim_F K.$$

The group of (field-)automorphisms of K which leave F elementwise invariant is called the *Galois group of K over F*, and is denoted by $\mathrm{Gal}(K/F)$,

$$\mathrm{Gal}(K/F) = \mathrm{Aut}_F K.$$

The extension K/F is called a *Galois* extension if F is the field of all elements which are invariant under some finite group of automorphisms of K. In other words, denoting by K^G the field of invariants under a group G of automorphisms of K, i.e.

$$K^G = \{x \in K \mid \sigma(x) = x \text{ for all } \sigma \in G\},$$

the extension K/F is Galois if and only if there exists a finite group G of automorphisms of K such that $F = K^G$.

For instance, if F is a field of characteristic zero and if $r_1, \ldots, r_n$ are the roots of a polynomial with coefficients in F, then Theorem 14.11, p. 250 (and Corollary 14.12, p. 251) show that $F(r_1, \ldots, r_n)$ is a Galois extension of F.

15.2. THEOREM (FUNDAMENTAL THEOREM OF GALOIS THEORY). *Let K be a field containing a subfield F. If $F = K^G$ for some finite group G of automorphisms of K, then*

$$[K : F] = |G| \quad \text{and} \quad G = \mathrm{Gal}(K/F).$$

The field K is then a Galois extension of every subfield containing F. Moreover, there is a 1–1 correspondence between the subfields of K containing F and the subgroups of G, which associates to any subfield L the Galois group $\mathrm{Gal}(K/L) \subset G$ and to any subgroup $H \subset G$ its field of invariants K^H.

Under this correspondence, the degree over F of a subfield of K corresponds to the index in G of the associated subgroup,

$$[L : F] = \big(G : \text{Gal}(K/L)\big) \quad \text{and} \quad (G : H) = [K^H : F].$$

Furthermore, a subfield L of K is Galois over F if and only if the corresponding subgroup $\text{Gal}(K/L)$ is normal in G. The Galois group $\text{Gal}(L/F)$ is obtained by restricting to L the automorphisms in G, and the restriction homomorphism induces an isomorphism $G/\text{Gal}(K/L) \xrightarrow{\sim} \text{Gal}(L/F)$.

Thus, it follows from Theorem 14.11 (p. 250) that the group $\text{Gal}(P/F)$ defined in §14.2 is the Galois group of $F(r_1, \ldots, r_n)$ over F, provided that its elements are considered as field-automorphisms of $F(r_1, \ldots, r_n)$ instead of permutations of $r_1, \ldots, r_n$.

The proof of this theorem requires some preparation. We start with a very simple observation which parallels Lemma 14.24, p. 266:

15.3. LEMMA. *Let $K \supset L \supset F$ be a tower of fields. Then*

$$[K : F] = [K : L][L : F].$$

Proof. Let $(k_i)_{i \in I}$ be a basis of K over L and $(\ell_j)_{j \in J}$ be a basis of L over F. If we prove $(k_i \ell_j)_{(i,j) \in I \times J}$ is a basis of K over F, the lemma readily follows.

The family $(k_i \ell_j)_{(i,j) \in I \times J}$ spans K since every element $x \in K$ can be written

$$x = \sum_{i \in I} k_i x_i$$

for some $x_i \in L$, and decomposing

$$x_i = \sum_{j \in J} \ell_j y_{ij}$$

with $y_{ij} \in F$, we end up with

$$x = \sum_{\substack{i \in I \\ j \in J}} (k_i \ell_j) y_{ij}.$$

To show that the family $(k_i \ell_j)_{(i,j) \in I \times J}$ is linearly independent over F, consider

$$\sum_{\substack{i \in I \\ j \in J}} k_i \ell_j y_{ij} = 0$$

for some $y_{ij} \in F$. Collecting terms which have the same index i, we get

$$\sum_{i \in I} k_i \left(\sum_{j \in J} \ell_j y_{ij} \right) = 0$$

hence

$$\sum_{j \in J} \ell_j y_{ij} = 0 \qquad \text{for all } i \in I,$$

since $(k_i)_{i \in I}$ is linearly independent over L, hence also

$$y_{ij} = 0 \qquad \text{for all } i \in I, j \in J,$$

since $(\ell_j)_{j \in J}$ is linearly independent over F. $\qquad\qquad\qquad\square$

The basic observation which lies at the heart of the proof of the fundamental theorem is known as the lemma *of linear independence of homomorphisms*. It is due to Artin, and generalizes an earlier result of Dedekind.

15.4. LEMMA. *Consider distinct homomorphisms $\sigma_1, \ldots, \sigma_n$ of a field L into a field K. Then $\sigma_1, \ldots, \sigma_n$, viewed as elements of the K-vector space $\mathcal{F}(L, K)$ of all maps from L to K, are linearly independent over K. In other words, if $a_1, \ldots, a_n \in K$ are such that*

$$a_1 \sigma_1(x) + \cdots + a_n \sigma_n(x) = 0 \qquad \text{for all } x \in L,$$

then $a_1 = \cdots = a_n = 0$.

Proof. Assume on the contrary that $\sigma_1, \ldots, \sigma_n$ are not independent, and choose $a_1, \ldots, a_n \in K$ such that

$$a_1 \sigma_1(x) + \cdots + a_n \sigma_n(x) = 0 \qquad \text{for all } x \in L \qquad (15.1)$$

with $a_1, \ldots, a_n$ not all zero, but such that the number of $a_i \neq 0$ be minimal. This number is at least equal to 2, otherwise one of the σ_i would map L to $\{0\}$. This is impossible since, by definition of homomorphisms of fields, $\sigma_i(1) = 1$ for all i. Changing the numbering of $\sigma_1, \ldots, \sigma_n$ if necessary, we may thus assume without loss of generality that $a_1 \neq 0$ and $a_2 \neq 0$.

Choose $\ell \in L$ such that $\sigma_1(\ell) \neq \sigma_2(\ell)$. (This is possible since $\sigma_1 \neq \sigma_2$.) Multiplying both sides of (15.1) by $\sigma_1(\ell)$, we get

$$a_1 \sigma_1(\ell) \sigma_1(x) + \cdots + a_n \sigma_1(\ell) \sigma_n(x) = 0 \qquad \text{for all } x \in L. \qquad (15.2)$$

On the other hand, substituting ℓx for x in equation (15.1), and using the multiplicative property of σ_i, we get

$$a_1\sigma_1(\ell)\sigma_1(x) + \cdots + a_n\sigma_n(\ell)\sigma_n(x) = 0 \qquad \text{for all } x \in L. \qquad (15.3)$$

Subtracting (15.3) from (15.2), the first terms of each equation cancel out and we obtain

$$a_2\big(\sigma_1(\ell) - \sigma_2(\ell)\big)\sigma_2(x) + \cdots + a_n\big(\sigma_1(\ell) - \sigma_n(\ell)\big)\sigma_n(x) = 0 \quad \text{for all } x \in L.$$

The coefficients are not all zero since $\sigma_1(\ell) \neq \sigma_2(\ell)$, but this linear combination has fewer non-zero terms than (15.1). This is a contradiction. $\square$

Remark. Only the multiplicative property of $\sigma_1, \ldots, \sigma_n$ has been used. The same proof thus establishes the linear independence of distinct homomorphisms from any group to the multiplicative group of a field.

15.5. COROLLARY. *Let $\sigma_1, \ldots, \sigma_n$ be as in Lemma 15.4 and let*

$$F = \{x \in L \mid \sigma_1(x) = \cdots = \sigma_n(x)\}.$$

Then $[L : F] \geq n$.

Proof. Suppose, by way of contradiction, $[L : F] < n$, and let $[L : F] = m$. Choose a basis $\ell_1, \ldots, \ell_m$ of L over F and consider the matrix $\big(\sigma_i(\ell_j)\big)_{\substack{1 \leq i \leq n \\ 1 \leq j \leq m}}$ with entries in K.

The rank of this matrix is at most m, since the number of its columns is m, hence its rows are linearly dependent over K. We can therefore find elements $a_1, \ldots, a_n \in K$, not all zero, such that

$$a_1\sigma_1(\ell_j) + \cdots + a_n\sigma_n(\ell_j) = 0 \qquad \text{for } j = 1, \ldots, m. \qquad (15.4)$$

Now, any $x \in L$ can be expressed as

$$x = \sum_{j=1}^{m} \ell_j x_j$$

for some $x_j \in F$. Multiplying equations (15.4) by $\sigma_1(x_j)$ (which is equal to $\sigma_i(x_j)$ for all i, since $x_j \in F$) and adding the equations thus obtained, we get

$$a_1\sigma_1(x) + \cdots + a_n\sigma_n(x) = 0.$$

This contradicts Lemma 15.4. $\square$

We are now ready for the proof of Theorem 15.2. We let $F = K^G$ for some finite group G of automorphisms of the field K.

Step 1: $[K : F] = |G|$.

Applying Corollary 15.5 with $L = K$ and $\{\sigma_1, \ldots, \sigma_n\} = G$, we already have

$$[K : F] \geq |G|.$$

If $[K : F] > |G|$, then for some $m > n \; (= |G|)$ we can find a sequence $k_1, \ldots, k_m$ of elements of K which are linearly independent over F. The matrix

$$\left(\sigma_i(k_j)\right)_{\substack{1 \leq i \leq n \\ 1 \leq j \leq m}}$$

has rank at most n, hence it columns are linearly dependent over K. Let $a_1, \ldots, a_m \in K$, not all zero, such that

$$\sigma_i(k_1)a_1 + \cdots + \sigma_i(k_m)a_m = 0 \qquad \text{for } i = 1, \ldots, n. \tag{15.5}$$

Changing the numbering of $k_1, \ldots, k_m$ if necessary, we may assume $a_1 \neq 0$. Moreover, multiplying $a_1, \ldots, a_m$ by a common non-zero element of K, we can transform a_1 into any other non-zero element of K, and we may therefore assume that a_1 has the following property:

$$\sigma_1^{-1}(a_1) + \cdots + \sigma_n^{-1}(a_1) \neq 0.$$

(Since $\sigma_1^{-1}, \ldots, \sigma_n^{-1}$ are linearly independent over K, by Lemma 15.4, we have $\sigma_1^{-1} + \cdots + \sigma_n^{-1} \neq 0$ in $\mathcal{F}(K, K)$, hence there exists $a_1 \in K$ for which the property above holds.)

Applying σ_i^{-1} to equations (15.5) and adding up the equations thus obtained, we get

$$k_1\left(\sigma_1^{-1}(a_1) + \cdots + \sigma_n^{-1}(a_1)\right) + \cdots + k_m\left(\sigma_1^{-1}(a_m) + \cdots + \sigma_n^{-1}(a_m)\right) = 0$$

i.e., since $G = \{\sigma_1, \ldots, \sigma_n\} = \{\sigma_1^{-1}, \ldots, \sigma_n^{-1}\}$,

$$k_1\left(\sum_{\sigma \in G} \sigma(a_1)\right) + \cdots + k_m\left(\sum_{\sigma \in G} \sigma(a_m)\right) = 0.$$

The coefficients of $k_1, \ldots, k_m$ are invariant under G and are not all zero, by the hypothesis on a_1. Therefore, this equation is in contradiction with the hypothesis that $k_1, \ldots, k_m$ are linearly independent over K. Thus,

$$[K : F] = |G|.$$

Step 2: $G = \text{Gal}(K/F)$.

Let $G = \{\sigma_1, \ldots, \sigma_n\}$. We already have $G \subseteq \text{Gal}(K/F)$, by definition. Suppose, by way of contradiction, that $\text{Gal}(K/F)$ contains an element τ which is not in G. Clearly,

$$\{x \in K \mid \sigma_1(x) = \cdots = \sigma_n(x)\} \supseteq \{x \in K \mid \sigma_1(x) = \cdots = \sigma_n(x) = \tau(x)\}.$$

But $F = \{x \in K \mid \sigma_1(x) = \cdots = \sigma_n(x)\}$ since $F = K^G$, and

$$\{x \in K \mid \sigma_1(x) = \cdots = \sigma_n(x) = \tau(x)\} \supseteq F$$

since $\sigma_1, \ldots, \sigma_n, \tau \in \text{Gal}(K/F) = \text{Aut}_F\, K$. Therefore, the last inclusion is an equality, and Corollary 15.5 yields

$$[K : F] \geq n + 1.$$

This is a contradiction, since it was seen in Step 1 that $[K : F] = n$.

Step 3: $\text{Gal}(K/K^H) = H$ and $(G : H) = [K^H : F]$ for any subgroup H of G.

The first equality follows from Step 2, with H instead of G. To obtain the second equality, we compare the following equalities which are derived from Step 1:

$$[K : F] = |G| \qquad \text{and} \qquad [K : K^H] = |H|.$$

Since the degrees of field extensions are multiplicative (by Lemma 15.3),

$$[K : F] = [K : K^H][K^H : F]$$

hence the preceding equalities yield

$$[K^H : F] = \frac{|G|}{|H|} = (G : H).$$

Step 4: $K^{\text{Gal}(K/L)} = L$ and $[L : F] = \big(G : \text{Gal}(K/L)\big)$ for any subfield L of K containing F.

By restriction to L, each $\sigma \in G$ induces a homomorphism of fields

$$\sigma|_L : L \to K.$$

If two such homomorphisms coincide, say

$$\sigma|_L = \tau|_L \qquad \text{for some } \sigma, \tau \in G,$$

then $\sigma(\ell) = \tau(\ell)$ for all $\ell \in L$, hence $\tau^{-1} \circ \sigma(\ell) = \ell$ for all $\ell \in L$. Therefore

$$\tau^{-1} \circ \sigma \in \text{Gal}(K/L),$$

which amounts, by Lemma 14.23, p. 266, to

$$\sigma \circ \text{Gal}(K/L) = \tau \circ \text{Gal}(K/L).$$

Therefore, if $\big(G : \text{Gal}(K/L)\big) = r$ and if $\sigma_1, \ldots, \sigma_r$ are elements of G in pairwise different cosets of $\text{Gal}(K/L)$, then the homomorphisms $\sigma_i|_L$ are pairwise different. Since $\sigma_1(x) = \cdots = \sigma_r(x)$ for any $x \in F$, Corollary 15.5 implies that

$$[L : F] \geq r \quad (= \big(G : \text{Gal}(K/L)\big)).$$

On the other hand, the inclusion $L \subseteq K^{\text{Gal}(K/L)}$ and Step 3 yield

$$[L : F] \leq [K^{\text{Gal}(K/L):F)} : F] = \big(G : \text{Gal}(K/L)\big),$$

hence, by the preceding inequality,

$$[L : F] = \big(G : \text{Gal}(K/L)\big).$$

Moreover, since $L \subseteq K^{\text{Gal}(K/L)}$ and since these fields both have the same (finite) dimension over F,

$$L = K^{\text{Gal}(K/L)}.$$

Steps 3 and 4 show that the maps $H \mapsto K^H$ and $L \mapsto \text{Gal}(K/L)$ are reciprocal bijections between the set of subgroups of G and the set of subfields of K containing F. These maps clearly reverse the inclusions,

$$H \subseteq J \Longrightarrow K^H \supseteq K^J \quad \text{and} \quad L \subseteq M \Longrightarrow \text{Gal}(K/L) \supseteq \text{Gal}(K/M).$$

Moreover, from Step 4 it also follows that each subfield L of K containing F is the field of invariants of some finite group of automorphisms (viz. of $\text{Gal}(K/L)$), hence K is Galois over L.

To complete the proof of the fundamental theorem, it now suffices to show that normal subgroups correspond to fields which are Galois over F.

Step 5: $\text{Gal}\big(K/\sigma(L)\big) = \sigma \circ \text{Gal}(K/L) \circ \sigma^{-1}$ for any $\sigma \in G$ and any subfield L of K containing F.

This readily follows from the inclusions

$$\sigma \circ \text{Gal}(K/L) \circ \sigma^{-1} \subseteq \text{Gal}\big(K/\sigma(L)\big)$$

and

$$\sigma^{-1} \circ \text{Gal}\big(K/\sigma(L)\big) \circ \sigma \subseteq \text{Gal}(K/L),$$

which are both obvious.

Step 6: If L is a subfield of K containing F such that $\mathrm{Gal}(K/L)$ is normal in G, then L is Galois over F, and there is an isomorphism $G/\mathrm{Gal}(K/L) \to \mathrm{Gal}(L/F)$ obtained by restricting to L the automorphisms in G.

By Step 5, the hypothesis that $\mathrm{Gal}(K/L)$ is normal in G implies that any $\sigma \in G$ induces by restriction to L an automorphism

$$\sigma|_L \colon L \to L$$

which leaves F elementwise invariant. We thus have a restriction map

$$\mathrm{res} \colon G \to \mathrm{Gal}(L/F).$$

Since the kernel of this map is $\mathrm{Gal}(K/L)$, there is an induced injective map

$$\overline{\mathrm{res}} \colon G/\mathrm{Gal}(K/L) \to \mathrm{Gal}(L/F).$$

Now, Steps 4 and 1 yield

$$\big(G : \mathrm{Gal}(K/L)\big) = [L : F] = |\mathrm{Gal}(L/F)|,$$

hence the groups $G/\mathrm{Gal}(K/L)$ and $\mathrm{Gal}(L/F)$ have the same finite order, and the injective map $\overline{\mathrm{res}}$ is therefore an isomorphism.

Step 7: If L is a Galois extension of F contained in K, then $\mathrm{Gal}(K/L)$ is a normal subgroup of G.

Let $\big(G : \mathrm{Gal}(K/L)\big) = r \; (= [L : F]$, by Step 4), and let $\sigma_1, \ldots, \sigma_r$ be elements of G in the r different cosets of $\mathrm{Gal}(K/L)$. We have shown in the proof of Step 4 that the restrictions $\sigma_i|_L \colon L \to K$ yield pairwise different homomorphisms of L in K which restrict to the identity on F. On the other hand, it follows from Corollary 15.5 that there are at most r homomorphisms of L in K which restrict to the identity on F. Therefore, any such homomorphism has the form $\sigma_i|_L$ for some $i = 1, \ldots, r$.

Now, since L is assumed to be Galois over F, we can find r automorphisms of L leaving F elementwise invariant. These automorphisms can be regarded as homomorphisms from L into K, since $L \subseteq K$, hence they have the form $\sigma_i|_L$. Thus,

$$\mathrm{Gal}(L/F) = \{\sigma_1|_L, \ldots, \sigma_r|_L\},$$

and consequently $\sigma_1, \ldots, \sigma_r$ map L into L,

$$\sigma_i(L) = L \qquad \text{for } i = 1, \ldots, r.$$

By Step 5, it follows that

$$\sigma_i \circ \mathrm{Gal}(K/L) \circ \sigma_i^{-1} = \mathrm{Gal}(K/L) \qquad \text{for } i = 1, \dots, r.$$

Since every element $\sigma \in G$ has the form $\sigma_i \circ \tau$ for some $i = 1, \dots, r$ and some $\tau \in \mathrm{Gal}(K/L)$ we also have

$$\sigma \circ \mathrm{Gal}(K/L) \circ \sigma^{-1} = \mathrm{Gal}(K/L) \qquad \text{for all } \sigma \in G,$$

hence $\mathrm{Gal}(K/L)$ is normal in G.

Exercises

1. Let $G = \{\sigma_1, \dots, \sigma_n\}$ be a group of automorphisms of a field K and let $e_1, \dots, e_n$ be a basis of K over K^G. Show that the matrix $\left(\sigma_i(e_j)\right)_{1 \le i,j \le n}$ is invertible in the ring of $n \times n$ matrices over K.

2. *The aim of the following exercise is to provide another approach to the proof of the fundamental correspondence (Theorem 15.2).* Let K/F be a Galois extension of fields with Galois group G and let $\mathcal{F}(G, K)$ be the K-vector space of all maps from G to K.

    (a) Show that there is a well-defined F-linear map

$$\varphi \colon K \otimes_F K \to \mathcal{F}(G, K)$$

    such that $\varphi(a \otimes b)(\sigma) = \sigma(a)b$ for $a, b \in K$.

    (b) Consider $K \otimes_F K$ as a vector space over K by $(a \otimes b) \cdot k = a \otimes bk$. Show that φ is K-linear and bijective.

    [Hint: show that the matrix of φ with respect to suitable bases of $K \otimes K$ and of $\mathcal{F}(G, K)$ over K is that of Exercise 1.]

    (c) Let G act on $K \otimes K$ by $\sigma(a \otimes b) = \sigma(a) \otimes b$. Show that the corresponding action on $\mathcal{F}(G, K)$ is such that $^\sigma f(\tau) = f(\tau\sigma)$.

    (d) Let G act on $K \otimes K$ by $\sigma(a \otimes b) = a \otimes \sigma(b)$. Show that the corresponding action on $\mathcal{F}(G, K)$ is such that $^\sigma f(\tau) = \sigma\big(f(\sigma^{-1}\tau)\big)$.

    (e) Show that a K-subalgebra of $K \otimes K$ is globally invariant under the action of G defined in (d) if and only if it has the form $L \otimes_F K$ for some subfield L of K containing F.

    [Hint: for any K-subalgebra A of $K \otimes K$, define $L = \{x \in K \mid x \otimes 1 \in A\}$. Show that every $\sum_{i=1}^r a_i \otimes b_i \in A$ is in $L \otimes K$ by induction on the number r of terms.]

(f) Establish a 1–1 correspondence between K-subalgebras of $\mathcal{F}(G, K)$ and partitions of G by mapping every partition $G = \bigcup_{i \in I} G_i$ to the subalgebra

$$\{f \colon G \to K \mid f(\sigma) = f(\tau) \text{ if } \sigma \text{ and } \tau \text{ belong to the same } G_i\}.$$

Show that a subalgebra of $\mathcal{F}(G, K)$ is globally invariant under the action of G in (d) if and only if the corresponding partition is a decomposition of G into left cosets of some subgroup.

(g) Use the bijection φ and parts (e) and (f) to set up a 1–1 correspondence between subfields of K containing F and subgroups of G. Use the action of G defined in (c) to show that this correspondence is the same as that in Theorem 15.2.

Selected Solutions

Chapter 10

Exercise 1. For $i = 1, 2, 3$, the permutations of $x_1, \ldots, x_4$ which leave u_i invariant are the same as those which leave v_i invariant. Therefore, v_i is a rational fraction in u_i with symmetric coefficients. Denoting by s_1, s_2 the first two elementary symmetric polynomials (see equation (8.2), p. 97), it is easily checked that $v_i = s_2 - u_i$ for $i = 1, 2, 3$, and

$$w_1 = s_1^2 - 4u_2, \qquad w_2 = s_1^2 - 4u_1, \qquad w_3 = s_1^2 - 4u_3.$$

Therefore, if $P(X)$ (resp. $Q(X)$, resp. $R(X)$) is the monic cubic polynomial with roots u_1, u_2, u_3 (resp. v_1, v_2, v_3, resp. w_1, w_2, w_3), then

$$P(X) = -Q(s_2 - X) = -\tfrac{1}{64}R(s_1^2 - 4X),$$

$$Q(X) = -P(s_2 - X), \qquad R(X) = -64P\Big(\frac{s_1^2 - X}{4}\Big).$$

Exercise 2. In order to reproduce the notation of Theorem 10.4, p. 142, let

$$g_1 = x_1 + x_2, \qquad g_2 = x_1 + x_3, \qquad g_3 = x_2 + x_3,$$
$$f_1 = x_1 x_2, \qquad f_2 = x_1 x_3, \qquad f_3 = x_2 x_3.$$

Then

$$a_0 = s_2, \qquad a_1 = s_1 s_2 - 3s_3, \qquad a_2 = s_1^2 s_2 - 5s_1 s_3,$$

$$\theta(Y) = Y^3 - 2s_1 Y^2 + (s_1^2 + s_2)Y - (s_1 s_2 - s_3),$$
$$\psi(Y) = Y^2 + (g_1 - 2s_1)Y + (g_1^2 - 2s_1 g_1 + s_1^2 + s_2),$$

317

and

$$f_1 = \frac{s_2 g_1^2 - s_1 s_2 g_1 - 3 s_3 g_1 + s_2^2 - s_1 s_3}{3 g_1^2 - 4 s_1 g_1 + s_1^2 + s_2}.$$

This expression is not unique. Indeed, it is clear that $f_1 = s_3/X_3$ and $g_1 = s_1 - X_3$, hence $f_1 = s_3(s_1 - g_1)^{-1}$. This non-uniqueness stems from the fact that $\theta(g_1) = 0$. Indeed, it is easy to check that

$$\frac{s_2 Y^2 - s_1 s_2 Y - 3 s_3 Y + s_2^2 + s_1 s_3}{3 Y^2 - 4 s_1 Y + s_1^2 + s_2} =$$

$$\frac{s_3}{s_1 - Y} - \frac{s_2 \theta(Y)}{(3 Y^2 - 4 s_1 Y + s_1^2 + s_2)(s_1 - Y)}.$$

Exercise 3. Let $\sigma \colon x_1 \mapsto x_2 \mapsto x_3 \mapsto x_1$. If $\sigma(f^3) \neq f^3$ then f^3, $\sigma(f^3)$ and $\sigma^2(f^3)$ are pairwise distinct. This contradicts the hypothesis that f^3 takes only two values. Therefore $\sigma(f^3) = f^3$ and it follows that $\sigma(f) = \omega f$ for some cube root of unity ω. Comparing coefficients in $\sigma(f)$ and f, we obtain $A = \omega B = \omega^2 C$, whence

$$f = A(x_1 + \omega^2 x_2 + \omega x_3).$$

Moreover, $\omega \neq 1$ since x_1, x_2 and x_3 can be rationally expressed from f.

Exercise 4. By Theorem 10.4 (p. 142), it suffices to prove that $t(\omega^k)t(\omega)^{-k}$ is invariant by the permutations which leave $t(\omega)^n$ invariant, i.e. by $\tau \colon x_1 \mapsto x_2 \mapsto \cdots \mapsto x_n \mapsto x_1$ (and its powers) (compare Proposition 10.8, p. 149). This is clear, since $\tau\big(t(\omega^k)\big) = \omega^{-k} t(\omega^k)$.

Exercise 5. Identifying $\{0, 1, \ldots, n-1\}$ with $\mathbb{F}_n$, we can represent σ_i and τ as follows:

$$\sigma_i(x) = ix, \qquad \tau(x) = x + 1 \qquad \text{for } x \in \mathbb{F}_n.$$

Then $\tau \circ \sigma_i(x) = ix + 1$ and $\sigma_i \circ \tau^k(x) = i(x + k)$, hence $\tau \circ \sigma_i = \sigma_i \circ \tau^k$ if $ik \equiv 1 \bmod n$. One easily checks that if $\sigma_i \circ \tau^j = \sigma_k \circ \tau^\ell$, then $i = k$ and $j = \ell$ (for $i = 1, \ldots, n-1$ and $j = 0, \ldots, n-1$), and it follows that $|GA(n)| = n(n-1)$.

Chapter 11

Exercise 1.

$$[\begin{array}{ccccc} \alpha & \beta & \gamma & \delta & \varepsilon \\ v & iii & iv & i & ii \end{array}] = a^\alpha b^\beta c^\gamma d^\delta e^\varepsilon + a^\delta b^\varepsilon c^\beta d^\gamma e^\alpha + a^\gamma b^\alpha c^\varepsilon d^\beta e^\delta + a^\beta b^\delta c^\alpha d^\varepsilon e^\gamma + a^\varepsilon b^\gamma c^\delta d^\alpha e^\beta,$$

$$[\begin{array}{ccccc} \alpha & \varepsilon & \delta & \beta & \gamma \\ v & iii & iv & i & ii \end{array}] = a^\alpha b^\varepsilon c^\delta d^\beta e^\gamma + a^\beta b^\gamma c^\varepsilon d^\delta e^\alpha + a^\delta b^\alpha c^\gamma d^\varepsilon e^\beta + a^\varepsilon b^\beta c^\alpha d^\gamma e^\delta + a^\gamma b^\delta c^\beta d^\alpha e^\varepsilon,$$

$$[\begin{array}{ccccc} \alpha & \gamma & \beta & \varepsilon & \delta \\ v & iii & iv & i & ii \end{array}] = a^\alpha b^\gamma c^\beta d^\varepsilon e^\delta + a^\varepsilon b^\delta c^\gamma d^\beta e^\alpha + a^\beta b^\alpha c^\delta d^\gamma e^\varepsilon + a^\gamma b^\varepsilon c^\alpha d^\delta e^\beta + a^\delta b^\beta c^\varepsilon d^\alpha e^\gamma,$$

$$[\begin{array}{ccccc} \alpha & \delta & \varepsilon & \gamma & \beta \\ v & iii & iv & i & ii \end{array}] = a^\alpha b^\delta c^\varepsilon d^\gamma e^\beta + a^\gamma b^\beta c^\delta d^\varepsilon e^\alpha + a^\varepsilon b^\alpha c^\beta d^\delta e^\gamma + a^\delta b^\gamma c^\alpha d^\beta e^\varepsilon + a^\beta b^\varepsilon c^\gamma d^\alpha e^\delta.$$

The sum of these four partial types is a partial type corresponding to the subgroup generated by the permutations $\tau\colon a \mapsto e \mapsto b \mapsto c \mapsto d \mapsto a$ and $\sigma\colon a \mapsto a$, $b \mapsto d \mapsto c \mapsto e \mapsto b$. If a, b, c, d, e are numbered as $a = 0, b = 2, c = 3, d = 4$, $e = 1$, then the subgroup is $\mathrm{GA}(5) \subset S_5$ (see p. 148).

Exercise 2. Applying $a \mapsto b \mapsto c \mapsto d \mapsto e \mapsto a$ to both sides of $a^2 = b + 2$ (for instance), one gets $b^2 = c + 2$, a relation which does not hold.

Exercise 3. The relations between a, b, c are the following:

$$a^2 = b + 2, \qquad b^2 = c + 2, \qquad c^2 = a + 2,$$
$$ab = a + c, \qquad bc = b + a, \qquad ca = c + b.$$

It is readily verified that these relations are preserved under $a \mapsto b \mapsto c \mapsto a$.

Exercise 4. Using the fact that 2 is a primitive root of 13, it follows from Proposition 12.18, p. 182 (see also Proposition 12.21, p. 187) that $2\cos\frac{2\pi}{13} \mapsto 2\cos\frac{4\pi}{13} \mapsto 2\cos\frac{8\pi}{13} \mapsto 2\cos\frac{10\pi}{13} \mapsto 2\cos\frac{6\pi}{13} \mapsto 2\cos\frac{12\pi}{13} \mapsto 2\cos\frac{2\pi}{13}$ preserves the relations among $2\cos\frac{2\pi}{13}, \ldots, 2\cos\frac{12\pi}{13}$.

Chapter 12

Exercise 3. Periods with an even number of terms are sums of periods of two terms, which are real numbers, see p. 184.

Exercise 4. Let $\zeta = \zeta'_k(= \zeta'^{g'^k})$ for some $k = 0, \ldots, p-2$ and let $g = g'^\ell$ for some interger ℓ, which is prime to $p-1$ since g is a primitive root of p (see Proposition 7.12, p. 89). A straightforward computation yields $\zeta_i = \zeta'_{\sigma(i)}$, where $\sigma: \{0, \ldots, p-2\} \to \{0, \ldots, p-2\}$ is defined by $\sigma(\alpha) \equiv \ell\alpha + k \bmod p-1$, for $\alpha = 0, \ldots, p-2$. The same arguments as in Proposition 10.6, p. 147, show that σ is a permutation. Moreover, $\alpha \equiv i \bmod p-1$ implies $\sigma(\alpha) \equiv \sigma(i) \bmod p-1$, hence

$$\sum_{\alpha \equiv i \bmod e} \zeta_\alpha = \sum_{\beta \equiv \sigma(i) \bmod e} \zeta'_\beta.$$

Exercise 5. Let $ef = gh = p-1$. If $K_g \subset K_f$, then

$$\sigma^e(\zeta_0 + \zeta_h + \cdots + \zeta_{h(g-1)}) = \zeta_0 + \zeta_h + \cdots + \zeta_{h(g-1)}.$$

In particular, $\sigma^e(\zeta_0) = \zeta_e$ is of the form $\zeta_{h\ell}$ for some ℓ. It follows that $e = h\ell$, hence h divides e and f divides g.

Let η be a period of f terms. From Proposition 12.23, p. 189, it follows that $K_f = K_g(\eta)$. Now, Proposition 12.25 (p. 191) shows that η is a root of a polynomial of degree $k = g/f$ over K_g. It is not a root of a polynomial of smaller degree, since if $P(\eta) = 0$ with $P \in K_g[X]$, then $P(\sigma^h(\eta)) = P(\sigma^{2h}(\eta)) = \cdots = P(\sigma^{h(k-1)}(\eta)) = 0$. Therefore, g/f is the degree of the minimum polynomial of η over K_g (see Remark 12.16, p. 180), and it follows from Proposition 12.15, p. 179, that $\dim_{K_g} K_f = g/f$.

Chapter 13

Exercise 1. The fundamental theorem of algebra (Theorem 9.1, p. 115) shows that all the roots of any polynomial equation $P(X) = 0$, with $P \in \mathbb{R}[X]$ (resp. $\mathbb{C}[X]$), are in $\mathbb{C}$, which is a radical extension of $\mathbb{R}$ (resp. $\mathbb{C}$).

Exercise 2. From Lagrange's result (Theorem 10.4, p. 142), it is known that x_1, x_2 and x_3 can be rationally expressed from $t = x_1 + \omega x_2 + \omega^2 x_3$, where $\omega = \frac{1}{2}(-1 + \sqrt{-3})$. Explicitly,

$$x_1 = \frac{1}{3}(s_1 + t + (s_1^2 - 3s_2)t^{-1}).$$

A straightforward computation yields

$$t^3 = s_1^3 - \frac{9}{2}(s_1 s_2 - 3s_3) + \frac{3\sqrt{-3}}{2}\Delta,$$

where $\Delta = (x_1 - x_2)(x_1 - x_3)(x_2 - x_3) = \sqrt{D(s_1, s_2, s_3)}$, where $D(s_1, s_2, s_3)$ is the discriminant (see §8.3). Therefore, a radical extension of $\mathbb{Q}(s_1, s_2, s_3)$ containing x_1 can be constructed as follows:

$$R_0 = \mathbb{Q}(s_1, s_2, s_3),$$

$$R_1 = R_0\left(\sqrt{-3D(s_1, s_2, s_3)}\right),$$

$$R_2 = R_1\left(\sqrt[3]{s_1^3 - \tfrac{9}{2}(s_1 s_2 - 3s_3) + \tfrac{3}{2}\sqrt{-3D(s_1, s_2, s_3)}}\right).$$

The field R_2 is not contained in $\mathbb{Q}(x_1, x_2, x_3)$ since $\sqrt{-3D(s_1, s_2, s_3)}$ is not in $\mathbb{Q}(x_1, x_2, x_3)$. In order to show that $\mathbb{Q}(x_1, x_2, x_3)$ does not contain any radical extension of $\mathbb{Q}(s_1, s_2, s_3)$ containing x_1 (or x_2 or x_3), one can argue as in §13.4: if $u \in \mathbb{Q}(x_1, x_2, x_3)$ has the property that some power u^p (with p prime) is invariant under $\sigma: x_1 \mapsto x_2 \mapsto x_3 \mapsto x_1$, then u is invariant under σ. Indeed, from $\sigma(u^p) = u^p$, it follows that $\sigma(u) = \omega u$ for some p-th root of unity ω. Since $\sigma^3 = \mathrm{Id}$, one has $u = \sigma^3(u) = \omega^3 u$, hence $p = 3$. But 1 is the only cube root of unity in $\mathbb{Q}(x_1, x_2, x_3)$, so $\sigma(u) = u$.

In order to obtain a solution of the general equation of degree 4, one first solves a resolvent cubic equation, for instance the equation with root $u = (x_1 + x_2)(x_3 + x_4)$, i.e.

$$X^3 - a_1 X^2 + a_2 X - a_3 = 0$$

with $a_1 = 2s_2$, $a_2 = s_2^2 + s_1 s_3 - 4s_4$, $a_3 = s_1 s_2 s_3 - s_1^2 s_4 - s_3^2$. (See Exercise 3 of Chapter 8.) Then, $v = x_1 + x_2$ is obtained as a root of the quadratic equation

$$X^2 - s_1 X + u = 0.$$

(The other root is $x_3 + x_4$.) Then x_1 and x_2 are obtained as roots of

$$X^2 - vX + \frac{(s_1 - 2v)s_4}{(s_1 - v)(s_2 - u) - s_3} = 0.$$

(Observe that the constant term is $x_1 x_2$, expressed as a function of v, u and the symmetric polynomials.) Therefore, a radical extension of $\mathbb{Q}(s_1, s_2, s_3, s_4)$ containing x_1 can be constructed as follows:

$$R_0 = \mathbb{Q}(s_1, s_2, s_3, s_4),$$

$$R_1 = R_0\left(\sqrt{-3D(a_1, a_2, a_3)}\right)$$

$$\left(= R_0\left(\sqrt{-3D(s_1, s_2, s_3, s_4)}\right), \quad \text{see Exercise 3 of Chapter 8, }\right)$$

$$R_2 = R_1(t)$$

where

$$t = \sqrt[3]{a_1^3 - \tfrac{9}{2}(a_1 a_2 - 3a_3) + \tfrac{3}{2}\sqrt{-3D(a_1, a_2, a_3)}}.$$

Then

$$u = \frac{1}{3}\left(a_1 + t + (a_1^2 - 3a_2)t^{-1}\right) \in R_2.$$

Let then

$$R_3 = R_2\left(\sqrt{s_1^2 - 4u}\right);$$

then $v = \tfrac{1}{2}(s_1 + \sqrt{s_1^2 - 4u}) \in R_3$ and

$$x_1 \in R_4 = R_3\left(\sqrt{v^2 - 4(s_1 - 2v)s_4\big((s_1 - v)(s_2 - u) - s_3\big)^{-1}}\right).$$

Exercise 3. Since 3 is a primitive root of 7, Proposition 12.21 (p. 187) shows that there is an automorphism σ of $\mathbb{Q}(\zeta_7)$ defined by $\sigma(\zeta_7) = \zeta_7^3$. Suppose $\mathbb{Q}(\zeta_7)$ is radical over $\mathbb{Q}$; then there is a tower of extensions

$$\mathbb{Q}(\zeta_7) = R_0 \supset R_1 \supset \cdots \supset R_h = \mathbb{Q}$$

where $R_i = R_{i+1}(u_i)$ with $u_i^{p_i} = a_i$ for some prime number p_i and some element $a_i \in R_{i+1}$ which is not a p_i-th power in R_{i+1}.

We are about to prove that if a_i is invariant under σ^2, then u_i is invariant under σ^2 too. Therefore every element in R_i is invariant under σ^2. By induction, it follows that every element in $R_0 = \mathbb{Q}(\zeta_7)$ is invariant under σ^2, a contradiction.

From $\sigma^2(a_i) = a_i$, it follows that $\sigma^2(u_i)^{p_i} = u_i^{p_i}$, whence $\sigma^2(u_i) = \omega u_i$ for some p_i-th root of unity $\omega \in \mathbb{Q}(\zeta_7)$. By Theorem 12.32, p. 199, the cyclotomic polynomial Φ_{p_i} is irreducible over $\mathbb{Q}(\zeta_7)$ if $p_i \neq 7$. Therefore, the only prime numbers p_i such that $\mathbb{Q}(\zeta_7)$ contains a p_i-th root of unity other than 1 are $p_i = 2$ or 7.

Now, Corollary 13.10 (p. 220) shows that $\dim_{R_{i+1}} R_i = p_i$. Therefore, it is impossible that $p_i = 7$, since $\dim_{\mathbb{Q}} \mathbb{Q}(\zeta_7) = 6$, by Theorem 12.13, p. 178. If

$p_i = 2$, then $\omega = 1$ or -1. However, it is impossible that $\sigma^2(u_i) = -u_i$, since by applying σ^2 twice to both sides of this equation one gets $\sigma^6(u_i) = -u_i$, a contradiction since $\sigma^6 = \mathrm{Id}$. Therefore, the only possibility is that $\sigma^2(u_i) = u_i$, and the claim is proved. On the other hand, $\mathbb{Q}(\zeta_7, \zeta_3)$ is radical over $\mathbb{Q}$. Indeed, let

$$\eta_0 = \zeta_7^{3^0} + \zeta_7^{3^2} + \zeta_7^{3^4} = \zeta_7 + \zeta_7^2 + \zeta_7^{-3},$$

$$\eta_1 = \zeta_7^{3^1} + \zeta_7^{3^3} + \zeta_7^{3^5} = \zeta_7^3 + \zeta_7^{-1} + \zeta_7^{-2}$$

be the periods of three terms of $\Phi_7 = 0$. It is readily checked that $\eta_0 + \eta_1 = -1$ and $\eta_0\eta_1 = 2$, hence $\eta_0, \eta_1 = \frac{1}{2}(1 \pm \sqrt{-7})$. Now, let $t = \zeta_7 + \zeta_3\zeta_7^2 + \zeta_3^2\zeta_7^{-3}$; then

$$\zeta_7 = \tfrac{1}{3}(\eta_0 + t + (2\eta_0 + 1)t^{-1})$$

and $t^3 = 8 + 2\eta_0 + 3\zeta_3 + 6\zeta_3\eta_0 \in \mathbb{Q}(\zeta_3, \eta_0) = \mathbb{Q}(\sqrt{-3}, \sqrt{-7})$, hence $\mathbb{Q}(\zeta_3, \zeta_7) = \mathbb{Q}(\zeta_3, \eta_0, t) = \mathbb{Q}(\sqrt{-3}, \sqrt{-7}, t)$ is a radical extension of $\mathbb{Q}$.

Exercise 4. If $R = F(u)$ with $u^p = a$, an isomorphism $f \colon F[X]/(X^p - a) \xrightarrow{\sim} R$ is given by $f(P(X) + (X^p - a)) = P(u)$.

Exercise 5. The cube roots of 2 in $\mathbb{C}$ are $\sqrt[3]{2} \in \mathbb{R}$, $\frac{1}{2}(-1 + i\sqrt{3})\sqrt[3]{2}$ and $\frac{1}{2}(-1 - i\sqrt{3})\sqrt[3]{2}$. Since the last two are not in $\mathbb{R}$, the field $\mathbb{Q}(\sqrt[3]{2}) \subset \mathbb{R}$ contains only one cube root of 2. Since, by the preceding exercise, all the fields obtained from $\mathbb{Q}$ by adjoining a cube root of 2 are isomorphic, they all contain only one cube root of 2. Therefore,

$$\mathbb{Q}(\sqrt[3]{2}), \quad \mathbb{Q}(\tfrac{1}{2}(-1 + i\sqrt{3})\sqrt[3]{2}) \quad \text{and} \quad \mathbb{Q}(\tfrac{1}{2}(-1 - i\sqrt{3})\sqrt[3]{2})$$

are pairwise distinct subfields of $\mathbb{C}$.

Chapter 14

Exercise 1. By Proposition 14.32, p. 279, $\mathrm{Gal}(P/F)$ acts transitively on the roots of P. By Theorem 14.35, p. 282, it follows that the number of roots of P divides $|\mathrm{Gal}(P/F)|$.

Exercise 2. Applying σ to the expressions $r_i = f_i(V)$, we get $\sigma(r_i) = f_i(\sigma(V))$, and since $\{\sigma(r_1), \ldots, \sigma(r_n)\} = \{r_1, \ldots, r_n\}$ it follows that $r_1, \ldots, r_n$ have a rational expression in $\sigma(V)$.

Exercise 3. This follows from Proposition 14.33, p. 280, by induction on the number of irreducible factors of P.

Exercise 4. Irreducibility of $(X - u_1) \cdots (X - u_r)$ over F readily follows from Proposition 14.32, p. 279.

Exercise 5. If T is a set of representatives of the right cosets of H in G, then

$$G(\alpha) = \bigcup_{\tau \in T} H_\tau(\alpha)$$

is a decomposition of $G(\alpha)$ into subgroups which have H as group of substitutions. If this decomposition is the same as $G(\alpha) = \bigcup_{\sigma \in R} \sigma(H(\alpha))$, then for each $\sigma \in R$ there is a $\tau \in T$ such that $\sigma \circ H = H \circ \tau$. In particular, $\sigma = \sigma \circ \mathrm{Id} = \eta \circ \tau$ for some $\eta \in H$. Therefore, for all $\xi \in H$,

$$\sigma \circ \xi \circ \tau^{-1} = \sigma \circ \xi \circ \sigma^{-1} \circ \eta \in H,$$

hence $\sigma \circ \xi \circ \sigma^{-1} \in H$. This proves that H is normal in G. The converse is clear, since if H is normal, then $\sigma \circ H = H \circ \sigma$ for all $\sigma \in G$.

Bibliography

[1] N.-H. Abel, *Œuvres complètes*, (2 vol.) (L. Sylow and S. Lie, eds.), Grøndahl & Søn, Christiania, 1881.

[2] E. Artin, *Galois Theory*, Notre Dame Math. Lectures, Notre Dame Univ. Press, Ind., 1948.

[3] ———, *Collected Papers* (S. Lang and J. Tate, eds.), Addison-Wesley, Reading, Mass., 1965.

[4] R.G. Ayoub, *Paolo Ruffini's contributions to the quintic*, Arch. Hist. Exact Sci., **23** (1980/81), 253–277.

[5] H. Bosmans, *Sur le "Libro de algebra" de Pedro Nuñez*, Biblioth. Mathem., (Sér. 3) **8** (1908), 154–169.

[6] N. Bourbaki, *Algèbre, chapitres 4 et 5*, Hermann, Paris, 1967.

[7] ———, *Algèbre, chapitres 4 à 7*, Masson, Paris, 1981.

[8] C.B. Boyer, *A History of Mathematics*, J. Wiley & sons, New York, N.Y., 1968.

[9] W.K. Bühler, *Gauss. A Biographical Study*, Springer, Berlin, 1981.

[10] F. Cajori, *A History of Mathematical Notations, vol 1: Notations in Elementary Mathematics*, Open Court, La Salle, Ill., 1974.

[11] G. Cardano, *The Great Art, or the Rules of Algebra*, translated and edited by T.R. Witmer, MIT Press, Cambridge, Mass., 1968.

[12] J.-C. Carrega, *Théorie des corps. La règle et le compas*, Coll. formation des enseignants et formation continue, Hermann, Paris, 1981.

[13] S.U. Chase, D.K. Harrison, A. Rosenberg, *Galois theory and Galois cohomology of commutative rings*, Mem. Amer. Math. Soc. **52** (1968), 1–19.

[14] P.M. Cohn, *Algebra, vol. 2*, J. Wiley & sons, London, 1977.

[15] R. Cotes, *Theoremata tum Logometrica tum Trigonometrica Datarum Flux-*

326 *Bibliography**Bibliography*

[16] R. Descartes, *The Geometry*, translated from the French and Latin by D.E. Smith and M.L. Latham, Dover, New York, 1954.

[17] ———, *Geometria* (2 vol.), trad. F. Van Schooten, Ex typographia Blaviana, Amstelodami, 1683 (ed. tertia).

[18] J. Dieudonné, *Abrégé d'histoire des mathématiques 1700–1900* (2 vol.), Hermann, Paris, 1978.

[19] R. Douady, A. Douady, *Algèbre et théories galoisiennes* (2 vol.) Cedic–Fernand Nathan, Paris, 1977, 1979.

[20] H.M. Edwards, *Galois Theory*, Graduate Texts in Math. **101**, Springer, New York, N.Y., 1984.

[21] É. Galois, *Écrits et mémoires mathématiques d'Évariste Galois*, (R. Bourgne et J.-P. Azra, éd.), Gauthier-Villars, Paris, 1962.

[22] S. Gandz, *The origin and development of the quadratic equations in Babylonian, Greek and early Arabic algebra*, Osiris **3** (1937), 405–557.

[23] C.F. Gauss, *Demonstratio nova theorematis omnem functionem algebraicam rationalem integram unius variabilis in factores reales primi vel secundi gradus resolvi posse*, Apud C.G. Fleckeisen, Helmstadii, 1799. (Werke Bd III, Georg Olms, Hildesheim, 1981, pp. 1–30.)

[24] ———, *Disquisitiones Arithmeticae*, Apud Gerh. Fleischer Iun. Lipsiae, 1801. (Werke Bd I, Herausg. König. Ges. Wiss. Göttingen, 1870.)

[25] ———, *Demonstratio nova altera theorematis omnem functionem algebraicam rationalem integram unius variabilis in factores reales primi vel secundi gradus resolvi posse*, Comm. soc. regiae scient. Gottingensis recentiores **3** (1816). (Werke Bd III, Georg Olms, Hildesheim, 1981, pp. 31–56.)

[26] A. Girard, *Invention Nouvelle en l'Algèbre*, réimpression par D. Bierens De Haan, Muré Frères, Leiden, 1884.

[27] N.H. Goldstine, *A History of Numerical Analysis from the 16th through the 19th century*, Studies in the History of Math. and Phys. Sciences **2**, Springer, New York, N.Y., 1977.

[28] H. Hankel, *Zur Geschichte der Mathematik in Alterthum und Mittelalter*, Teubner, Leipzig, 1874.

[29] G.H. Hardy, E.M. Wright, *An Introduction to the Theory of Numbers*, Clarendon Press, Oxford, 1979.

[30] T.L. Heath, *The Thirteen Books of Euclid's Elements*, Cambridge Univ. Press, Dover, New York, N.Y., 1956.

[31] J. Hudde, *Epistola prima, de Reductione Aequationum*, in [17, vol. 1], pp. 406–506.

[32] ———, *Epistola secunda, de Maximis et Minimis*, in [17, vol. 1], pp. 507–516.

[33] N. Jacobson, *Lectures in Abstract Algebra, vol. III*, Van Nostrand, New York, N.Y., 1964.

[34] ———, *Basic Algebra I*, Freeman, San Francisco, Ca., 1974.

[35] I. Kaplansky, *Fields and Rings*, Chicago Lectures in Math., Univ. Chicago Press, Chicago, Ill., 1972.

[36] L.C. Karpinski, *Robert of Chester's Latin Translation of the Algebra of Al-Khowarizmi*, Univ. of Michigan Studies, Humanistic series **11**, MacMillan, New York, N.Y., 1915.

[37] B.M. Kiernan, *The development of Galois Theory from Lagrange to Artin*, Arch. Hist. Exact Sci., **8** (1971), 40–154.

[38] M. Kline, *Mathematical Thought from Ancient to Modern Times*, Oxford Univ. Press, New York, N.Y., 1972.

[39] W.R. Knorr, *The Evolution of the Euclidean Elements*, Synthese Historical Lib. **15**, D. Reidel, Dordrecht, 1975.

[40] J.-L. Lagrange, *Réflexions sur la résolution algébrique des équations*, Nouveaux Mémoires de l'Acad. Royale des sciences et belles-lettres, avec l'histoire pour la même année, **1** (1770), 134–215; **2** (1771), 138–253. (Œuvres de Lagrange, vol. 3 (J.-A. Serret, éd.) Gauthier-Villars, Paris, 1869, pp. 203–421.)

[41] H. Lebesgue, *L'œuvre mathématique de Vandermonde*, Enseignement Math. Sér. II, **1** (1955), 201–223.

[42] G.W. Leibniz, *Mathematische Schriften, Bd V*, herausg. von C.I. Gerhardt, Georg Olms, Hildesheim, 1962.

[43] ———, *Der Briefwechsel von Gottfried Wilhelm Leibniz mit Mathematikern*, herausg. von C.I. Gerhardt, Georg Olms, Hildesheim, 1962.

[44] P. Morandi, *Field and Galois Theory*, Graduate Texts in Math. **167**, Springer, New York, N.Y., 1996.

[45] I. Newton, *The Mathematical Papers of Isaac Newton*, ed. by D.T. Whiteside, vol. I: 1664–1666, Cambridge Univ. Press, Cambridge, 1967; vol. IV: 1674–1684, Cambridge Univ. Press, Cambridge, 1971; vol. V: 1683–1684, Cambridge Univ. Press, Cambridge, 1972.

[46] ———, *The Mathematical Works of Isaac Newton, vol. 2*, assembled with

an introduction by D.T. Whiteside, The sources of science, Johnson Reprint Corp., New York, London, 1967.

[47] L. Nový, *Origins of Modern Algebra*, Noordhoff, Leyden, 1973.

[48] A. Romanus, *Ideae Mathematicae Pars Prima, sive Methodus Polygonorum*, apud Ioannem Masium, Lovanii, 1593.

[49] M. Rosen, *Abel's theorem on the lemniscate*, Amer. Math. Monthly **88** (1981), 387–395.

[50] J. Rotman, *Galois Theory* (2d edition), Universitext, Springer, New York, N.Y., 1998.

[51] P. Ruffini, *Opere Matematiche* (3 vol.), E. Bortolotti, ed., Ed. Cremonese della Casa Editrice Perrella, Roma, 1953–1954.

[52] P. Samuel, *Théorie algébrique des nombres*, Coll. Méthodes, Hermann, Paris, 1971.

[53] J.-A. Serret, *Cours d'algèbre supérieure* (2 vol.), Gauthier-Villars, Paris, 1866 (3ème éd.).

[54] D.E. Smith, *A Source Book in Mathematics* (2 vol.), Dover, New York, N.Y., 1959.

[55] S. Stevin, *The Principal Works of Simon Stevin. Volume II B: Mathematics*, D.J. Struik, ed., C.V. Swets en Zeitlinger, Amsterdam, 1958.

[56] I. Stewart, *Galois Theory*, Chapman and Hall, London, 1973.

[57] R. Taton, *Les relations d'Évariste Galois avec les mathématiciens de son temps*, Rev. Hist. Sc. **1** (1948), 114–130.

[58] E.W. Tschirnhaus, *Methodus Anferendi Omnes Terminos intermedios ex data aequatione*, Acta Eruditorum (Leipzig) (1683), 204–207.

[59] A.T. Vandermonde, *Mémoire sur la résolution des équations*, Histoire de l'Acad. Royale des Sciences (avec les mémoires de Math. & de Phys. pour la même année, tirés des registres de cette Acad.) (1771), 365–416.

[60] A. Van der Poorten, *A proof that Euler missed ... Apéry's proof of the irrationality of* $\zeta(3)$, Math. Intel. **1** (1979), 195–203.

[61] B.L. Van der Waerden, *Algebra* (2 vol.), (7th ed. of *Moderne Algebra*), Heidelberger Taschenbücher 12 & 23, Springer, Berlin, 1966 & 1967.

[62] _____, *Science Awakening I*, Noordhoff, Leyden, 1975.

[63] _____, *Die Galoissche Theorie von Heinrich Weber bis Emil Artin*, Arch. Hist. Exact Sci. **9** (1972), 240–248.

[64] F. Vieta, *Ad Problema quod omnibus Mathematicis totius orbis construendum proposuit Adrianus Romanus Francisci Vietae Responsum*, apud Iametium Mettayer, Parisiis, 1595.

[65] ———, *The Analytic Art*, translated by T.R. Witmer, Kent State Univ. Press, Kent, Ohio, 1983.

[66] E. Waring, *Meditationes Algebraicae*, translated by D. Weeks, Amer. Math. Soc., Providence, RI, 1991.

[67] H. Weber, *Lehrbuch der Algebra, Bd. I*, F. Vieweg u. Sohn, Braunschweig, 1898.

[68] A. Weil, *De la métaphysique aux mathématiques*, Sciences (1960), 52–56. (Œuvres Scientifiques–Collected Papers, vol. 2, Springer, New York, N.Y., 1979, pp. 408–412.)

[69] ———, *Two lectures on number theory, past and present*, Enseignement Math. **20** (1974), 87–110. (Œuvres Scientifiques–Collected Papers, vol. 3, Springer, New York, N.Y., 1979, pp. 279–302.)

Index

Abel, Niels-Henrik, 210–212
  Abel's condition, 231–232, 242–243,
    293–294
  theorem on natural irrationalities,
    219–225
abelian group, 242
al-Khowarizmi, Mohammed ibn Musa,
  9–11, 22
Alembert, Jean Le Rond d', 115
algebra
  Arabic, 9–12, 22
  Babylonian, 2–5, 7, 22
  Greek, 5–9, 21
alternating group, 228, 276
Apéry, Roger, 112
Artin, Emil, 305–306, 309

Bachet de Méziriac, Claude, 87
Bernoulli, Jacques, 110
Bernoulli, Nicholas, 75
Bezout, Etienne
  Bezout's method, 123–126, 134–137
  Bezout's theorem, 87
  elimination theory, 69
Bolzano, Bernhard, 121
Bombelli, Rafaele, 20, 26
Bourbaki, Nicolas, 306

Cardano, Girolamo, 13–15, 21–22, 25–26,
  40

Cardano's formula, 15–20, 78,
  128–132, 277
casus irreducibilis, 19, 109
Cauchy, Augustin-Louis, 117, 210–212,
  228
complete solvability (by radicals), 264
congruence (modulo an integer), 168
constructible point, 201
coset, 141, 255
Cotes, Roger, 75–77
Cotes–de Moivre formula, 73, 76–81,
  83, 94–95
cyclic group, 89
cyclotomic polynomial (or equation),
  90–92, 231
  Galois group, 241–242
  irreducibility, 175–178, 188, 196–200
  solvability by radicals, 83, 158–164,
    192–196
cyclotomy, 83

Dedekind, Richard, 196, 305, 309
degree (of a field extension), 307
Delambre, Jean-Baptiste, 209
derivative (of a polynomial), 53
Descartes, René, 22, 28–30, 37
  Descartes' method for quartic
    equations, 64–65
Diophantus of Alexandria, 8, 29
discriminant, 106, 112–114, 276–278, 292